Advances in Atom and Single Molecule Machines

Advances in Atom and Single Molecule Machines is the first comprehensive series of books dealing specifically with single atom and molecule machines. Derived from a number of long-term European Commission Future and Emerging Technologies (FET) projects including AtMol, Elfos, Focus, Diamant, Artist, PAMS and MEMO, volumes in this series comprise topical reviews, lecture-course derived textbooks and re-worked proceedings of workshops.

More information about this series at http://www.springer.com/series/10425

Christian Joachim
Editor

Building and Probing Small for Mechanics

Springer

Editor
Christian Joachim
CEMES-CNRS
Toulouse, France

ISSN 2193-9691 ISSN 2193-9705 (electronic)
Advances in Atom and Single Molecule Machines
ISBN 978-3-030-56776-7 ISBN 978-3-030-56777-4 (eBook)
https://doi.org/10.1007/978-3-030-56777-4

This Springer imprint is published by the registered company Springer Nature Switzerland AG
The registered company address is: Gewerbestrasse 11, 6330 Cham, Switzerland

Preface

This book was constructed to gather in one unique volume all the current mechanical machinery miniaturization trends using the gear artifact invented long ago as the basic elementary device in mechanics. It takes benefit from the research activities of the partners of the H2020 FET Open European project "Mechanics with Molecule(s)" (MEMO, 2018–2021) and also from the first MEMO workshop "Building & Probing Small" held Place of the Academies in Brussels from March 25–27, 2019.

One technology node after the other, MEMO is exploring the miniaturization roadmap of mechanics beneficiating from machining tools, micro-lithography, nano-lithography and certainly synthetic chemistry. The forefront of the MEMO research is the border between the solid state nanogears down to a few tens of nanometer in diameter and molecule-gears with diameters up to a few nanometers. A molecule-gear is a single molecule made exclusively of covalently bound chemical groups. Here, the scanning tunneling microscope and its atomic scale manipulation capabilities are playing an essential role is exploring, for example, the functioning of a train of molecule-gears as reported in this book. However, not all the small organic molecules can function in this way, since many molecules rapidly diffuse laterally on a surface rather than rotating. Moreover, and under excitations, a molecule can also be easily detached from its rotational axle. In solution, a "molecular rotor" is constituted of the chemical groups (or molecule) stabilized together by weak interactions like van der Waals or hydrogen bond. This was the main scope of the Brussels March 2019 workshop as exemplified also in this volume.

Along the volume, we pay attention to provide a good balance between history of technology, advanced micro & nano-lithography, organic and organo-metallic chemistry together with experimental surface science and theoretical approach. Importantly, chapters concerning quantum chemistry of molecule-gears mechanics open the design of new molecule machinery in the techno-mimetic approach, in the nano-architectronics approach or simply to confirm intuitive approaches of designing a solid-state nanogear, a molecular gear or a molecule-gear.

The main part of the research works presented in this volume was supported by the European Union Horizon 3792020 FET open project "Mechanics with Molecule(s)" (MEMO, Grant 766864). We thank deeply the European commission for its full support. In addition, Springer-Nature and its staff are also thankful for the publication of this volume. The editor of this volume n°13 of the series deeply relies on the MEMO consortium for the finalization of this volume. Thanks to all of the MEMO partners. A special thanks to Pr. Anne-Sophie Duwez from University of Liege to have organized the Brussels March 2019 workshop which also leads to the two very nice conclusive chapters.

Toulouse, France

Christian Joachim

Contents

From the Anthycytera Astronomical Clock to Single Molecule Scale Machinery 1
Christian Joachim

From the Pascaline to a 5-Digits Metallic and Planar Miniature Mechanical Calculator 1 cm in Lateral Size 9
Pierre Abheillou, Bertrand Gatti, Kevin Froissard, Nicolas Joachim, and Christian Joachim

Photo-Microlithography Fabrication of the Parts of a Micro-Mechanical Calculator 21
Christian Bourgerette, Laure Noé, Sebastien Pinaud, and Christian Joachim

Fabricating Solid State Gears at the Nanoscale: A Top–Down Approach 41
D. Mailly and G. Faini

Prototypes of Molecular Gears with an Organometallic Piano-Stool Architecture 65
Seifallah Abid, Guillaume Erbland, Claire Kammerer, and Gwénaël Rapenne

Design and Synthesis of a Nano-winch 81
Yohan Gisbert, Agnès M. Sirven, Gwénaël Rapenne, and Claire Kammerer

Chemical Anchoring of Molecular Rotors 99
Oumaima Aiboudi and Franziska Lissel

Anchoring Molecular Rotors by On-Surface Synthesis 117
Kwan Ho Au Yeung, Tim Kühne, Frank Eisenhut, and Francesca Moresco

Transmission of Rotational Motion Between Molecule-Gears 131
W.-H. Soe, S. Srivastava, and C. Joachim

A Simple Example of a Molecule-Gear Train: PF_3 Molecules on a Cu(111) Surface 143
S. Srivastava, W.-H. Soe, and C. Joachim

Mechanical Transmission of Rotation for Molecule Gears and Solid-State Gears 165
Huang-Hsiang Lin, Jonathan Heinze, Alexander Croy, Rafael Gutierrez, and Gianaurelio Cuniberti

Rotations of Adsorbed Molecules Induced by Tunneling Electrons 181
N. Lorente and C. Joachim

Motion and Nanomechanical Effects in Supramolecular Catalysts 195
Michael Schmittel, Abir Goswami, Indrajit Paul, and Pronay Kumar Biswas

Five Minutes in the Life of a Molecular Shuttle: Near-Equilibrium Measurements of Shuttling Dynamics Using Optical Tweezers 219
Kateryna M. Lemishko, Teresa Naranjo, Emilio M. Pérez, and Borja Ibarra

From the Anthycytera Astronomical Clock to Single Molecule Scale Machinery

Christian Joachim

Abstract The historical background of this volume "Building and Probing small for mechanics" of the Springer-Nature series "Advances in Atom and Single Molecule Machines" is presented to put in context the actual effort towards the construction of functioning molecular machinery, one molecule per one molecule or in a supramolecular assemblage. The accent is made on 3 different roots: surface science, organic chemistry and supramolecular chemistry supported by the recent progresses in nanolithography and by the quantum chemistry approaches and their semi-classical trends for mechanics in designing and interpreting the functioning of the first molecule-machines.

Keywords Top-down · Bottom-up · Molecular machinery · Molecule-machines

1 Introduction

Mechanical elementary devices like gears, worms, rods, sliders, pull-tabs, pendulums, ratchets, springs and motors are found in many modern machines functioning in transferring a movement, producing work or creating information. Some of them have been in competition for performances, sustainability, robustness and portability with electrical machines constructed using electronic devices since the invention of the vacuum tubes by J. A Fleming in 1904 and of the solid-state transistor by W. Shockley, J. Bardeen and W. Brattain in 1947. Well before the starting of the *miniaturization* race in 1950s for the electronic devices, miniaturization have already render mechanical machinery more and more portable, faster in their functioning and

---

C. Joachim (✉)
Centre D'Elaboration de Matériaux et D'Études Structurales (CEMES), Centre National de La Recherche Scientifique (CNRS), Université de Toulouse, 29 Rue J. Marvig, BP 4347, Toulouse Cedex 31055, France
e-mail: christian.joachim@cemes.fr

International Center for Materials Nanoarchitectonics (WPI-MANA), National Institute for Material Sciences (NIMS), 1-1 Namiki, Tsukuba, Ibaraki 305-0044, Japan

C. Joachim (ed.), *Building and Probing Small for Mechanics*, Advances in Atom and Single Molecule Machines, https://doi.org/10.1007/978-3-030-56777-4_1

better in energy consumption during their operation. A watch is nothing more than a miniature of our solar system.

Constructed around 200 BC, the Antikythera astronomical clock is 21 cm in lateral dimension with a functioning mechanism of about 30 bronze 10 cm in diameter gears [1]. Centuries after centuries, handicraft, know-how and the introduction of new mechanical devices other than gears and rigid rods for the transmission of motion leads to the invention of a water free mechanical clock by Gerbert d'Aurillac in 994, of a portable watch using a spiral spring and 1 cm in diameter metallic gears by P. Henlein in 1510 and of the first mechanical calculator using at the begining wooden and centimeter in size gears by Blaise Pascal in 1642 (see also chapter "From the Pascaline to a 5-Digits Metallic and Planar Miniature Mechanical Calculator 1 cm in Lateral Size" of this volume). Taking a gear and its diameter as the representative of all elementary mechanical devices, miniaturization of mechanical artifacts like gears accelerates first the diffusion of small watches. Gears were also essentially for the construction of the first self-functioning heat engines by James Watt in 1763. The miniaturization and planarization of the Pascaline by Jacob Auch in 1790 (see chapter "From the Pascaline to a 5-Digits Metallic and Planar Miniature Mechanical Calculator 1 cm in Lateral Size" of this volume) and its large complexification by Charles Babbage in 1834 leads to the construction of modern mechanical calculators that were still active during the first half of the 20th century with metallic gears going down to a few millimeters in diameter.

2 The Top-Down Approach for Mechanic

The semiconductor planar technology invented in 1958 by Jack Kilby from Texas Instruments and Robert Noyce from Fairchild Semiconductor was also the occasion to boost mechanical devices miniaturization with the invention of surface micromachining by K. E. Peterson from IBM in 1977 [2]. It drives an explosion of technological researches for miniaturizing all possible mechanical devices reaching elementary gear diameters down to a few hundred microns using photolithography and the sacrificial layer technology [3] (see also chapter "Photo-Microlithography Fabrication of the Parts of a Micro-Mechanical Calculator" of this volume). It was completed in 1980 by the invention of the LIGA technique with its companion molding technique to fabricate robust metallic micro-gears [4].

The next miniaturization era started using electron and focused ion beam nanolithography (see chapter "Fabricating Solid State Gears at the Nanoscale: A Top–Down Approach" of this volume) to reach in our days gear diameters down to 30 nm [5]. This opens the race for the nanofabrication of ultra-miniaturized mechanical calculators able to function at very high temperature and under intense neutrons and cosmic ray radiations better than their nanoelectronics solid-state counterparts [6, 7].

Continuing this miniaturization trend for mechanical artifacts, engineers may reach one day the limit where there will be not enough atoms in a gear for this gear to function. This limit also applies to our mechanical series: gears, worms, rods, sliders, pull-tabs, pendulum, ratchet, springs and motors. For physicists, a gearing effect would require some cohesion between the atoms entering in the composition of the gear. All those atoms are supposed to rotate together in a collective manner because they are supposed to belong to the same piece of matter called a gear [8, 9].

For physicists, there must be also some quantum decoherence for this minimum gear set of atoms for them together to abandon their quantum mechanics mechanical properties. This leads to a reversal of the miniaturization trend when starting from the atoms themselves and going up in the number of atoms entering in the composition (in the fabrication) of a mechanical artifact. One may simply ask the question which related to the so-called *monumentalization* approach: From the bottom, what is the starting number of atoms required to construct for example a functioning gear that is gears that are able to be assembled in a chain and on a surface to transfer a rotation along this gear train in a gearing like operation [10].

3 The Bottom-up Approach for Mechanics

This so-called monumentalization trend i.e. the finding along a bottom-up approach of the minimum number of atoms required for example to observed a gearing effect appears already in 1987 at the single molecule level [11]. At that time, single molecule to molecule gearing effects were called on board to explain why for example in an NH_3 molecular monolayer chemisorbed on a metallic surface like Ag(111), Cu(111), Ni(111) or Ru(0001), single NH_3 molecules ae still randomly rotating at low surface temperature and why it is not the case for example for a PF_3 molecule in a PF_3 monolayer [12]. The answer is that a molecular rotation mechanical hindering effect comes into play when increasing the molecular coverage density for PF_3 and not for NH_3 on those metallic surfaces [11]. It also creates mechanical disorders at the boundaries between well-ordered extended molecular domains [13, 14].

The invention of the scanning tunneling microscope (STM) in 1981 by G. Binning and H. Rohrer and the first STM image of an isolated molecule on Ag(111) recorded in 1987 by J. K. Gimzewki [15] opens the observation and the drive of the rotation of a single molecule alone. In 1998, a single O_2 molecule was step by step randomly rotated on an Ni(111) surface by inelastic tunneling effect [16]. Also in 1998, a larger decacyclene molecule with legs and pro-chiral on a surface was at wishes put in rotation driven by the heat of its Cu(111) surface [17]. A rack and pinion molecular machinery was soon after constructed in 2005 [18] and the first 1.2 nm in diameter functioning molecule-gear experimented in 2009 [19]. It brings a first answer to the above *monumentalization* question: a mechanical gearing effect (see chapter "Anchoring Molecular Rotors by on-Surface Synthesis and Transmission of Rotational Motion Between Molecule-Gears" of this volume) and by extension a single molecule-motor (see chapter "Design and Synthesis of a Nano-winch and Rotations of Adsorbed Molecules Induced by Tunneling Electrons" of this volume) do not need a lot of atoms to operate: the 2009 molecule-gear has less than 50 atoms

[18, 19]. Its quantum mechanics mechanical decoherence is induced by its adsorption on the supporting surface. The molecule-gear cohesion of its chemical structure is ensured by the covalent bonds because even on the surface, the electronic part of the complete molecule-gear quantum wave function is still coherent. This avoids the atoms of a molecule-gears to fall apart. Still, intramolecular flexibility brings a limit to the gearing effect like observed for a PF_3 molecule (see for example chapters "A Simple Example of a Molecule-Gear Train: PF3 Molecules on a Cu(111) Surface and Mechanical Transmission of Rotation for Molecule Gears and Solid-State Gears" of this volume).

Between the 1 nm range of the molecule-gears [19] and the 10 nm range of the smallest solid state nanogears [5], there is a large spectrum of macromolecular machinery long explored and optimized by Nature like macromolecular protein motors [20]. Starting in the mid 60's, it took about 25 years for molecular biochemists like Boyer [21], Mitchell [22] and Walker [23] (to name a few) to demonstrate that the ATP synthase enzyme is a marvelous macromolecular motor about 10 nm in diameter [24]. K. Kinoshita succeed in 1997 to observed the rotation of a single ATP synthase macromolecular assemblage [25] and C. Montemagno started in 2000 the engineering part by grafting a rigid rod on a genetically modified ATP motor [26]. Natural evolution had even found a way to incorporate in insects solid state micro-gears with a diameter of about 200 microns [27].

Coming from the end of the 70's there is the in-solution molecular mechanics approach where the molecular machinery cohesion is ensured by covalent bond. Here, the mechanical quantum decoherence is caused by the billions of solvent molecules where the active molecular machines are diluted in. Gearing effect were first explored for example at the end of the 1970's by Oki [28] and Mitsov [29] in their famous triptycene dimers propeller studies opening the path for the 1999 R. T. Kelly molecular ratchet effect studied in solution [30]. Functioning in solution, the 1999 B. Feringa molecular motors and its improvements [31] are also finding their origin in the Oki and Mitsov chemistry atropisomerism studies [28, 29] (see also chapters "Prototypes of Molecular Gears with an Organo-Metallic Piano-Stool Architecture and Chemical Anchoring of Molecular Rotors" of this volume).

For insuring the stability and also the cohesion of a molecular machinery in solution, weak electrostatic, hydrogen bond and/or van der Walls forces can also be used. Those forces are at the basis of rotaxane and catenate based supramolecules invented in the 90s notably Stoddart [32] and Sauvage [33] (see chapters "Motion and Nanomechanical Effects in Supramolecular Catalysts and Five Minutes in the Life of a Molecular Shuttle: Near-Equilibrium Measurements of Shuttling Dynamics Using Optical Tweezers" of this volume). Those supramolecular machinery were developed in parallel with the covalent approach mentioned just above. They found their origin in the search on one hand for molecular hysteresis with applications in molecular information processing [30] and on the other hand on the mastering of a linear mechanical motion in a supramolecular assemblage [32, 34]. In a bottom-up perspective, they can be positioned between the 1 nm single molecule and the 10 nm in scale bio-macromolecular machinery. At the end of the 90s, Leigh comes into play to fill up this 1 nm and 10 nm gap with more and more complex supramolecular machinery working in solution [35].

4 Conclusion

All scales up and down have now been explored to construct miniature and functioning mechanical devices. The top-down approach is coming a long way from the Greek civilization and their 30 cm gears to our actual nanolithography fabricated 30 nm solid state nanogears. Moving up along the bottom-up approach, one can stop exactly at the bottom playing with 1 nm in size single molecule-gears and motors on a surface. In solution, molecular motors are functioning but statistically. Going up in scale, the 10 nm scale has its biomolecular machinery of all kinds particularly well adapted to sustain living organisms from the molecular scale. After this introductory chapter, this volume 13 of the series "Advances in Atom and Single Molecule machines" is presenting in the first 3 chapters how the top down approach is looking at the molecular scale. Then, the reader will enter in the 8 chapters of the single molecule approach diversity overlapping in the last 2 chapters with the molecular machinery approach for mechanics and catalysis.

Acknowledgements We thank the European Union Horizon 2020 FET open project "Mechanics with Molecule(s)" (MEMO, grant 766864) for financial support.

References

1. Freeth, T., Jones, A., Steele, J.M., Bitsakis, Y.: Calendars with Olympiad display and eclipse prediction on the Antikythera mechanism. Nature **454**, 614–617 (2008)
2. Petersen, K.E.: Micromechanical light modulator array fabricated on silicon. Appl. Phys. Lett. **31**, 521–524 (1977)
3. Kovacs, G.T.A., Maluf, N.I., Petersen, K.E.: Bulk micromachining of silicon. Proc. IEEE **86**, 1530–1551 (1998)
4. Becker, E. W.; Ehrfeld, W.; Münchmeyer, D.; Betz, H.; Heuberger, A.; Pongratz, S.; Glashaser, W., Michel, H.J., Siemens, R.: Production of separation-nozzle systems for uranium enrichment by a combination of x-ray lithography and galvanoplastics. Naturwissenschaften **69**, 520–523 (1982)
5. Yang, J., Deng, J., Troadec, C., Ondarcuhu, T., Joachim, C.: Solid-state SiO_2 nano-gears AFM tip manipulation on HOPG. Nanotechnology **25**, 465305 (2014)
6. Bradley, K.C.: Mechanical computing in microelectromechanical systems (MEMS). Ph.D. Thesis AFIT/GE/ENG/03-04, Department of the Air Force, Air University
7. Merkle, R.C.: Two types of mechanical reversible logic. Nanotechnology **4**, 114–131 (1993)
8. Deng, J., Troadec, C. Ample F., Joachim, C.: Fabrication and manipulation of solid state SiO_2 nano-gears on a gold surface. Nanotechnology **22**, 275307 (2011)
9. Lin, H.H., Croy, A., Gutierrez, R., Joachim, C., Cuniberti, G.: Mechanical transmission of rotational motion between molecular-scale gears. Phys. Rev. Appl. **13**, 034024 (2020)

10. Zhao, R., Qi, F., Zhao, Y.-L., Hermann, K.E., Zhang, R.-Q., Van Hove, M.A.: Interlocking molecular gear chains built on surface. J. Phys. Chem. Lett. **9**, 2611–2619 (2018)
11. Alvey, M.D., Yates Jr., J.T., Uram, K.J.: The direct observation of hindered rotation of a chemisorbed molecule: PF3 on Ni(111). J. Chem. Phys. **87**, 7221 (1987)
12. Zwanzig, R.: Rotational relaxation in a gear network. J. Chem. Phys. **87**, 4870 (1987)
13. Kerkart, M., Woodrufft, D.P., Avilat, J., Asensioi, M.C., Fernindez-Garcia, M., Conesas, J.C.: Structure determination for PF3 absorption on Ni(111). J. Phys.: Condens. Matter **4**, 6509–6522 (1992)
14. Jones, R.G., Abrams, N.E., Jackson, G.J., Booth, N.A., Butterfield, M.T., Cowi, B.C.C., Woodruff, D.P., Crapper, M.D.: The structure of PF3 adsorbed on Cu(111). Surf. Sci. **414**, 396–408 (1998)
15. Gimzewski, J.K., Stoll, E.P., Schlittler, R.R.: Scanning tunnelling microscopy on individual molecules of copper phthalocyanine adsorbed on polycrystalline silver surfaces. Surf. Sci. **181**, 267 (1987)
16. Stipe, B.C., Rezai, M.A., Ho, W.: Inducing and viewing the rotational motion of a single molecule. Science **279**, 1907–1909 (1998)
17. Gimzewski, J.K., Joachim, C., Schlittler, R.R., Langlais, V., Tang, H., Johanson, J.: Rotation of a single molecule within a supramolecular bearing. Science **281**, 531–533 (1998)
18. Chiaravalloti, F., Gross, L., Rieder, K.H., Stojkovic, S., Gourdon, A., Joachim, C., Moresco, F.: A rack and pinion device at the molecular scale. Nat. Mater. **6**, 30–33 (2007)
19. Manzano, C., Soe, W.-H., Wong, H.S.J., Ample, F., Gourdon, A., Chandrasekhar, N., Joachim, C.: Step by step rotation of a molecule-gear mounted on an atomic-scale axis. Nat. Mater. **8**, 576–579 (2009)
20. Kozlov, I.A., Skulachev, V.P.: H + -Adenosine triphospatase and membrane energy coupling. Biochem. Biophys. Acta. **463**, 29–89 (1977)
21. Boyer, P.D., Cross, R.L., Momsen, W.: A new concept for energy coupling in oxidative phosphorylation based on a molecular explanation of the oxygen exchange reactions. Proc. Nat. Acad. Sci. USA **70**, 2837–2839 (1973)
22. Mitchell, P.: Coupling of phosphorylation to electron and hydrogen transfer by a chemical-osmotic type of mechanism. Nature **191**, 144–148 (1961)
23. Walker, J.E., Saraste, M., Gay, N.J.: The UNC operon: nucleotide sequence, regulation and structure of ATP-synthase. Biochem. Biophys. Acta. **768**, 164–200 (1894)
24. Boyer, P.D.: The ATP synthase— a splendid molecular machine. Annu. Rev. Biochem. **66**, 717–749 (1997)
25. Noji, H., Yasudat, R., Yoshida, M., Kinosita, K.: Direct observation of the rotation of F1-ATPase. Nature **386**, 299–302 (1997)
26. Soong, R.K., Bachand, G.D., Neves, H.P., Olkhovets, A.G., Craighead, H.G., Montemagno, C.D.: Powering an inorganic nanodevice with a biomolecular motor. Science **290**, 1555–1558 (2000)
27. Malcolm, Burrows M., Sutton, G.: Interacting gears synchronize propulsive leg movements in a jumping insect. Science **341**, 1254–1256 (2013)
28. Oki, M.: Unusually high barriers to rotation involving the tetrahedral carbon atom. Angew. Chem. Int. Ed. Engl. **15**, 87–93 (1976)
29. Iwamura, H., Mislow, K.: Stereomechanical consequence of dynamic gearing. Acc. Chem. Res. **21**, 175–182 (1988)
30. Kelly, T.R., De Silva, H., Silva, R.A.: Unidirectional rotary motion in a molecular system. Nature **401**, 150–152 (1999)
31. Koumura, N., Zijlstra, R.W.J., van Delden, R.A., Harada, N., Feringa, B.L.: Light-driven monodirectional molecular rotor. Nature **401**, 152–155 (1999)
32. Anelli, P.L., Spencer, N., Stoddart, J.F.: A molecular shuttle. J. Am. Chem. Soc. **113**, 5131–5133 (1991)
33. Livoreil, A., Dietrich-Buchecker, C.O., Sauvage, J.P.: Electrochemically triggerred swinging of a [2]-catenate. J. Am. Chem. Soc. **116**, 9399–9401 (1994)

34. Ballardini, R., Balzani, V., Gandolfi, M.T., Prodi, L., Venturi, M., Philp, D., Ricketts, G.H., Stoddart, F.: A photochemically driven molecular machine. Angew. Chem. Int. Ed. Engl. **32**, 1301–1303 (1993)
35. Erbas-Cakmak, S., Leigh, D.A., McTernan, C.T., Nussbaumer, A.L.: Artificial molecular machines. Chem. Rev. **115**, 10081–10206 (2015)

From the Pascaline to a 5-Digits Metallic and Planar Miniature Mechanical Calculator 1 cm in Lateral Size

Pierre Abheillou, Bertrand Gatti, Kevin Froissard, Nicolas Joachim, and Christian Joachim

Abstract It is demonstrated how the mechanism of a mechanical calculator can be reduced to a 2 levels planar machinery. The initial 1790 J. Auch planar design was transformed and the mechanical parts of this new design were 3D printed reducing 1:5 the active surface occupied by the 1790 original. To avoid back-carry propagations, the ratchet springs are still metallic. In a next miniaturization step, modern metal machining techniques were used to reach a 4 mm thick planar version, 1:100 in surface of the 1790 original. This open the way to further miniaturization steps of our unique design using modern micro and nanolithography techniques.

Keywords Mechanical calculator · Planar mechanics · 3D printing · Electro-erosion

1 Introduction

Building small for mechanics started its top down approach more than 2000 years ago, its bottom-up approach about 30 years ago and Nature had succeeded to explore the small for mechanics starting at the 10 nm scale. In this chapter, we will respect the spirit of the top-down miniaturization approach by giving the example of the step by step miniaturization of a mechanical calculator using tools that one can almost found at home from Lego bricks to 3D printing and machining techniques. The Hublot clock maker had demonstrated how to faithfully miniaturized 1/10 the Antikythera analogue clock with its 495 mechanical pieces astronomical clock [1]. We are following here the same experimental path with the miniaturization of the

P. Abheillou · B. Gatti · K. Froissard · N. Joachim · C. Joachim (✉)
Centre d'Elaboration de Matériaux et d'Études Structurales (CEMES), Centre National de la Recherche Scientifique (CNRS), Université de Toulouse, 29 Rue J. Marvig, BP 4347, 31055 Toulouse Cedex, France
e-mail: christian.joachim@cemes.fr

C. Joachim
International Center for Materials Nanoarchitectonics (WPI-MANA), National Institute for Material Sciences (NIMS), 1-1 Namiki, Tsukuba, Ibaraki 305-0044, Japan

C. Joachim (ed.), *Building and Probing Small for Mechanics*, Advances in Atom and Single Molecule Machines, https://doi.org/10.1007/978-3-030-56777-4_2

1642 Blaise Pascal mechanical calculator. In Sect. 2, we will used 1230 Lego bricks to first construct for training an exact replica of this calculator. In Sect. 3, we will start with the 1790 Jacob Auch planar version of the Pascaline to step by step miniaturize it 1:5 with a new design using 3D printing technology. In Sect. 4, our new design of the Auch calculator will be constructed 1:100 fully in metal using modern electro-erosion and laser machining technologies with mechanical beams 100 μm in width. It calculates with 10 teeth gears down to 2.8 mm in diameter. In conclusion, we open the next step of miniaturization of a mechanical calculator using optical, e-beams and focused ions beam lithography techniques described in the chapters "Photo-Microlithography Fabrication of the Parts of a Micro-Mechanical Calculator" and "Fabricating Solid State Gears at the Nanoscale: A Top–Down Approach" of this volume before going to the molecular scale for the 10 remaining chapters.

2 The Pascaline in Lego Bricks

A good example of a complex mechanical machine, possibly to be miniaturized down to molecular scale is a mechanical calculator such as the Pascaline. This volume describes the use of molecule-gears and motors in line with the French philosopher Blaise Pascal work in 1642: the first ever mechanical calculator which can be operated many times (Fig. 1a). The Pascaline is a true calculator meaning that it can perform a

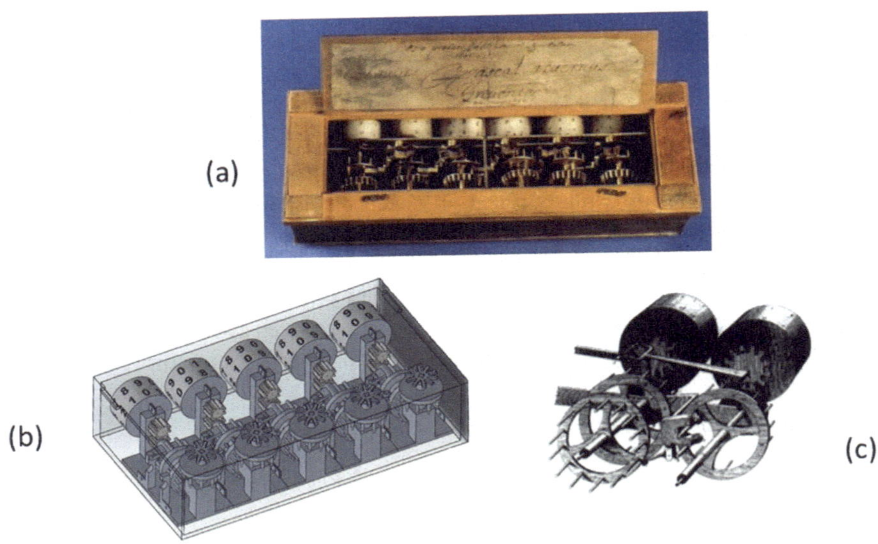

Fig. 1 **a** The 6-digits 13 cm × 36 cm × 80 cm original Blaise Pascal calculating machine. **b** Details of the mechanism for 5-digits. **c** Per digit: 5 toothed wheels, 3 rotation axles mechanical system made of a swinging fork, a jumper and a pawl leading to 56 pieces per digit (**c** from Robert Bénard "*Machine arithmétique de Pascal*", Encyclopedia Diderot and d'Alembert)

carry operation automatically and propagate the carry along the total number of digits cascaded in the machine. The data are input using a mini stylus. Along this mechanism and a few centimeters in diameter, each toothed wheel moved only forward by one unit when the previous wheel had completed its own revolution of ten (Fig. 1b). The Pascaline is essentially a 3-dimensional machine with at least 4 levels of mechanical machinery. It is using the gravity phenomenon to propagate the carry using a swinging fork system per digit to arm the carry completed by a jumper and a pawl (Fig. 1c).

To appreciate the complexity of this machinery and prepare its miniaturization, we have first constructed a Lego brick version of the Pascaline (Fig. 2a). The interesting point about this Lego construction is the possibility to study the carry propagation with the objective to design a planar version of this mechanical calculator. Planarization is one condition to be able to miniaturize a mechanical calculator down to the molecular scale. The other intriguing property of the Blaise Pascal design is the use of gravity to shape and propagate the carry. Lego bricks are so light that we were confronted by the inefficiency of the push-down for the carry per digit after the engagement of the swinging fork (Fig. 2b). At the macroscopic scale, this was easy compensated but gravity cannot be used at the molecular scale and even at the micron scale. Another obstacle of the Blaise Pascal design is the number of active mechanical layers per digit in its mechanism. Depending on how the motion for the

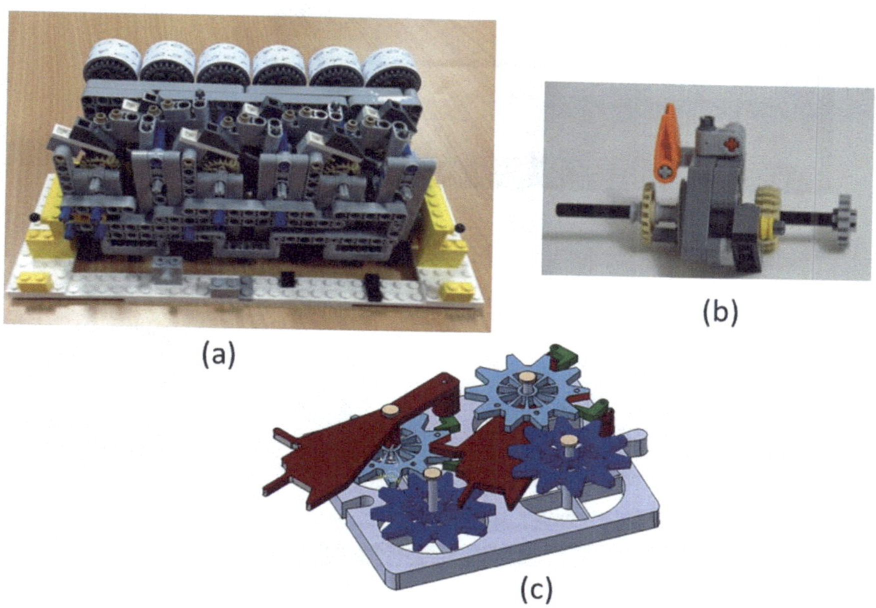

Fig. 2 **a** The original 6-digits Pascaline (Fig. 1a) constructed with more than 1000 Lego bricks. **b** the detail of the carry mechanism. Adapted for this Leo construction. **c** A CAD tentative of the planarization a single digit stage with a 4 mechanical levels version of the Pascaline mechanics. This single stage can be cascaded as presented with the lateral left pending brown ratchet. The metallic springs are not represented

carry is decomposed, it is between 4 and 5 levels of mechanics. This is really too large for a miniaturization in a planar form even if such a planarization can be computer aid designed (CAD) with modern software tools (Fig. 2c). Gravity is not used in this planarization which requires the positioning of 2 springs per digit.

For building a miniature mechanical calculator, it is therefore necessary to reduce the number of mechanical layers of this starting Pascaline design. After the Blaise Pascal mechanical calculator came the famous 1685 G. Leibnitz and the less known 1774 P.M. Hahn designs. Both are quite massive with also a large number of active mechanical layers for their operation because the multiplication and the division function were latter added to the functionality of those calculators. It turns out that Jacob Auch, a formal assistant of the P.M. Hahn in its clock maker workshop, proposed in 1790 a planar version of the Blaise Pascal calculator (Fig. 3a). Jacob Auch succeeded to reduce the number of active mechanical layers to 3 and to avoid gravity using instead springs and a unique slider to propagate the carry from digit to digit (Fig. 3b). The overall machine is 1.9 cm thick (Fig. 3c), a quite remarkable result in term of clock maker know-how. This was our starting point for mechanical calculator planarization and miniaturization.

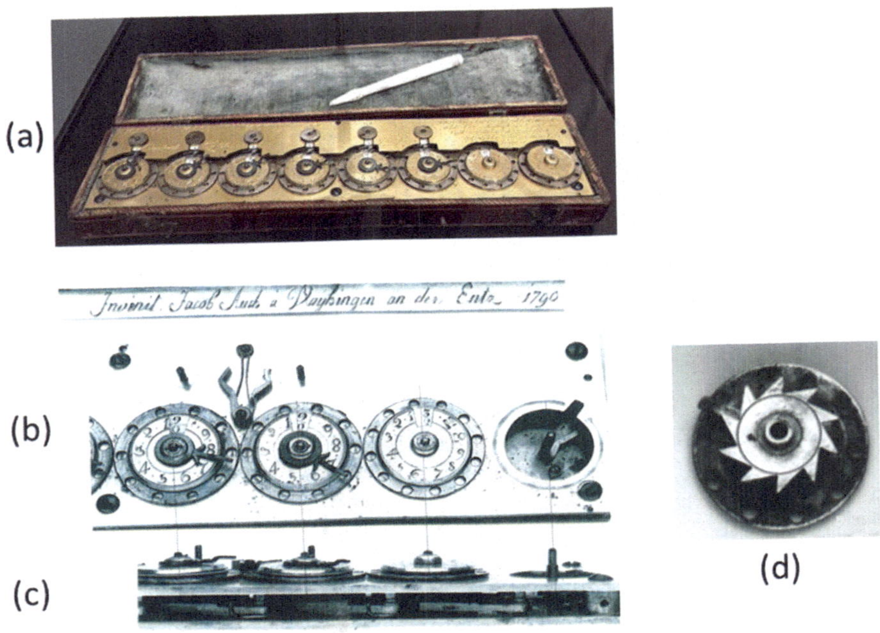

Fig. 3 **a** The 8-digits 1.9 cm × 6.1 cm × 22.8 cm Jacob Auch calculating machine. **b** A detail top view of 4 digits with one dismantled. **c** A cross-section of the machine with its 3 levels of mechanics, 1.9 cm in thickness plate to plate. **d** The dismantled toothed wheel with 10 asymmetric teeth (Thanks to the Dresden Science Museum)

3 A Very Planar and Miniaturized 3D Printed Version of the Jacob Auch Calculator

The miniaturization of a mechanical calculator down to the molecular scale imposes a lot of constrains: the planarity of the machinery support, the positioning and anchoring of the rotation axles of the molecule-gears, the distribution of the carry among the molecular digits, the elasticity for internal molecular degrees of freedom, and the softness of a molecule. All those concerns are also true at the nanoscale where solid state nanogears (for example 100 nm in diameter, see chapter "Fabricating Solid State Gears at the Nanoscale: A Top–Down Approach" of this volume) must be made of a very cohesive material for the classical laws of mechanics be applicable to the ensemble of atoms constitutive of this 100 nm nanogear. For example, a minimum of atoms is required for a mechanical nanobeam to respect in torsion the classical Hookes law [2].

For miniaturization and planarization, important lessons from the past have to be learnt. There were many attempts along the 18th century and after to planarize and miniaturize the Pascaline. The Fig. 3 Auch design was our starting point because all the rotation axles of its mechanical machinery are oriented in the same direction, perpendicular to the calculator base support. This is exactly the configuration which can be mastered at the molecular scale where each atomic scale rotation axle is made of a single metallic ad-atom (see chapter "Transmission of Rotational Motion Between Molecule-Gears" of this volume) or of a single dangling bond (see chapters "Prototypes of Molecular Gears with an Organo-Metallic Piano-Stool Architecture, "Design and Synthesis of a Nano-winch", "Chemical Anchoring of Molecular Rotors" and Anchoring Molecular Rotors by on-Surface Synthesis" of this volume) also perpendicular to the supporting surface.

Previous to the 1790 Jacob Auch flat mechanical calculator, other planar mechanical calculators were proposed to reach portability and because the miniaturization of clocks and watches was progressing quite fast during this period. The 1664 S. Morland mechanical adder was only 0.8 cm thick as compared to the 1.9 cm thickness of the Auch calculator. But its mechanics doesn't allow an automatic distribution of the carry which must be performed by hand like with an abacus. The 1780 C. Mahon (C. Stanhop) mechanical calculator was 2.2 cm thick with toothed wheels and a single cog per digit. But its carry operation was not robust because of the large force required to pass the carry from one digit to the next. Jumping over in time, the US Air Force attempts in 2003 to planarize a mechanical calculator had followed the same historical path than the one followed in this chapter [3]. In its nice PhD Thesis research work, Lieutenant K. C. Bradley jumps directly from the Pascaline to the Charles Babbage large mechanical calculator and then to the optical lithography sacrificial layer technology (see chapter "Fabricating Solid State Gears at the Nanoscale: A Top–Down Approach" of this volume) without trying first to improve the past planarization and miniaturization attempts started by Jacob Auch.

To start with and thanks to the Bonn Arithneum museum, we have first CAD designed and 3D printed on scale and in plastic all the mechanical pieces of the

1790 Auch mechanical calculator. We have assembled them and added the required metallic springs to learn how to add 2 numbers with this mechanism (Fig. 4). At each complete revolution of a toothed wheel, a rod presses on the rocker (in black in Fig. 4) which pivots on its axis while coming to bear on the built-in stopper. Continuing to rotate with the same wheel and the rod pushes on the entire slider (in red in Fig. 4). Since the rocker pivots to ensure a carry propagation, Jacob Auch placed it in front of the next tooth of the second toothed wheel. It allows to propagate the carry to the next digit. As a consequence, for each revolution of a wheel (10 teeth), the next wheel passes one unit more. This carry system is achieved by means of pawls and springs. For each unit passed, the tooth pushes the pawl. The spring then brings the pawl against to the next tooth. It is the mechanism that allows an efficient carry propagation.

By assembling all the digits one after the other along the common slider, Jacob Auch made it possible to perform calculations with 7–8 digits numbers in base 10. The only limit being the mechanical resistance of the springs because the slider friction becomes very large and prevents the carry to propagates over. For our 3D printed

Fig. 4 A photography of our 3D printed J. Auch mechanical calculator respecting exactly the dimensions and the design of the 1790 J. Auch original machine presented in Fig. 3. Only 3 digits have been assembled for this version (Thanks to Ina Prinz, Arithmeum museum, Bonn)

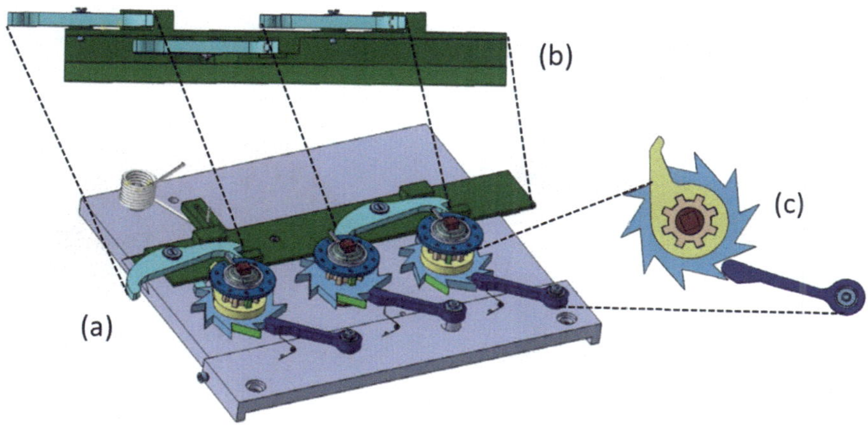

Fig. 5 **a** The CAD details of the 3D printed Auch machine assembled as presented in Fig. 4. All the plastic parts have been 3D printed and mounted one after the other. Only the 3 digits set-up with the corresponding 2 carry flip-flop mechanism is presented. **b** A cross-section of the carry slider with the exact positioning of the carry bascule (in light blue) per digit. **c** The counting toothed wheel and its ratchet system (dark blue). The spring per ratchet is visible on **a** but not represented on **c**. Notice on (**a**) top left, the very large spring to compensate for the friction of the dark green slider

polymer version, the resistance of the plastic materials to bending and wear come also into play. This first step towards miniaturization was essentially to appreciate the crucial role of the long slider to distribute the carry among the digits.

With its 4 levels of mechanics, with the long slider common to all the digits, and with the large number of springs (Fig. 5), this initial design is not very adapted for miniaturization down to the molecular scale and even to the intermediates solid state micro and nano scales. Molecular rods are very flexible and a common molecular rode distributing the carry among different molecular digits will not function properly unless the section of this rode is enlarged to approach the solid-state behaviors (see for example [4]). Therefore, and step by step, we have improved the Jacob Auch planar design to be compatible with the micro and nanolithography processes (see chapters "Photo-Microlithography Fabrication of the Parts of a Micro-Mechanical Calculator" and "Fabricating Solid State Gears at the Nanoscale: A Top–Down Approach" of this volume) and with the molecular scale mechanics know-how of our time (see all the chapters after chapter "Fabricating Solid State Gears at the Nanoscale: A Top–Down Approach").

The next step was to eliminate the slider and to rely only on rotations with the objective to go from many to 2 levels of mechanics per digit. As illustrated in Fig. 6, we were first forced to stage up and down the toothed wheels to be able to propagate the carry. By turning, each long tooth (yellow in Fig. 6) attached to its corresponding toothed-wheel causes the next toothed-wheel to rotate by one unit. At each complete revolution (corresponding to a 10), the yellow tooth carries one unit more the next digit and thus allows the carry to propagate. Notice that as compared to the Fig. 5

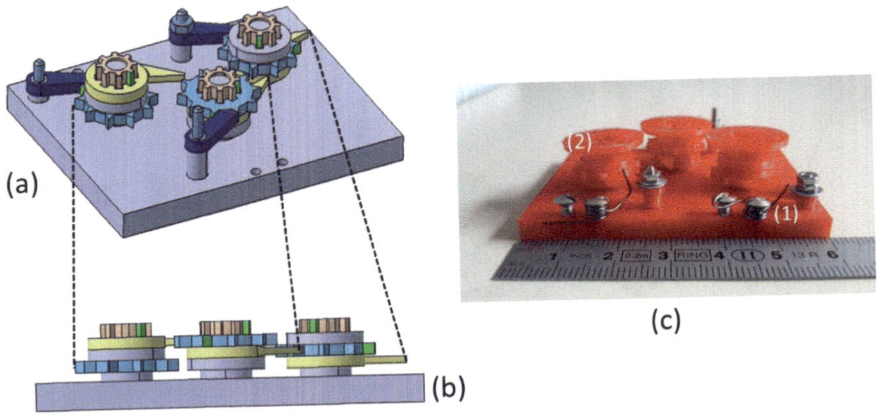

Fig. 6 **a** The CAD details of our new 3D printed mechanical calculator assembled as presented in **c**. All the plastic parts have been 3D printed and mounted one after the other. Only a 3 digits set-up with the corresponding 2 carry flip-flop mechanism is presented but can be cascaded further. **b** A cross-section of our mechanism showing how the carry can propagate by the different levels assume by the yellow carry teeth attached to the counting toothed-wheel per digit (in light blue). A grey circular spacer per digit avoid the interaction of the yellow carry teeth with the next counting toothed well. **c** The photography of the assembled 1 cm × 5 cm × 6 cm 3-digits calculator in base 10 with its metallic springs (1)

design, the toothed-wheels have all their teeth symmetric. It is now the long carry tooth which is providing the ratchet effect essential for the carry operation to function. As in Fig. 5, the pawls are always there to avoid back-rotations. But their geometries have been modified so that they match the contour of the teeth and the space between them. The level alternation presented in Fig. 6b between the toothed-wheel, the rod and the spacer (in gray per digit) allows the rod to touch only the next toothed-wheel to propagate the carry. Indeed, when the yellow long carry tooth passes the next digit toothed-wheel, it can continue its rotation without touching it. It only faces the grey spacer which lets it pass without mechanical interaction. The 3D printed Fig. 6c mechanical calculator is perfectly functional.

The Fig. 6 design is very specific because each digit is positioned at different levels relative to the calculator base to avoid the blocking of the machine by a back-carry effect. Nevertheless, it can be adapted to atomic scale stepped surface with for example one molecule-gear positioned at the border of each terrace interacting mechanically (Van der Waals or repulsion forces) with the molecule-gears of the next and previous top and down terraces using molecule-gears with one tooth longer than the others. In our Fig. 6 calculator, each gear is 15 mm in diameter (external diameter) and some parts are still metallic to insure a reliable mechanical functioning and a long operation duration as compared to plastic beams. On the way to planarization, there are only 3 levels of active mechanics in this machinery and 10 mechanical pieces per digit. Compared to the Fig. 4 Jacob Auch calculator, our calculator is 4 times smaller in lateral size.

The final step of our design was to reduce to 2 the number of active mechanical levels. This reduction is very well adapted to miniaturization down to the molecular scale because of the well-known double decker molecules having potentially 2 active molecular levels (See chapter "Prototypes of Molecular Gears with an Organo-Metallic Piano-Stool Architecture" of this volume). As presented in Fig. 7, a 5-digits mechanical calculator can be designed and assembled part by part with only 2 active levels of mechanics. Having also the mechanical beams made of the same material than the other parts of the machine is a very important indication that all the pieces of a miniature mechanical calculator can be fabricated out of the same material. This is crucial in the next two chapters for micro and nanolithography where mounting by hand or under a microscope for example a miniature metallic spring becomes impossible as soon as miniaturization is going on.

Our Fig. 7 calculator is strictly a 2 levels mechanics machinery. There is no need here to stage up and down the different digits. For this purpose, the long orange carry tooth per digit in Fig. 7a finds its length progressively reduced from digit to digit with a simple up and down alternation in its height. This design was not explored in the 19th century because the spatial extension of a digit was a concern for portability. With miniaturization, this is no more a limitation.

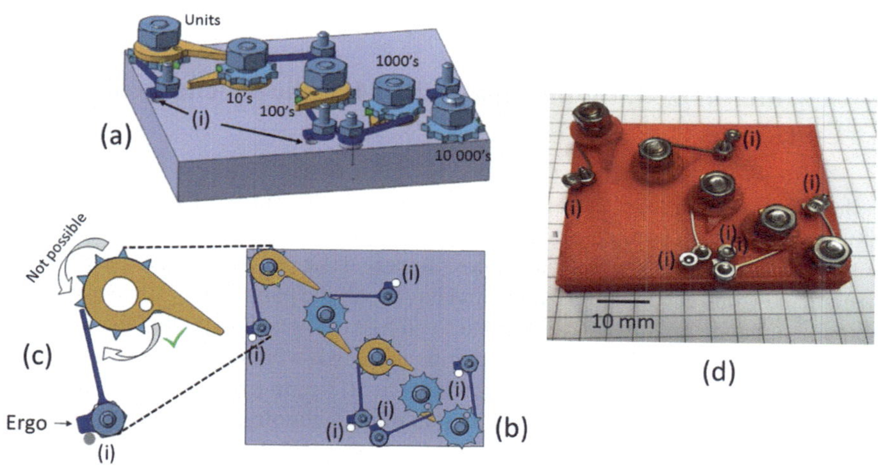

Fig. 7 **a** The CAD details of our final 3D printed mechanical calculator assembled as presented in **d**. All the plastic parts have been 3D printed and mounted one after the other. A 5 digits set-up with the corresponding 4 carry mechanism is presented. **b** A top view of our mechanism showing how the carry can propagate by progressively reducing the distance between the digits and keeping an alternating height of the orange long carry teeth per digit attached to the counting toothed-wheel per digit (in light blue). **c** An enlargement of a single digit. An ergo was added (fixed by a metallic screw (**i**) on the (**d**) final set-up) to block a possible back rotation of the carry. **d** The photography of the assembled 0.8 cm × 5.5 cm × 6 cm 5-digits calculator in base 10 with its metallic beams. **i** Are the little screws blocking the back rotation of the beams which are now in a planar configuration. To compare with the Fig. 6 3-digits calculator, a 3-digits calculator will 0.8 cm × 2.5 cm × 3 cm in dimension

All the mechanical pieces of this miniature and planar mechanical calculator where 3D printed like in the above previous presented designs. Only the springs remain metallic because of the too large flexibility of the plastic mechanical beams when reducing the over size of the machine. With its 10 holes, a small input disk was also mounted on the rotation axle of each digit to allow the input of the numbers to be added using a simple sharp metallic pin. Even if our about 10 times smaller than the Jacob Auch design is working perfectly, it is facing large friction effects. The diameter of each toothed-wheel per digit is becoming so small that the torque developed by one under an ultrathin pin input becomes too small to propagate the carry and to fight against the surface friction. One has to de-tight the holding screw of each digit to ease this propagation. It often leads to a slippery of the carry and also to a backward rotation of the calculating toothed-wheel.

The Fig. 7 design is at the limit of plastic 3D printing technology. Furthermore, the tip end of the orange long carry-tooth per digit and also the tip end of each teeth of the calculating toothed-wheels are becoming so sharp by design (ideally less than a few hundred-microns radius of curvature) that after a few operations, they are becoming severely round and not operative any more. This also limits the number of digits that our Fig. 7 plastic machine can handle by a systematic cascading of the digits after a rescaling of their spacing on the calculator base for each new digit to be added.

4 A Miniaturized, Planar and Fully Metallic Mechanical Calculator

In the previous section, planarization and miniaturization of the Jacob Auch calculator was based on a complete step by step redesign of the elementary digits leading to the Fig. 7 very pocket like calculator, already 1:10 of the original Jacob Auch machine. To continue miniaturization and having all the parts of the calculator made from the same material, we have decided to go from 3D printing to the limit of metal machining technology. In our days, machine tools in a workshop complemented by electro-erosion and laser etching can go down to 10 μm in precision. Further down, the LIGA (photo-lithography plus metal moulding) technology [5] will take the lead for the next miniaturization step (see the next chapter of this volume). In a last step, we have kept the Fig. 7 design and machine tooled all the parts. In the Fig. 8 metallic calculator, the calculating toothed-wheels were fabricated by the electro-erosion process and the mechanical beams (with their ergo) were laser machined. The two essential parts of our design are presented in Fig. 9.

As presented in Fig. 8, we have decided to assembled the calculator on a standard UHV Omicron sample holder used in all the LT-UHV STM experiments presented in this volume. This miniature mechanical calculator is 1:100 the surface area of the 1790 Jacob Auch calculator. Its thickness is only 4 mm, 1:2 of the 1780 C. Mahon

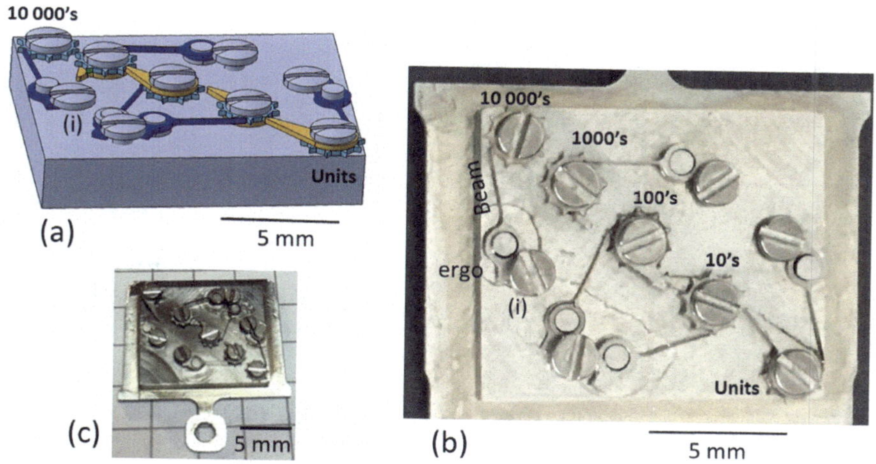

Fig. 8 **a** The CAD details of our full metallic mechanical calculator assembled as presented in **b**. All the parts have machined and mounted one after the other on a standard UHV omicron STM sample holder. **c** A 5 digits set-up with the corresponding 4 carry mechanism is presented. **b** A top view of our 0.4 cm × 1 cm × 1.4 cm 5-digits calculator mechanism showing how the carry can propagate by progressively reducing the distance between the digits and keeping an alternating height of the orange long carry teeth per digit (orange in **a**) attached to a counting toothed-wheel per digit (light blue in **a**). (Laser micro-machining by Micro Tolerie Dallard (France))

in thickness design. It is working perfectly with its 2.8 mm metal toothed-wheels. A very sharp needle operated under a binocular optical microscope is now required to input the numbers.

5 Conclusion

In this chapter, we have demonstrated how mechanical calculators can be reduced to 2 levels of mechanics with the rotation axle of the calculating toothed-wheels oriented perpendicular to the base of the calculator. The adaptation of the initial 1790 Jacob Auch planar design to the modern 3D printing technology opens the way to reduce by 1:5 the active mechanics surface area of the original Auch machine. After having tested the functioning of our miniature and planar mechanical calculator, modern metal machining fabrication techniques were used to furthermore miniaturized our calculator 1:100 of the original Auch which is working perfectly in its full metallic and planar version. This open the way to a further miniaturization of our unique design using modern micro and nanolithography techniques as proposed in the next 2 chapters of the volume before entering in the molecular scale for the rest of this volume.

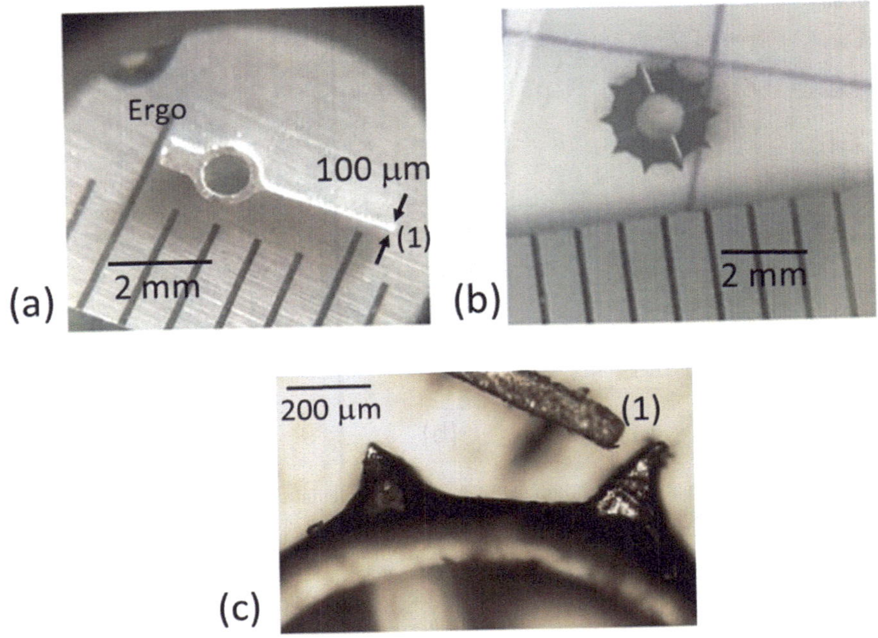

Fig. 9 The photography of **a** a metallic beam with its co-linear ergo and **b** a metallic toothed-wheel 2.8 mm in diameter (including the teeth length). **c** An optical microscope image of the contact between the end of the beam in **a** and one teeth of the toothed-wheel in **b**. Notice the irregularity of the teeth in **c** with an end tip radius of curvature approaching 10 µm

Acknowledgements We thank the European Union Horizon 2020 FET open project "Mechanics with Molecule(s)" (MEMO, grant 766864) for financial support.

References

1. See Hublot watch maker. www.hublot.com, Calibre-Hublot-Anticythere-2033-CH01 (2011)
2. Bombis, C., Ample, F., Meilke, J., Mannsburger, M., Villagomez, C., Roth, Ch., Joachim, C., Grill, L.: Mechanical behaviour of nanocrystalline NaCl islands on Cu(111). Phys. Rev. Lett. **104**, 185502 (2010)
3. Bradley, K.C.: Mechanical Computing in Microelectromechanical Systems (MEMS). Ph.D. Thesis AFIT/GE/ENG/03-04, Department of the Air Force, Air University (2003)
4. Merkle, R.C.: Two types of mechanical reversible logic. Nanotechnology **4**, 114–131 (1993)
5. Becker, E.W., Ehrfeld, W., Münchmeyer, D., Betz, H., Heuberger, A., Pongratz, S., Glasha ser, W., Michel, H.J., Siemens R.: Production of separation-nozzle systems for uranium enrichment by a combination of x-ray lithography and galvanoplastics. Naturwissenschaften 520–523 (1982)

Photo-Microlithography Fabrication of the Parts of a Micro-Mechanical Calculator

Christian Bourgerette, Laure Noé, Sebastien Pinaud, and Christian Joachim

Abstract A scanning photo-lithography process is developed to miniaturize mechanical calculators down to 40 μm in diameter for their calculating micro-gears. Our first moulding process span the dimensions from 1 mm to 60 μm for the micro-gears. Down to 100 μm, the planar calculator construction can still be based on a micro-manipulation of the moving parts under an optical microscope. Below and to reach the 10 μm, a double photo-lithography process was developed on a specific graphite/SiO_2/Si wafer for mastering surface frictions. After a baking at 120 °C, the photo-resist becomes the material constitutive of all the moving micromechanical pieces. Only the rotation micro-axles remain metallic to ensure their good anchoring to the surface. A 2-digits micro-calculator is fabricated. In base 10, the carry propagation is demonstrated.

Keywords Mechanical calculator · Microlithography · Micro-gears · Planar mechanics

1 Introduction

Miniaturizing a mechanical calculator below the millimetre scale requires to abandon the standard machining fabrication i.e. the mechanical removal of materials using miniature tools and go for lithography techniques event if micromachining [1] and nanomachining are progressing [2]. Following the previous chapter "From the Pascaline to a 5-Digits Metallic and Planar Miniature Mechanical Calculator 1 cm in Lateral Size" of this volume, gears rotating each on a robust axle, ratchets, pawls

C. Bourgerette · L. Noé · S. Pinaud · C. Joachim (✉)
Centre d'Elaboration de Matériaux et d'Études Structurales (CEMES), Centre National de la Recherche Scientifique (CNRS), Université de Toulouse, 29 Rue J. Marvig, BP 4347, 31055 Toulouse Cedex, France
e-mail: christian.joachim@cemes.fr

C. Joachim
International Center for Materials Nanoarchitectonics (WPI-MANA), National Institute for Material Sciences (NIMS), 1-1 Namiki, Tsukuba, Ibaraki 305-0044, Japan

C. Joachim (ed.), *Building and Probing Small for Mechanics*, Advances in Atom and Single Molecule Machines, https://doi.org/10.1007/978-3-030-56777-4_3

and long beams are all required at the micron scale to calculate mechanically and distribute automatically the carry among the digits encoded by the toothed-wheels rotation. Micromachining by photolithography is the actual technique able to explore micromechanics from the millimetre to the 10 μm scale [1]. After the deposition of a photo-resist on a support, this resit can be either decomposed or polymerized locally using UV light passing through a mask or scanned with an ultra-sharp beam light on the surface of the photo-resist. In the next step of the fabrication process, the decomposed or non-polymerized photo-resit is diluted and the support rinsed. The resulting on-surface patterns can be filled up with a material (generally a metal) or the resist-free part of the surface can also be etched (see chapter "Fabricating Solid State Gears at the Nanoscale: A Top–Down Approach" of this volume for more details). As presented in this chapter, the remaining polymerized photo-resist can also be baked to become the active material constituting the microfabricated mechanical device.

In Sect. 2, the photo-lithography and moulding techniques are used to fabricated micro-gears, micro-ratchets and pawls and micro-beams between 500 and 50 μm in scale. Thanks to our moulding technique, all those mechanical devices can be picked up from the fabrication support and positioned with a micrometre precision under an optical microscope. Below the 10 μm scale, it become difficult to mount those devices one after the others on their holding pads and rotation axles using for example small tweezers or magnetic tip under an optical microscope. Here, photolithography processes can be used to fabricate all the elementary mechanical devices directly in place on the surface of the support, almost ready to be activated mechanically for example by a very sharp glass micropipette. This process is presented in Sect. 3 where UV scanning photo-lithography is used to directly microfabricate by photo-polymerization all the required mechanical parts of an elementary mechanical calculator. The scanning photo-lithography is very important here to align the micro-gears with their on-surface pre-fabricated rotation axles without the use of micro-manipulation tools. This scanning photo-lithography technique was also selected because its scanning mask software is compatible with our 3D printer scanning software used in chapter "From the Pascaline to a 5-Digits Metallic and Planar Miniature Mechanical Calculator 1 cm in Lateral Size". This scanning technique has also the advantage to be more flexible than a static physical mask used for example in Sect. 3 because a soft-mask is stored in the beam light scan controlling computer and do not need to be physically re-fabricated after a new modification of the mechanical machinery design.

In Sects. 2 and 3, we have simplified the design of our mechanical calculator. As presented in Fig. 1, the ratchet effect is now occurring at the end of the elastic micro-beams. The carry propagates by using one tooth longer than the other per digit. It is pushing on the next micro-gear after a complete turn (by unit of tens) as presented in Fig. 1. It is a complete planar design with only one level of mechanics. Following the chapter "From the Pascaline to a 5-Digits Metallic and Planar Miniature Mechanical Calculator 1 cm in Lateral Size" results presented in this volume, this new design was tested using our 3D printed technology before entering in the photo-lithography processes presented in Sects. 2 and 3. In Sect. 3, we have even simplified again our

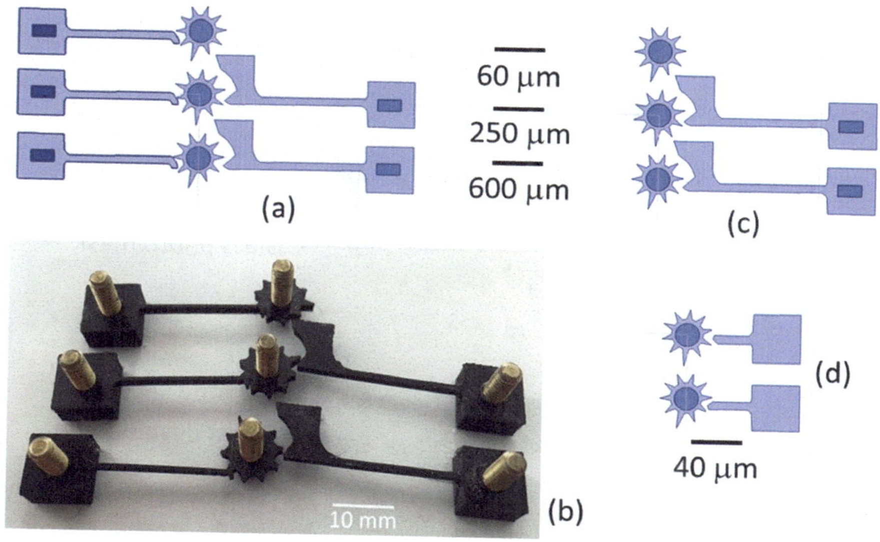

Fig. 1 The different one level of mechanics photo-lithographed in this chapter. **a** is the complete 3 digits mechanical calculator design with **b** the perfectly working millimetre scale 3D plastic printed machinery. **c** A simplified version of **a** with the 3 left ratchets suppressed as photo-lithographed and assembled under an optical microscope as presented in Sect. 2 supposing that the surface friction will guaranty a non-back rotation of the micro-gear after a carry operation. **d** An even over simplified version of **c** as photo-lithographed in Sect. 2. A first tentative of **c** is also presented in Sect. 2. The 3 different scale bars for microfabricated (**c**) are given above and the 40 μm scale is also indicated for (**d**). Those scales are corresponding to the micro-gear diameters not considering the long teeth per toothed wheel in charge of triggering the carry propagation

design of the ratchet with the prospect to test how a simple asymmetric tip at the end of each micro-beam will function.

In Sect. 4, a first study is presented on micro-beam elasticity in charge of creating the ratchet effect per digit. As already met in chapter "From the Pascaline to a 5-Digits Metallic and Planar Miniature Mechanical Calculator 1 cm in Lateral Size" with plastic 3D printed beams, we show how mechanical cracks rapidly occur after a few operations using the photo-resits as the micro-beam material even after a hard baking. In conclusion, we open the next chapter of this volume concerning nanolithography to fabricate the part of a meso-calculator with 10 teeth solid state nanogears with a diameter reaching 100 nm.

2 Micro-Molding of Gears, Ratchets and Planar Beams

The microfabrication process of the Fig. 1c mechanical parts is presented in Fig. 2. It is a standard moulding process involving an electro-deposition of nickel, a demoulding and the one by one deposition of all the mechanical pieces on another part of surface

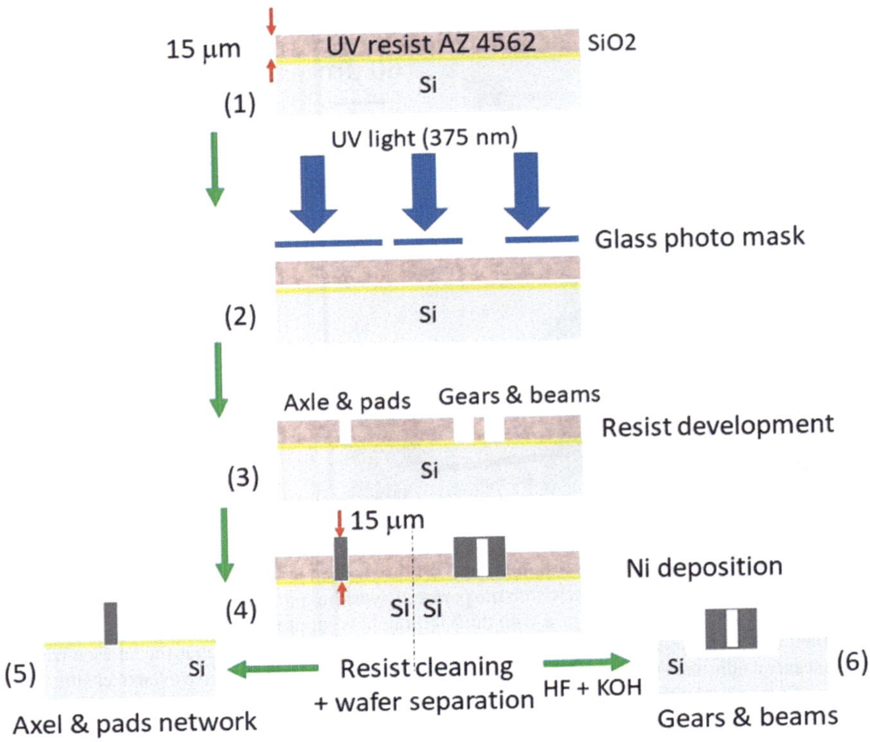

Fig. 2 Our micro-fabrication process from (1) the AZ 4562 UV photo-resist deposition to (5) and (6) the cutting of the 4 in. wafer in 2 cm in lateral size chips with the axles and clipping pads for the ratchets on one side (5) and on the other side (6) the 2 cm chips having all the Ni mechanical pieces which are detached from the SiO_2 chip by HF and deposited in a Petri box (see Figure below). The axles and pads in (5) are deeply anchored inside the Si surface

where the rotation axles and the holding pads are microfabricated in parallel. All those pieces were first release from their native SiO_2 surface by an active under etching of the Si surface [1] and deposited in a Petri box. In parallel, the other part of the native SiO_2 wafer surface is prepared where the micro-axles and the micro-pads are micro-fabricated well anchored on the wafer surface to avoid the problem faced by the moving axles as discussed in Sect. 3. This wafer can be very easily UHV clean for example by annealing at 120 °C during a week at 10–9 mbar if a compatibility with molecular scale UHV processes is required.

For the Fig. 2(2) step, the glass UV photo-mask Cr covered was designed and fabricated on purpose and is presented in Fig. 3. It is a standard glass UV mask where we have included all the micro-gears and ratchet-beams dimension of interest in this chapter as indicated in Fig. 1c. The rotation axles and the holding pads for the micro-beams are also part of this mask to UV exposed the photo-resist only once.

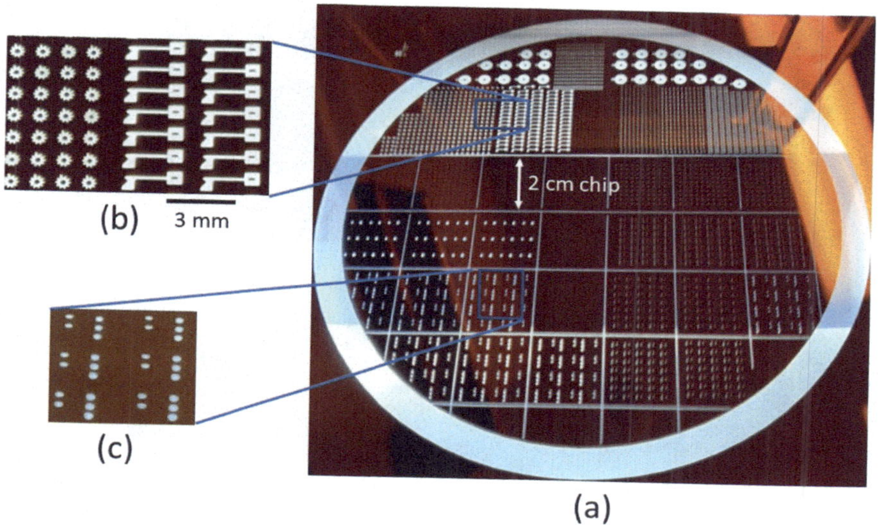

Fig. 3 **a** A photography of the 4 in. glass optical mask on purpose fabricated for all the mechanical pieces required for the 3 scales indicated in Fig. 1c: 60 μm, 250 μm and 600 μm for the micro-gears with respectively the corresponding 20 μm, 80 μm and 250 μm axle diameters and the corresponding micro-ratchet beams. **b** A zoom-in on the large gears and beams area. **c** A zoom-in on 6 networks of pads and axles where those large-scale micro-beams and toothed wheels will be mounted. The clipping micro-pads of those ratchet-beams are also defined according to the Fig. 1c design

To ease the micro-assembly of all the mechanical parts under an optical microscope and using a micro-manipulator, all the parts (including the rotation axle and the blocking micro-pads of the ratchet-beams) were microfabricated in nickel according the step (4) in the Fig. 2 process. Notice that all the nickel mechanical pieces moulded can also be UHV cleaned after their fabrication.

The Fig. 4 is presenting a series of optical microscopy photos of the micro-fabricated by moulding Ni mechanical pieces. According to the Fig. 1c design, three diameters of 10 teeth nickel micro-gears were fabricated: 60, 250 μm and 600 μm with respectively the corresponding 20, 80 and 250 μm axle diameters and the corresponding ratchet-beams (see Fig. 5).

In Fig. 5, a given network of micro-axles and micro-pads is presented ready to accept their micro-gears and ratchet-beams. There were also nickel micro-fabricated with a 15 μm height corresponding to the 15 μm micro-gears and ratchet-beams nickel material thickness. Their positioning on the surface is respecting the Fig. 1c design and are indicated in dark blue.

The picking and pre-positioning of the mechanical pieces on the axles and clipping pads SiO_2 wafer are presented in the Fig. 6 after the release of each of them from their original chip and storage in the Petri box. To be more precise in the final positioning and testing, one can beneficiate from used ultra-sharp STM tips of our LT-UHV STM

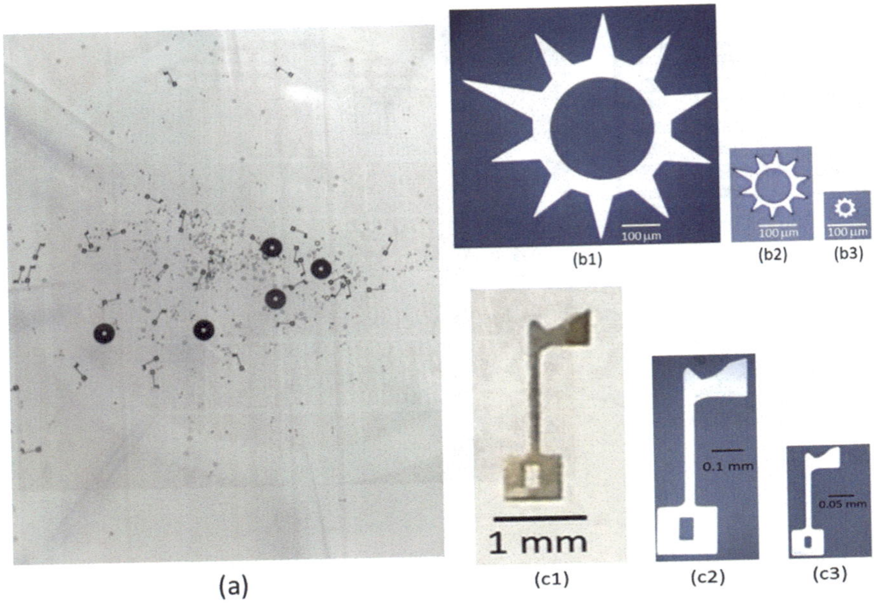

Fig. 4 **a** An optical photo of a many mechanical pieces of different size dispersed in a Petry box after their microfabrication and detachment from the Si wafer surface. **b** a zoom-in using an optical microscope showing at the same scale: (b1) a large toothed wheel (600 μm), (b2) a medium (250 μm) and (b3) a small one (60 μm), all with a single long carry tooth. **c** a zoom-in showing the corresponding ratchet-beam with their respective anchoring squared hole

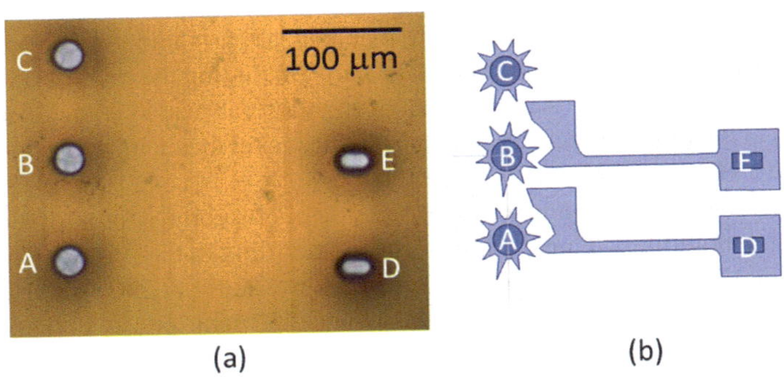

Fig. 5 **a** The optical photo of a network of 3 axles (A, B, C) and 2 pads (D, E) for mounting respectively the micro-gears and the ratchet-beams after their demoulding as presented in Fig. 4. **b** Reproduced here, the Fig. 1c design to clarify the positioning of the axles and pads. This network is for three 60 μm toothed wheels (Fig. 4b3) and their 2 companions (Fig. 4c3) ratchet beams

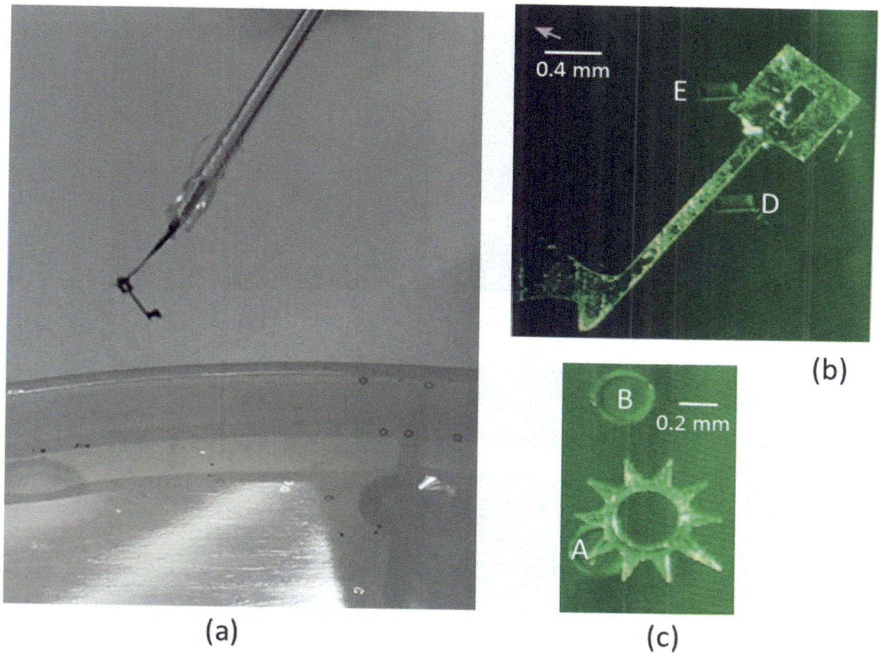

Fig. 6 **a** The picking up from the Petri box of a 2.3 mm long single Ni ratchet-beam piece using a micropipette equipped at the end by a metallic tip where a little magnet had been glued. Notice the five small 250 μm gears and a few small ratchet-beams remaining in the Petri box. **b** The (**a**) ratchet-beam pre-positioned near two clipping micro-pads. **c** A 600 μm micro-gear also prepositioned using the same technique near its Ni rotation axle A following the Fig. 5b design

instrument easy to be glued at the end of a glass micro-pipette. The micro-pipette was navigated on the wafer surface using a standard micro-manipulator.

Starting from Fig. 6c, the final positioning and rotation of this 600 μm micro-gear is presented in Fig. 7. Its locking on the A rotation axle was obtained by step by step manipulating the micro-gear with the tip. The Ni axle is very robust and the teeth of the micro-gear are not deformed during the rotation. A complete 5-stage planar analogue calculator is now under its final construction respecting the Fig. 1c design.

3 Direct Photo-Lithography of Micro-Gears and Ratchets

Presented in Fig. 8, the microfabrication process of the Fig. 1d mechanical parts has the objective to avoid the de-moulding and micro-manipulations steps of assembling the micro-mechanical calculator piece by piece when those pieces are miniaturized below about 50 μm but still resulting from a photo-lithography process. The idea developed in this section is that while polymerized under UV light, the photo-resist

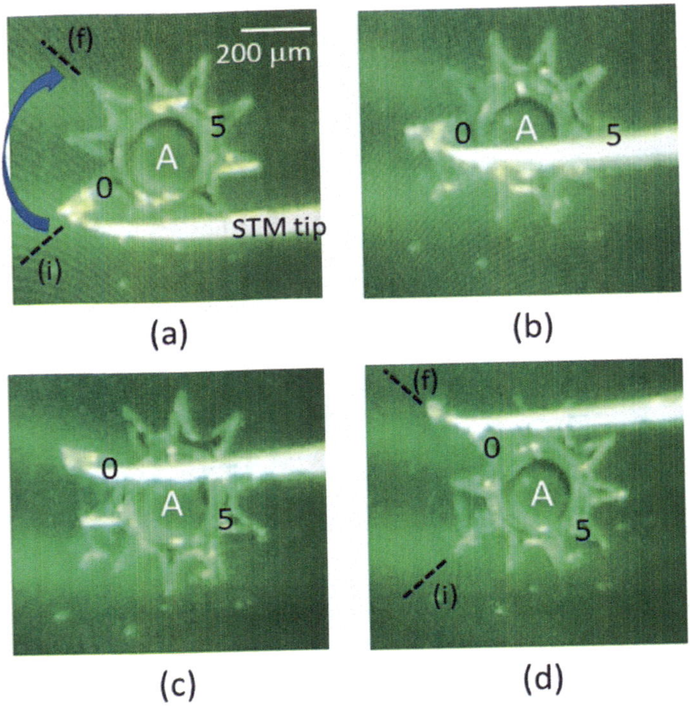

Fig. 7 Starting from the pre-positioning presented in Fig. 6, this 600 μm micro-gear was first **a** exactly engaged on it's a Ni axle using a used ultra-sharp STM tip coming from an LT-UHV STM instrument. Then, it was step by step rotated from **a** to **d** using the same used STM tip. The rotation angle from **i** to **f** is about 70° corresponding to entering the number "2" on a mechanical calculator

itself is robust enough and becomes the only materials of the toothed wheels and of the ratchet-beam. As compared to the Fig. 1c design, we have avoided the complex ratchet shape at the end of each micro-beam. We have also suppressed the holding pads of those beams supposing that a lateral squared extension at the opposite end of a micro-beam is enough, after polymerization to hold the beam in place.

Different molecular UV photo-resists were used to test their material cohesion after UV polymerization and to test the resulting elasticity of the micro-beams. They were all spin coated on the wafer (SiO_2 surface and after Graphene/SiO_2 surface) whose lateral size are then cut in chips whose lateral size varied from 1 to 2 cm. Hereafter, V2 is the acceleration (rpb/s), V the rotation speed (rmp) and T the complete rotation time of the spin coating process. To prepare the anchoring of the micro-gear axles, a positive AZ 5214 photo-resist was first used with V2 = 2000, V = 4000, T = 30 s and a 105 °C anealing temperature during 1 min leading to a 1.4 μm resist thickness. For the fabrication of micro-gears, the first negative photo-resist was AZnLoF with V2 = 3000, V = 3000, T = 30 s and a 100 °C annealing temperature during 1 mn

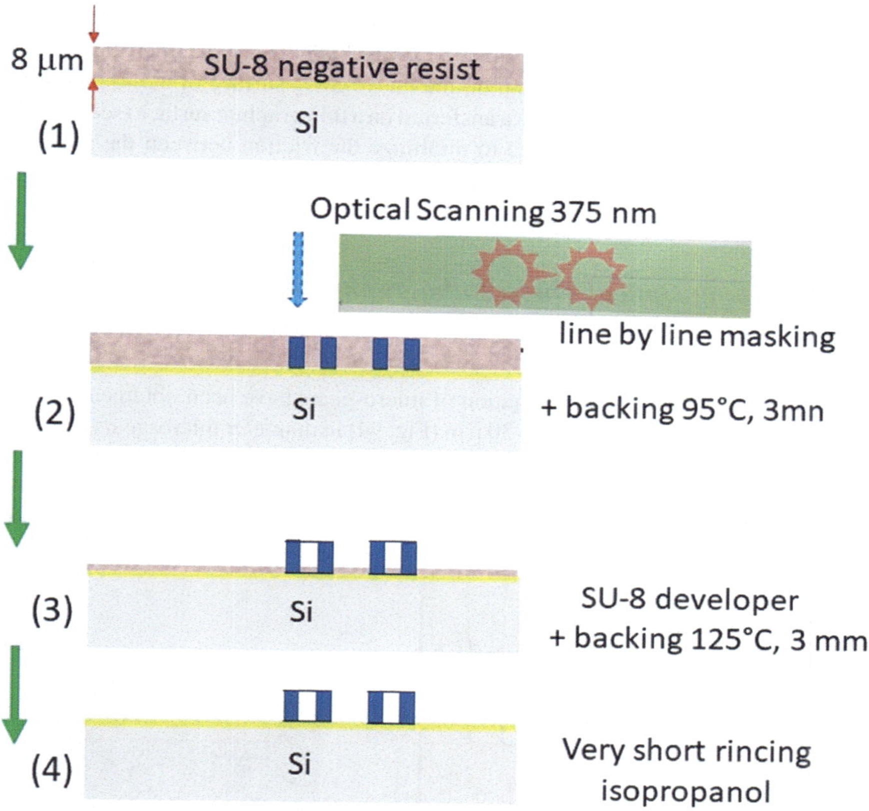

Fig. 8 The general steps of our photo-lithography process leading to transform the polymerized photo-resist at step (2) in the materials of the toothed wheels of the beams and of its holding micro-pads after step (3). The surface used was either SiO_2 on a Si wafer or the graphene surface transfer printed on this Si wafer (see below). The line by line scanning is per step on 2 mm on the DILASE 650 from KLOE

leading to a 6 µm resist thickness. Important to transform this polymerized resist in a robust material, the post-annealing temperature was 110 °C during 1 mn. The other photo-resist was the well known SU-8 [3] with V2 = 3500, V = 3000, T = 30 s and a 95 °C annealing temperature during 3 mn leading to a 8 µm resist thickness. The post-annealing temperature was here 95 °C during 3 mn. After revelation in acetone, the remaining part of the resist is also hard baken at 125 °C during 3 mm to insure the material cohesion of the micro-gears and of the micro-beams.

To avoid any assembling step under an optical microscope as demonstrated in the previous section, we have used a scanning optical lithography technique. This permits to perform a series of two consecutive insulations: one for the micro-fabrication of the rotation axles and one for the microfabrication of the micro-gears and of the micro-beams. We have use the DILASE 650 scanning photo-lithography instrument from KLOE. It is a direct laser writer with a 2 μm in diameter optical beam and with two possible working UV wavelength: 375 nm and 405 nm. All the microfabrication tests were performed in a cleanroom on the native SiO_2 surface of a 500 μm thick silicon wafer. Then, our process was transferred on a thin graphite surface (see below), contact printed on the SiO_2 surface to minimize the friction between the movable mechanical parts and the supporting surface.

3.1 Scanning Optical Micro-Lithography for Micro-Gears on SiO_2

Different sizes, shapes and combination of micro-gears have been obtained starting from very large 0.9 mm (Fig. 9a) to 30 μm (Fig. 9d) in diameter micro-gears overlapping well in size for comparison with the Fig. 4b micro-gears. One awaited difference is that after the final rinsing, it is very difficult on the SiO_2 surface to rotate, displace or picking up those micro-gears using for example a glass micropipette as a pusher.

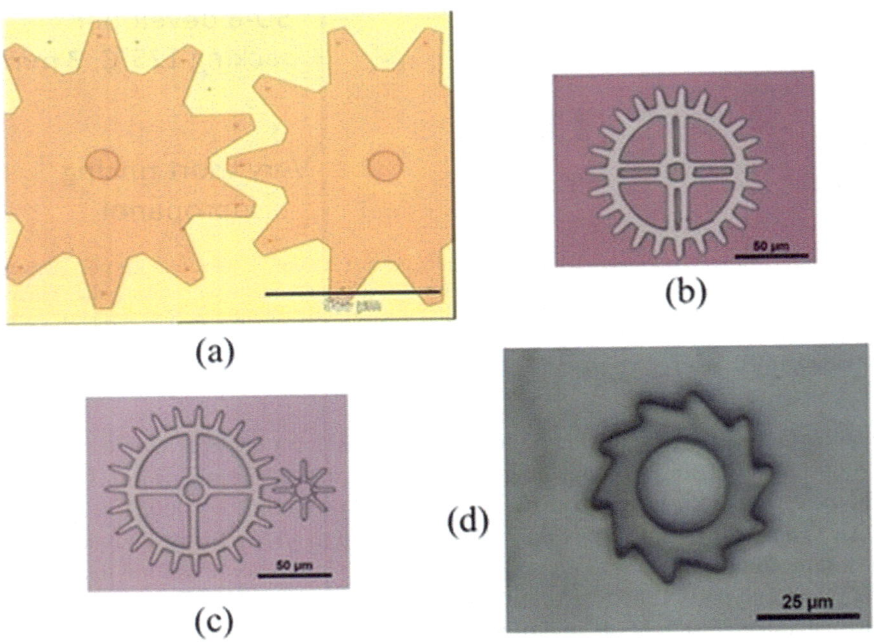

Fig. 9 Different kinds of micro-gears micro-fabricated on a SiO_2 surface using the AZnLoF resist. **a** A very large almost milli-gear with the axle hole not free. **b** A simple micro-gear with 23 teeth and an empty central part and rigidification bars. **c** A train of 2 micro-gears, the second one having a 40 μm diameter. **d** A small 30 μm micro-gear with 10 asymmetric teeth and a large central hole to accommodate an axle

3.2 Fabrication of the Micro-Gears with Their Central Axle

An axle of rotation is usually one of the most difficult part to fabricate and this at any scale since whatever sequence of fabrication (the axle being fabricated before, at the same time or after the miniature gear), either the axle or this gear can be damaged or moved away during the process. Furthermore, an axle of rotation must be very robustly anchored to the supporting surface and this is not always possible depending on the material. The previous section had presented a solution working well down to the 100 μm scale. But this becomes not very practicable below 50 μm for photo-lithography. The same difficulty will be discussed in the next chapter "Fabricating Solid State Gears at the Nanoscale: A Top–Down Approach" for the 100 nm scale and in many chapters of this volume for the molecular scale.

For fabricating a micro-gear and its rotation axle at the same time, our initial design was to define a computerized mask where during the UV beam scanning, the centre of the micro-gear is first filled up by its axle made of the same material that the micro-gear itself. We have tried different axle shapes and the best one to avoid lateral adhesion between the interior of the gear and the axle was a cross as presented in the Fig. 11a. The problem with this cross like axle is that its height is the same than the thickness of the micro-gears because coming from the same resist thickness. Therefore, there is no holding effect. When a micro-gear is rotated using for example a glass micropipette in a pushing mode of manipulation, it is easily dismounted from its axle. Furthermore, if this micro-gear rotates (which is not often the case on SiO_2 but which will be the case on a supported graphite surface (see below)), this cross like axle is moving too with the micro-gear.

Therefore, and following our Sect. 2, we have selected another material for the micro-axle fabrication. As demonstrated in Fig. 7, metallic axles are very robust for example either Al or Ni pillar normally 10–20 μm in diameter and 8–10 μm in height. A first photo-lithography test was performed using a positive AZ5214 resist, followed by the UV 375 nm insulation, the resit development and the growth-anchoring of the metal pillar by their electrolytic deposition. The resist is then completely washed out in isopropanol for the supporting surface to be ready for the second micro-gear scanning photo-lithography step. An example of such a network of micro-axles is presented in Fig. 10.

After the metal micro-axle fabrication, a new resist can be spin coated to perform the micro-gear insulation also using the scanning photo-lithography. The lithography process is here the same than the one described above including a final baking to ensure the material cohesion of each micro-gear now self-mounted on its metallic axle (Fig. 11). Here the interest of our scanning photo-lithography technique is that the re-alignment of the centre of the micro-gears on the already fabricated micro-pillar for the axles is quite simple and very reproducible. This leads to a very reproducible assemblage of micro-gears and axles as presented in Fig. 12.

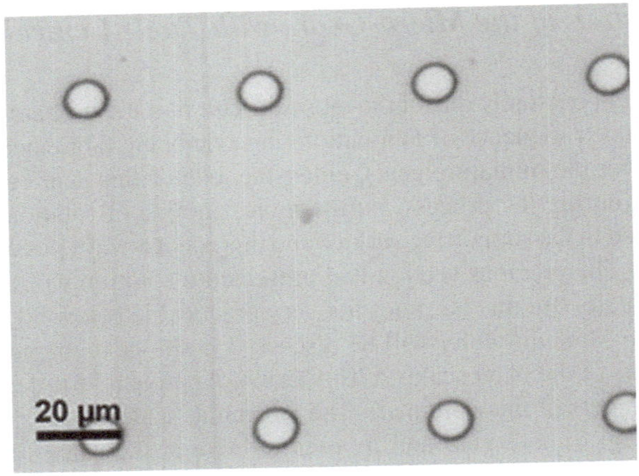

Fig. 10 An optical microscope photo of a typical series of 8 Al pillars, 10 µm in diameter and 10 µm high micro-fabricated by photo-lithography, anchored well on the SiO_2 surface using a pre-etching before the metal deposition and used in Fig. 12 for positioning micro-gears using a second micro-lithography process optically aligned on the first one

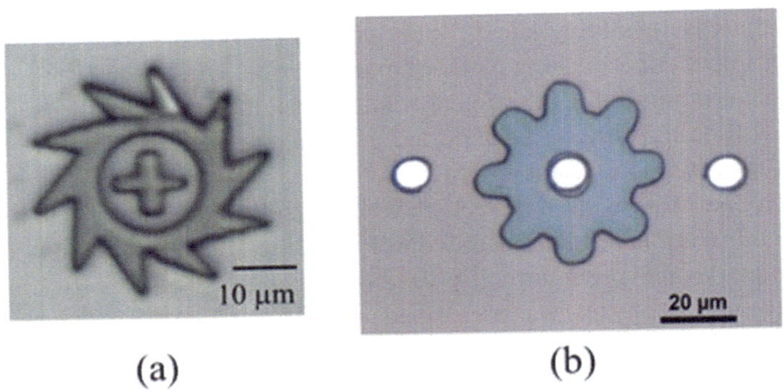

Fig. 11 Two example of micro-gears with their axle. **a** The cross like axle was fabricated at the same time than the micro-gear and in **b** the Al pillar axle was fabricated before in a first photo-lithography step and the micro-gear after in a second step and using another resist

3.3 Fabrication of the Micro-Beams for a Ratchet Effect and the Anchoring Pads

The essential part of our Fig. 1 mechanical calculator design is the beam supposed first to play the role of a ratchet for the calculating micro-gear to rotate in only one direction and second to propagate the carry from digit to digit. As presented in Fig. 1, those beams must also be anchored at one end to the supporting surface. At the millimetre scale, this anchoring is ensured by screws which is impossible here.

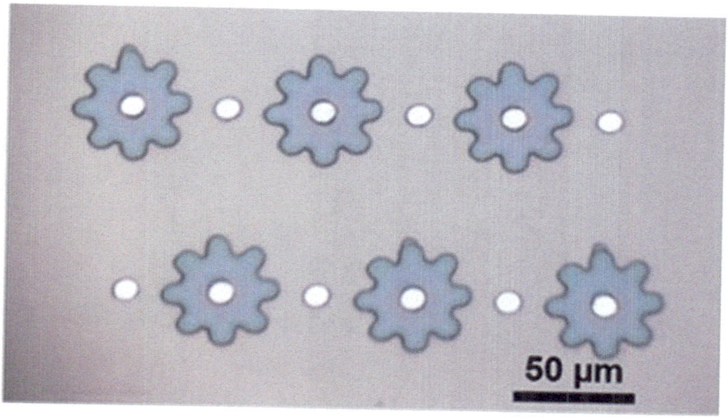

Fig. 12 A network of 12 Al micro-axles with its 6 non-interdigitated micro-gears showing the precision of our double scanning photo-lithography process

As presented in Fig. 1d, we limit here the microfabrication to a 2-digits mechanical calculator having all the required mechanical pieces: two calculating micro-gears with one gear having a longer tooth to transfer the carry, two rotation axles, two ratchet-like beams with their characteristic asymmetric end and two large anchoring pads to anchor each beam. This planar design is inspired by the mechanical machinery fabricated in chapter "From the Pascaline to a 5-Digits Metallic and Planar Miniature Mechanical Calculator 1 cm in Lateral Size". As indicated in Fig. 1d, it was decided to use only one raw of ratchets considering that, due to surface friction, the calculating micro-gears would not turn back after a rotation is actuated by a glass micro-pipette.

One important point of the Fig. 1d design recalled in Fig. 13 is the thickness of the beams and the lateral size of their anchoring pads. In absence of a precise measurement of beam elasticity and spring constant when fabricated with an 8 μm initial thickness polymerized and baked SU-8 resist, we have tried different lengths, widths and anchoring shapes of the end pad. After a few trials and errors, a 4 μm in width, 45 μm in length and a curved attachment root to the anchoring pad was design (see Fig. 13). For this first complete design a very simple curvature was adopted assuming that the longitudinal rigidity of each beam will stabilize the ratchet effect.

3.4 *The Graphite on SiO_2 Supporting Surface*

After a few mechanical manipulations using glass micro-pipette mounted on a micro-manipulator under an optical microscope, it became clear that the SU-8 material micro-gears and beams were too adherent to the native SiO_2 surface for reliable rotations of a micro-gear. Some micro-gears were broken and it was rather difficult to beneficiate from the lateral elasticity of the micro-beams. Therefore, we have explored another surface. The MoS_2 and the graphite surface are known to be quite

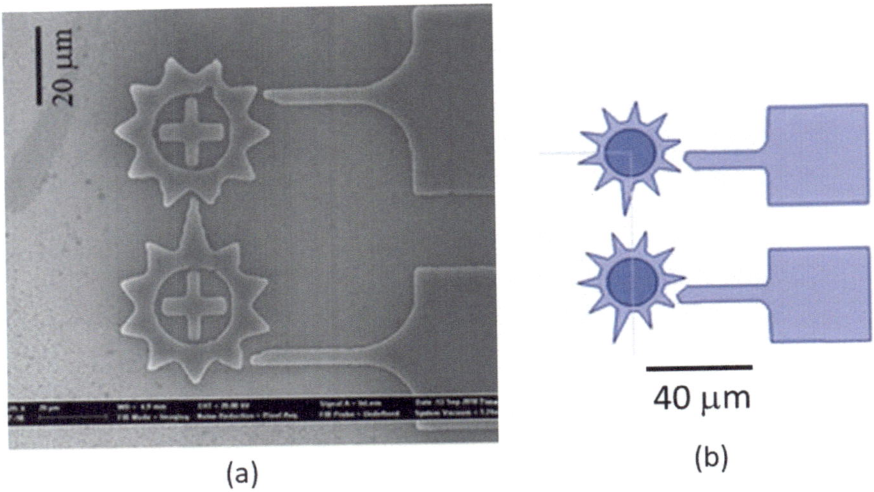

Fig. 13 A scanning electron microscope image after the complete fabrication of a 2 stages mechanical calculator respecting the Fig. 1d design on an SiO_2 native surface. An 8 μm thick SU-8 resist was used with a 375 nm UV light. It is very difficult to activate the rotation of a calculating micro-gear on this surface for example using a glass micro-pipette. For this run, the central axle is still a cross made of the same material than the micro-gears

friction-less [4]. MoS_2 is known to be a nice lubricant and the graphite surface was already used for HSQ resist meso-wheel manipulation with a AFM tip [5]. The graphite surface was selected with a new surface preparation process leading to the production of a very fresh planar graphite surface deposited on a standard SiO_2/Si chip. The preparation process of this surface is presented the Fig. 14. The advantage of this process is that the as-prepared surface is fresh with very large flat thick graphene areas as presented in Fig. 15.

All the scanning photo-lithography processes developed above for the SiO_2 surface were adapted to this graphite surface. The SU-8 resist was kept while changing only the development and rising duration not to destabilize the graphite over layer. The micro-axles are made of electro-deposited Ni micro-pillars as in Sect. 2 to anchor them inside the graphite surface. A complete elementary mechanical machinery with its two calculating stages was then micro-fabricated as presented in Fig. 16 respecting the simplified Fig. 1d design. Like in Fig. 13 on the SiO_2 surface, we first start with cross like micro-axles made also in SU-8 and then use the Ni micro-pillar for the axles.

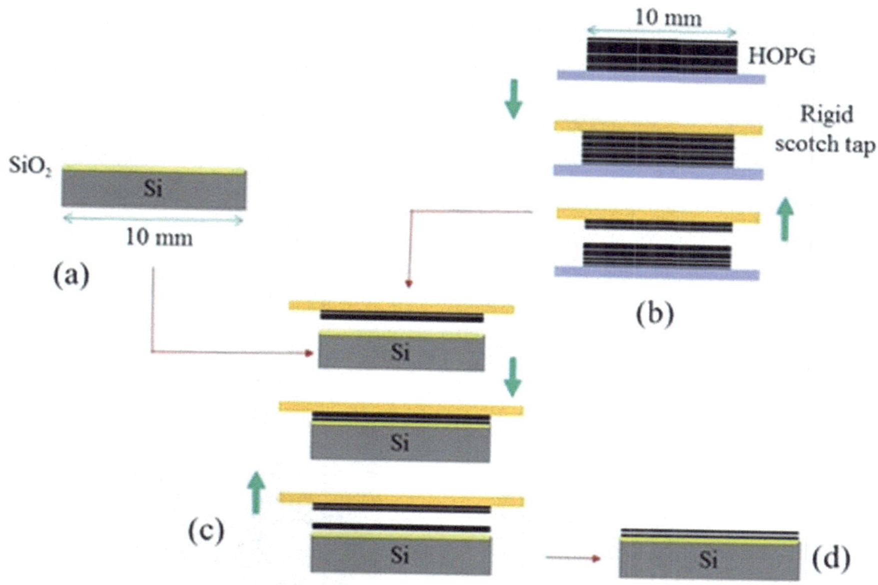

Fig. 14 The transfer printing process used to deposit a fresh graphite surface on a SiO_2/Si wafer. **a** The native SiO_2/Si surface wafer is cleaned in acetone and hardly dry. **b** The 3 steps sequence of picking up a layer of graphite from a commercial HOPG bulk sample using a rigid scotch tape. **c** The graphite layer is transfer printed on the decontaminated SiO_2/Si surface. Part of the layer remains on the scotch tape and about 8 μm in thickness are deposited on the SiO_2/Si surface. **d** This graphite/SiO_2/Si surface is ready for scanning photo-lithography

Fig. 15 An 8 mm × 8 mm Graphite/SiO_2/Si wafer in its plastic protection box. The average thickness of this fresh graphite surface is about 8 μm measured by ellipsometry. Lateral left and bottom right are graphite layer going out of the silicon chip after the transfer printing

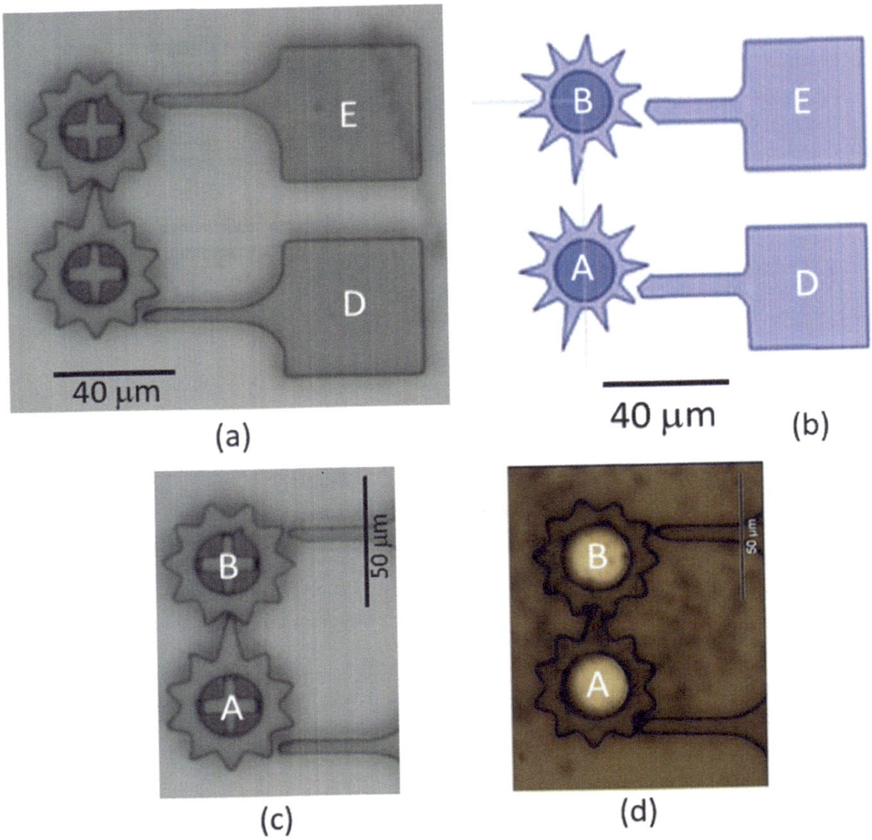

Fig. 16 **a** An optical microscope image of a 2 stages mechanical calculator still crossed axles micro-lithographed on the graphite/SiO_2/Si surface. **b** The recall of the Fig. 1d design on scale with **a** and the E and D holding micro-pads indicated. **c** A slight zoom-in of **a** to compare with the same 2 stages mechanical device but now with the Ni micro-axle also micro-lithographed on the graphite/SiO_2/Si surface. The cleanness of the graphite surface has still to be improved for example in **d**

4 Flexibility of a Planar Beam at the Micron Scale

After the micro-fabrication presented in Fig. 16, we have tested the lateral elasticity of the micro-beams and the transmission of a carry that is to propagate the rotation of a micro-gear after a complete revolution of the micro-gear having one elongated tooth in charge of the carry. Both were performed under an optical microscope using a 2 μm apex glass micro-pipette positioned and manipulated under this microscope using a standard micromanipulator.

For the micro-beam elasticity, the main question was to test the lateral elasticity of a 4 μm in width, 8 μm in height and 45 μm in length micro-beam made in SU-8 and baked at 125 °C during 3 min. Lateral deformation test were performed on many of those micro-beams. First, the sharp end of a given micro-beam needs generally to be detached from the surface because the last baking step of our process is re-enforcing surface adhesion even on the graphite surface. Normally a gentle mechanical push is enough and the optical contrast atop the micro-beam is changing from grey to white as demonstrated in passing from Fig. 17(1–3). Once this end is detached, it generally remains like this during micropipette manipulations (see for example from Fig. 17(4, 5)). A lateral curvature up to 40° is not breaking a micro-beam which is enough for a ratchet operation. Notice also that the large squared anchoring micro-pad is very robust and was not moving during the Fig. 17 complete manipulation sequence. As presented in Fig. 17(2), an already moderate 5° push is inducing some cracks along the body of a micro-beam.

Fortunately, up to 30°–40°, those cracks are stabilized and seems to give more flexibility to the micro-beam structure with no more damage. Figure 18 provides a zoom of the 4 cracks observed in Fig. 17. They are nearly equally spaced and are blocked at the same position during the deformation sequence shown in Fig. 17(2–8).

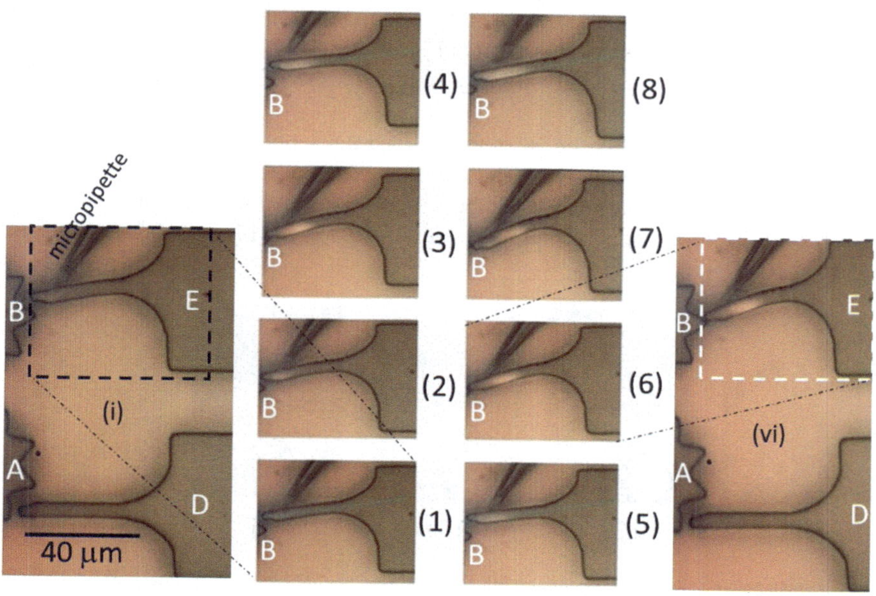

Fig. 17 One micro-beam lateral elasticity being tested by pushing with the 2 μm end apex of a glass micro-pipette. Two sequence are illustrated here. From (1) to (4) a moderate push leading to 10° of tilt and from (5) to (8) a stronger sequence with a 30° tilt. A zoom on the 4 cracks appearing already on (1) is given in Fig. 18. (i) A large image of (1) and (vi) of (6) showing the 2 stages machinery according to the Fig. 1d design

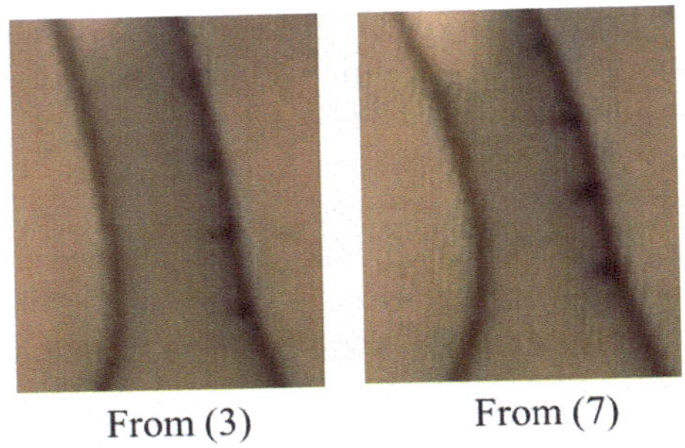

Fig. 18 A optical microscope zoom in of the 4 cracks appearing already in Fig. 17(1). The beam width is 4 μm

A more detail and systematic study is now required to optimize the lateral elasticity of the micro-beams as a function of the beam size and of the final baking temperature. Notice that such experiments were not possible on an SiO_2 surface where the adhesion of the baked SU-8 is very strong.

To test a carry propagation, it was important to use micro-gears anchored on Ni axles and to work on the graphite surface. The carry transmission was study on a two-stages mechanism by first fully rotating the micro-gear with the appropriate long carry tooth to reach the position needed to transmit the carry to the next micro-gear. Then, the micro-pipette was used again as the stylus used in a macroscopic mechanical calculator to add a "one" to this micro-gear and follow the rotation of the next stage micro-gear. As presented in Fig. 19, this carry is working properly i.e. the Ni micro axles are anchored strongly enough on the graphene surface to let the carry transmission to operate.

More important and after the 36° rotation, the micro-gear in charge of the carry is not going backward thanks to the surface friction. This indicates that in the design of the machinery of a planar micro-mechanical calculator as fabricated here only one set of lateral beam-ratchets is required as anticipated in the Fig. 1c, d designs.

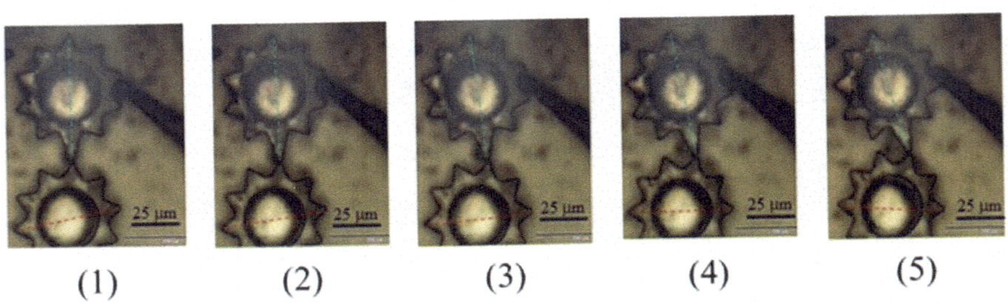

Fig. 19 The demonstration that the carry transmission is working from one stage to the other on a graphite surface with Ni micro-axle. (1), the top toothed wheel with its long carry teeth what rotated first step by step in the position to transfer a carry if a 1 is added to the top gear (a 36° rotation to the right). From (2) to (4) the 1 is entered using the micro-pipette. In (5) the transmission of the carry almost proceeds; the down gear had rotated by 36° and the top toothed wheel can continue to rotate for 9 steps before transmitting again a 1

5 Conclusion

Starting from the mechanical design of a planar one level millimetre size calculator in chapter "From the Pascaline to a 5-Digits Metallic and Planar Miniature Mechanical Calculator 1 cm in Lateral Size", we have developed a scanning photo-lithography process to miniaturize the structure of a mechanical micro-calculator down to 40 μm in diameter micro-gears. The first moulding process span dimension from 1 mm to 60 μm and the assemblage of the micro-calculator by micro-manipulation under an optical microscope is practicable down to 100 μm. Below and to reach the 40 μm scale, a double photo-lithography process was used and a specific graphene/SiO_2/Si surface was developed for mastering surface friction in particular for the ratchet-beam effect. Another originality of our process is that the resist for the photo-lithography is used as material for all the micromechanical pieces. Only the micro-axles remain metallic to ensure their good anchoring to the graphene/SiO_2/Si supporting surface. Not only the micro-gears, but also well anchored axles, quite elastic beam-ratchets and lateral pads fixing those beams have been design, fabricated, and tested. A two stages calculator was fabricated and the carry propagation in base 10 was demonstrated, which can now be optimized for more calculating mechanical stages. The next miniaturization steps down to the 100 nm are described in chapter "Fabricating Solid State Gears at the Nanoscale: A Top–Down Approach".

Acknowledgements We thank the European Union Horizon 2020 FET open project "Mechanics with Molecule(s)" (MEMO, grant 766864) for financial support. This work was supported by LAAS-CNRS Technology platform, a member of the French RENATECH network.

References

1. Bustillo, J.M., Howe, R.T., Muller, R.S.: Surface micromachining for microelectromechanical systems. Proc. IEEE. **86**, 1552–1574 (1998)
2. Cui, D.D., Zhang, L.C.: Nano-machining of materials: understanding the process through molecular dynamics simulation. Adv. Manuf. **5**, 20–34 (2017)
3. Keller, S., Blagoi, G., Lillemose, M., Haefliger, D., Boisen, A.: Processing of thin SU-8 films. J. Micromech. Microeng. **18**, 125020 (2008)
4. Deng, J., Troadec, C., Kim, H.K., Joachim, C.: Direct transfer of Au nano-islands from a MoS2 stamp to an SiH surface. J. Vac. Sci. Tech. B. **28**, 484 (2010)
5. Yang, J., Deng, J., Troadec, C., Ondarcuhu, T., Joachim, C.: Solid-state SiO_2 nano-gears AFM tip manipulation on HOPG. Nanotechnology **25**, 465305 (2014)

Fabricating Solid State Gears at the Nanoscale: A Top–Down Approach

D. Mailly and G. Faini

Abstract A rapid overview on the main lithographic tools used in the top–down approach in nanotechnology including their intrinsic and extrinsic resolution is given in this chapter. Moreover, we discuss the advantages of the use of focused ion beam for very specific applications, complementary to the more standard ones. Finally, we will describe, using focused He ions, how to fabricate solid state gears at the nanoscale, in the range 200 nm down to 50 nm.

Keywords Nanotechnology · Nanofabrication · Lithography · Helium ion beam

1 Introduction

1959 Richard Feynman's famed talk during at the annual American Physical Society meeting, "There's Plenty of Room at the Bottom: An Invitation to Enter a New Field of Physics" [1] is considered as the date of birth of nanoscience. End of 50s corresponds also to the work of Martin M. Atalla and Dawon Kahng at Bell Labs demonstrating the first realization of a full planar technology to fabricate field effect devices [2, 3] and announced in 1960. They proposed the concept of Metal Oxide Semiconductor (MOS) integrated circuits (IC) using this planar technology approach. More than sixty years later we can assert that this new approach was the "kick-off" of the microtechnology era with the advent of the microelectronics industry first, of the nanotechnology era second.

To realize objects at the nanoscale range, two mains routes are used in nanotechnology (see Fig. 1). The first one, named "top–down" approach, is directly related to the developments issues from the planar technology mentioned above and the related improvement of the techniques used in the microelectronics industry. Starting from

D. Mailly · G. Faini (✉)
CNRS, Centre de Nanosciences et de Nanotechnologies (C2N), Université Paris-Saclay, 10 Bd Thomas Gobert, 91120 Palaiseau, France
e-mail: giancarlo.faini@c2n.upsaclay.fr

D. Mailly
e-mail: dominique.mailly@c2n.upsaclay.fr

C. Joachim (ed.), *Building and Probing Small for Mechanics*, Advances in Atom and Single Molecule Machines, https://doi.org/10.1007/978-3-030-56777-4_4

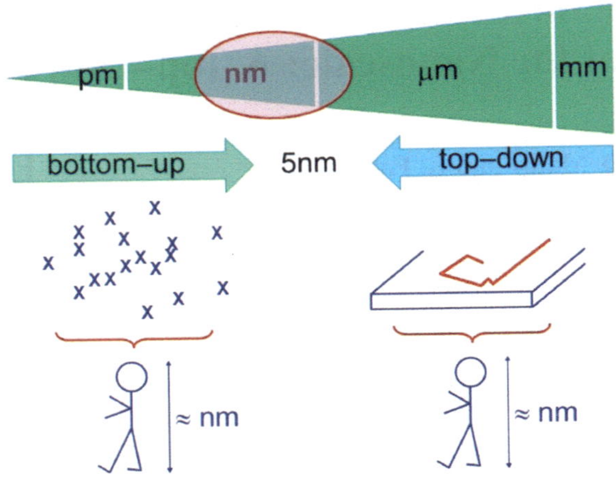

Fig. 1 Nanofabrication techniques: top-down and bottom-up routes

a macroscopic material (semiconductor crystal, metal, superconductor, functional oxide…), the device is tailored at the nanometer scale using different tools: lithography, etching, deposition, implantation. The opposite route, named "bottom–up" approach, is that directly issued from the chemistry and the material growth engineering sciences. Here it concerns the assembly of elementary bricks, atoms and molecules, to build more complex entities at the nanoscale range. As shown in this book, both routes are complementary and are together used in nanotechnology.

In this chapter we will focus on some tools using the "top–down" approach. The flowchart for a typical fabrication process used using these techniques is shown in Fig. 2, summarizing the different type of tools necessary to meet the objectives: lithography, deposition, etching, implantation or irradiation.

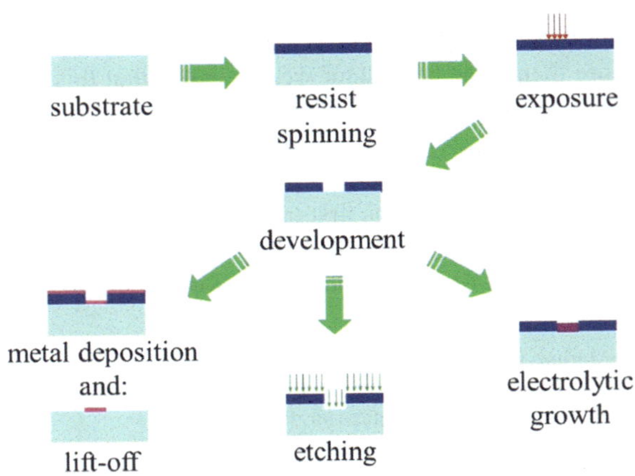

Fig. 2 Flowchart of a typical top-down fabrication process

2 Lithographic Tools

The word lithography comes from the Greeks λιτηοσ (stone) and γραφοσ (writing) and it is a well-known very old techniques used even nowadays mainly in arts. The planar process invented by Atalla and Kahng at Bell Labs for the first MOS devices was in fact a kind of rediscovery of the masking effect used in this technique. Moreover, the origin of the lithographic process is also intimate linked to the original process invented by Nicéphore Niépce in 1826 for the photography and for this reason considered by many as the very first photograph. In that case, Niépce used a natural mineral resin, the "bitumen of Giudea" or asphalt, to record the optical image of the object he wished to photograph. After the development in a diluted solution of essence of lavender, he used the residual resin (negative image of the subject) as an etching mask to the chemical acid attack of copper plates at the beginning and tin plates later. Actually, the term "resist" used in micro- and nano-technology derives from this method since the resin resists against the aggression of the chemical bath protecting the wanted portions of the underlying material.

The lithographic techniques using a resist are characterized by the radiation used to modify the properties of the resist. It is mainly the nature of the radiation that determine the resolution that can be achieved using a given technique, i.e. the smallest feature one could achieve. The Fig. 3 summarizes the typical range of pattern sizes as a function of the lithographic used tool.

As shown in Fig. 4 the lithographic methods can be roughly divided into two categories according to the writing strategy chosen:

- Parallel writing strategy like those used in optical, x-ray and, more recently nano-imprint lithography
- Sequential writing strategy like those used by the controlled deflection of a focused beam (electron or ion beams) or the mechanical displacement of a probe (near field approaches).

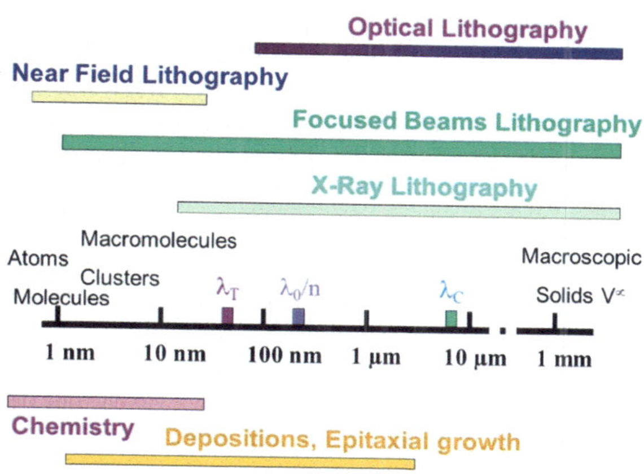

Fig. 3 Range of resolutions achievable by the different techniques, both for the top–down and the bottom–up approaches

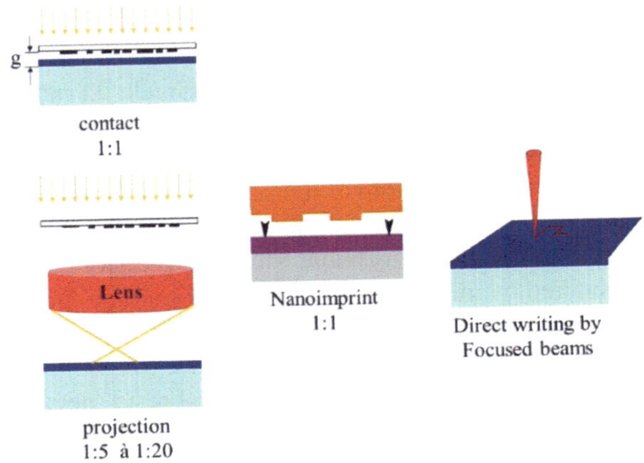

Fig. 4 Parallel and sequential lithographic writing strategies

In the first category, all the patterns are generated at the same time, in a unique exposing step. This is possible by interposing between the resist and the radiation source, a mask imprinted by the patterns to be exposed. The mask is nothing else than a transparent material to the radiation (e.g. glass or quartz in UV lithography) where the wanted pattern is printed in opaque area (metals like Cr in the same example) protecting the resist from the radiation. It is then easy to understand that these methods are the quickest since in a single run, large area and a very high number of devices can be produced. These methods are of course preferred by the industry in order to conserve a large as possible throughput in term of mass production. The counterpart of these techniques is that the mask design is fixed once and for all. There are no possibilities to slightly modify the geometry, the location or the size of the devices: it is a method of a simple duplication of the master mask. The second requirement to be fulfilled using this approach is the possibility to use the same master mask a huge number of times, without any degradation of the mask itself that could be detrimental to the final realization.

In the sequential lithography methods, the patterns are written point by point at the surface of the resist. The elementary tool is the focussed beam that can by displaced pixel after pixel in a very well controlled manner. It is obvious that these methods are time consuming and are certainly inappropriate for productions purposes (except in some very particularly very delicate and high resolution steps and for very particular and rare applications). But the advantage here is that these methods don't need the use of any mask: there is direct writing of the resist. For this reason, the shape, the location, the size, can be changed at will since the only modifications to be done are in the software used for the pattern design. Moreover, these approaches give the higher resolutions as we will describe below in this chapter. As we will show for instance in the case of ion beam lithography, in some applications the lithography can be performed without any use of a resist, but directly modifying the physical or/and chemical properties of the patterned material. It is also worth emphasising that, generally, the sequential tools of lithography are used to produce the master masks for the parallel lithography industrial needs.

3 Optical Lithography

The optical lithography by contact or proximity is one of the oldest methods used to reproduce patterns by means of ultraviolet (UV) radiation. After the spinning of the resist on top of the surface of the sample, the ensemble is exposed to the UV radiation through a mask that is in intimate contact or in proximity to the resist surface. The mask comprises transparent and opaque regions defining the pattern design one wants to reproduce on the wafer. This technique is quite easy to carry out and, indeed, it was the technical hinge of the whole microelectronics industrial production till the middle of the seventies.

The fundamental limitation of this technique is related to the phenomenon of the light diffraction at the edges of the opaque portions of the mask. The correct description of the light intensity distribution at the coming out of the mask is given by the Fresnel diffraction theory (involving spherical waves) which is much more complex and needs the use of complex calculations.

As an example, in Fig. 5 we have represented a typical optical lithographic setup in a contact or proximity configuration, where t is the resist thickness, s the separation between mask and resist ($s = 0$ in contact mode), $s + t = g$ the total gap between the mask and the sample surface. In this simple illustration the pattern design is made of metallic lines equally separated by a distance b. In the lower part of the Fig. 5 we have drawn the light intensity distribution on the resist surface along the transverse direction of the lines. Due to the diffraction phenomenon, the real light intensity distribution, responsible for the exposure of the resist, drawn in red, is deformed from the expected square one. The higher is the distance s between mask

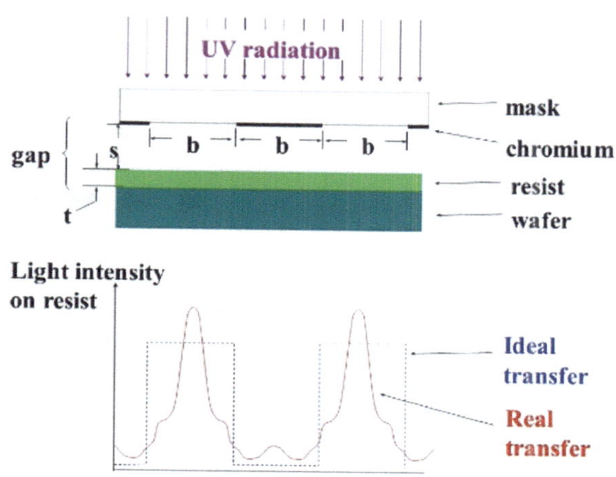

Fig. 5 Light distribution on the resist

and resist, the more is the deformation from the ideal case. The more the distance b approaches the wavelength λ, the more is the deviation from the ideal square profile. The consequence of the diffraction phenomenon is that one collects an important light contribution in the area of the resist corresponding to the opaque chromium line with a corresponding loss of the critical dimensions of the design pattern. In a first approximation, the resolution, i.e. the minimal feature size achievable, is given by $R \approx \sqrt{g\lambda}$. It is possible to achieve resolutions of the order 200 nm or less using thinner photoresists or shorter exposing wavelengths in contact mode: as an example $R = 150$ nm are achievable using the $\lambda = 193$ nm wavelength supplied by an ArF excimer laser. It is worth noting that the decrease of the used optical wavelength is limited by the absorption of the light by the glass or the quartz of the mask. Using shorter wavelengths needs more complex and expensive reflective optical setups working in vacuum chambers.

In the middle of the 80s, the improvement of the production for industrial purposes led to the development of the optical lithography using projective setups as sketched in Fig. 6a. Here the sample wafer and the mask are separated one from each other and the image of the mask is projected and focussed on the resist surface deposited on the top the wafer. Using this approach, it is possible to reduce the pattern designed on the mask by a factor ranging from 5 to 20 times. The consequence of this is that it is not possible to cover all the wafer surface with the image with a unique exposure step, the projection setup reducing the dimensions of the focalised field. The most used writing strategy is thus a technique named "step and repeat", the pattern designed is thus separated by a set of identical exposing fields. After a step of exposure of the projected mask patterns on the surface of the wafer this latter is mechanically displaced to repeat the exposure process in the neighbouring field and so on until the entire resist surface is covered.

The limiting resolution factor here also is the diffraction of the light at the edges of the metallic portions of the mask. Since we are dealing with an optical projection setup, the wave involved here are plane waves and the correct description of the

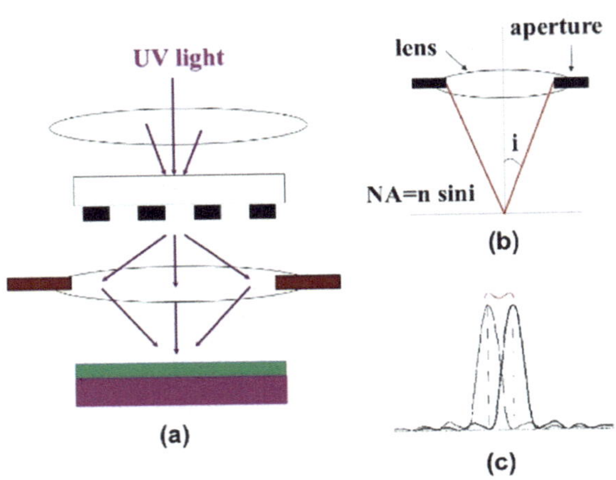

Fig. 6 **a** scheme of the setup. **b** Numerical aperture definition. **c** Rayleigh criteria illustration

physical mechanisms is given by the Fraunhofer diffraction theory. Here the very important parameter is the numerical angular aperture $N.A.$ of the setup. Indeed, the greater is the angular aperture, the larger is the number of diffracted light order collected and the better is the resolution of the projected mask pattern image. As sketched in Fig. 6b, the numerical angular aperture is defined by $N.A. = n \sin i$, where n is the refractive index of the propagating medium and i the maximal angle of collection. The minimal separation between two objects projected by a lens is given by the Rayleigh criteria as shown in Fig. 6c, $R = 0.61\lambda/N.A.$, valid in the ideal case of no mask defects, no lens aberration. The general formula to define the final resolution achievable with an optical projection setup is then $R = k\lambda/2(N.A.)^2$, where k is a technological parameter considering all the deviations from the ideal case and related to the particular setup and process used for the lithographic operation (mask, optical setup, resist…).

The other parameter to play with, is the angular numerical aperture. Progress in optical setup design have increases this value from 0.3 to 0.9. But there is another limitation since this value cannot be increased at will because of the value of the depth of focus that determine the total resist thickness exposed by the light. The depth of focus is given by $DoF = \lambda/2(N.A.)^2$. As an example, using typical values like $\lambda = 365$ nm, $k = 0.8$, $N.A. = 0.48$ yields to $R = 0.6\ \mu$m and $DoF = 0.8\ \mu$m. Typical resist thicknesses are larger than the micrometre leading to a poor transfer of the pattern design through the resist during the development and a consequent failure of the lithographic process. But as mentioned above different strategies have been developed to push the resolution well upwards the diffraction limits. The limitations described above in reducing the light wavelength can be overcome by reducing the k technological parameter playing with, for instance, the mask or the resist properties: phase shifting masks, top surface imaging, multiple exposures or resist recess techniques. The k factor can be reduced down to 0.3. Even if these techniques permit to overcome the diffraction limit, they are restricted to very simple pattern geometries. Indeed, they are not adapted to more complex designs as, for instance, the realisation of solid state gears at the nanoscale.

Nowadays, immersion lithography, where the NA is increased by using a liquid between the final lens and the resist to improve the n index up to 1.35, is commonly used and reach 38 nm resolution using a 193 nm light [4]. Diffraction limit is also push away with EUV lithography at a wavelength of 13.5 nm. Since there is not any material transparent at this wavelength, precise mirrors are used instead of lens to image the mask. An ultimate resolution of 13 nm is achieved in the last generation lithographic systems [5].

4 Near Field Techniques

A scanning tunneling microscope is able to reach the ultimate resolution by the manipulation of single atoms. A remarkable example has been achieved by the group of IBM Almaden in '93 who realize a quantum corral made of 48 Fe atoms on a copper surface. But the drastic experimental conditions needed to perform this feat e.g. very low temperature, ultra-high vacuum and very low mechanical vibrations, preclude any use of this technique to produce realistic devices. Later in the beginning of 2000, IBM launch the Millipede project. It is now a silicon cantilever AFM tip that is heated in order to indent 10 nm holes in a thin polymer film. An array of 1024 of such tips fabricated using NEM's technologies was produced to achieve a data storage density of a trillion bits per square inch [6]. Unfortunately, this technique has still a poor throughput and the progress of optical lithography buried the project.

A commercial equipment based on this idea with a single AFM tip to fabricate prototype or research samples with a 10 nm resolution is available [7]. Scanning probe techniques based on local oxidation, material deposition, magnetic reversal can also be used to produced nanostructures.

5 Electron Beam Lithography

Since very long time we know how to focus charge particle beams: the electron beam microscope was invented in 1931 by E. Ruska (1986 Physics Nobel Price) and M. Knoll, two engineers of the Technical University of Berlin. Control of the displacement of the electron beam permitted few years later, in 1935, to M. Knoll to invent the scanning electron microscope. Improvement in the electron optics has continued to be made in order to obtain a good control on the positioning of the beam spot, the stability of the beam and the correction of the aberrations due to the optics.

In Electron Beam Lithography (EBL) electrons are used instead of photons to irradiate the resist. The electron beam is sequentially moved in a controlled way along the resist surface, pixel per pixel, in order to reproduce the desired pattern. The control of the dwell time spent by the beam on each pixel will define locally the dose of radiation on the resist. This dwell time can be adjusted for each pixel permitting to vary the exposure dose all along the pattern realization. The final resolution obtained by this technique is directly related to the final spot size achievable at the surface of the sample and the interaction of the electron with matter. The size of the final electron spot is a complex quantity to calculate and to predict since it is linked to several factors, most of them interconnected. Indeed, the spot shape and size depend on the electron gun source, i.e. on the virtual source size, and on the chromatic and spherical aberrations of the forming-beam optic lenses, giving rise to an energy spreading of the beam. Overall, the final spot size depends on:

- The accelerating voltage: the higher the voltage, the smaller the spot size. Typically, on the range of 1–5 nm for V = 100 kV
- The current of the spot probe: the larger the current, the larger the spot size

The energy of the accelerated electrons is $E = eV = p^2/2m$, where $e = 1.6 \times 10^{-19}\,C$ is the charge and $m = 9.1 \times 10^{-31}$ kg the mass of electrons, V the accelerating voltage in the electronic column. Since the De Broglie wavelength is $\lambda = h/p = h/\sqrt{2mE} \approx 3.8 \times 10^{-1}/\sqrt{V}$ nm (here V is expressed in kV), this

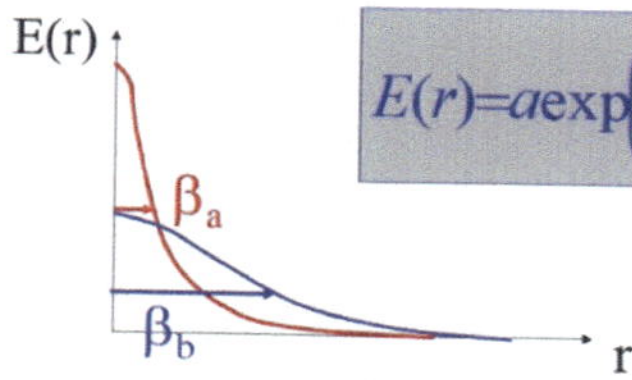

$$E(r)=a\exp\left(-\frac{r}{\beta_a}\right)^2+b\exp\left(-\frac{r}{\beta_b}\right)^2$$

Voltage (kV)	β_a(mm)	β_b(mm)
20	0.08	2
50	0.04	9
60	-	13
120	-	43

Substrate: Si

Fig. 7 Simple double Gaussian model for the calculation of the electron-solid interaction contributions to forward and backward scattering mechanisms

yields for instance that at V = 100 kV the electron wavelength is $\lambda \approx 4 \times 10^{-2}$ nm. EBL is thus not limited by the diffraction as in the optical lithography case. The only limitations on the on the achievable resolution are thus the aberrations in the electronic optics as mentioned above and the physical mechanisms of the electron-solid matter interactions.

As electrons enter in a solid (resist and sample wafer material), the interaction with the atoms composing the solid cause a scattering of the primary electron yielding to a broadening of the pattern design. It is worth to note that resist materials are sensitive to very small energy (typically in the eV range) whereas the incident electron are in the kV range. A detailed knowledge of the electron energy loss is needed to understand the exposure process. A simple double Gaussian model is sketched in Fig. 7. This model describes the two scattering phenomena undergone by the incident electron while losing its energy as far as it enters and collides with the solid atoms:

- A forward scattering: beam broadens when entering the resist. As indicated in the table value of the Fig. 7, this mechanism essentially depends on the voltage and the resist: minimizes for high energy beams and thinner resists
- A backward scattering: essentially depends on the sample substrate and the voltage: increases as the voltage increases.

The consequence of this backscattering mechanism is that not only the beam broadens much more in the sample substrate than in the resist but, worse, some electrons return to the resist. This is illustrated in the Monte Carlo simulation shown in Fig. 8 [8].

The electrons that are backscattered (in red in the Fig. 7) to the resist cause additional exposure and originate the so named proximity effects, giving serious limitations on the respect of the critical dimensions of the designed patterns or even on the shape itself. A direct consequence of the proximity effects is that one encounters a real difficulty to draw patterns very close each other without losing the critical dimensions or the separation between the objects themselves that could be thus spuriously connected. In Fig. 8 we can easily see that this effect is higher as the

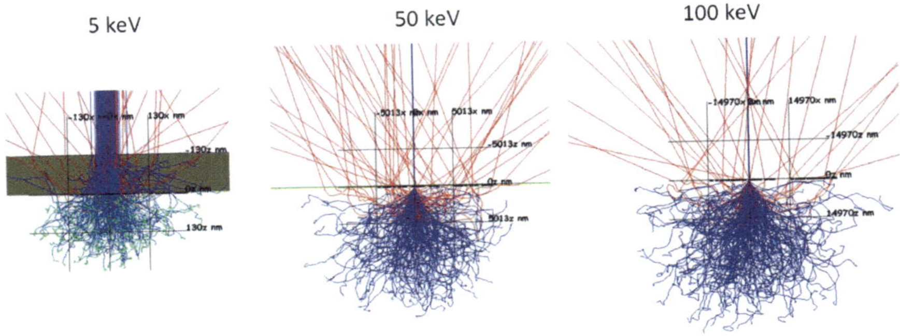

Fig. 8 Monte Carlo simulation simulations of the electron trajectories into the Si stack (Si wafer covered by 130 nm of resist) for an ebeam energy of 5, 50 and 100 keV (simulated with the CASINO software). Forward scattered trajectories are in blue, backscattered ones in red. Reproduced with permissions from the authors [8] and from SPI (International Society for Optics and Photonics 2020)

incident electron energy increases: backscattered electrons penetrate the resist at distance up to 500 nm from the incident beam spot at 5 keV, up to 30 μm at 100 keV. It results also that, as the accelerating voltage is higher, the proximity effects are diluted since the backscattered electrons are more disperse and the resulting energy density deposited by them in the resist is then lowered.

Several strategies are then used to beat these proximity effects. The simplest, but computing time consuming for very complex designs, is to vary the dose (deposited charge per pixel) as a function of the pattern design. Computing programs have been developed to perform the calculation for the dose correction. Figure 9 is a simple example for Al crossing lines 22 nm wide: at the crossing point the critical dimensions are respected.

These programs permit also to take into account pattern designs with a very complex environment. Figure 10 is an example of realization in the framework of science-art collaborative projects [9], it represents the realization by EBL and Au deposition on Si substrate reproducing a winged genius with the head of a bird from the relief engraved on an Assyrian palace of the first millennium B.C. The smallest details of this lithography are in the range of 30 nm.

Other strategies to overcome the proximity effects limitations are to use higher energy electron beam in order to dilute the effects (see Fig. 8) since the local density of backscattered electrons is in this case low, or on the contrary to use very low energy (example with near field probes) but in this case the forward scattering is large. Other possibilities are to use resists sensitive to very high energy like inorganic resists or, finally, write on membranes ruling out backscattering effect due to the absence of a solid substrate.

As mentioned above, the drawback of EBL is related to the sequentially scanning of the beam on the resist surface, precluding high throughput requirements mandatory for industrial production purposes. To overcome this limitation a multi-beam

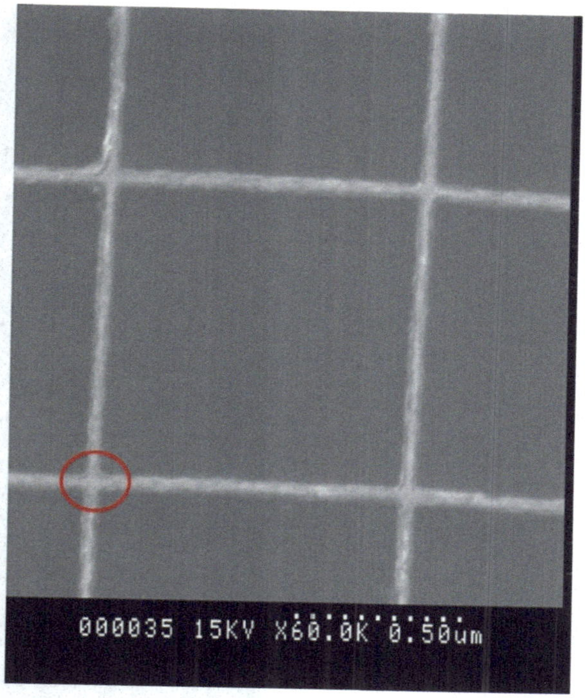

Fig. 9 22 nm Al crossing lines

approach was proposed and realized by some companies like MAPPER LITHOGRAPHY located in Delft, founded in 2000 and now part of the ASML company. Here the EBL system [10] is composed by 13,260 parallel beams able to write on the wafer sample with a resolution in the range of 10–14 nm. The 13,260 beams are generated by splitting a single electron beam issued by an electron gun. The 13,260 beams are focused on the resist surface by a series of electrostatic lenses. Each beam has its own optics avoiding any mixing problem. In that way, it is possible to reach a high throughput of 10 wafers per hour with a standard resolution of 25 nm. Nevertheless, this technique didn't satisfy the market requirement and the production of the machines was interrupted in 2018.

6 Focused Helium Ions

Focused gallium sources are widely used for nanofabrication processes utilizing Focused Ion Beam (FIB). This one of the most popular Liquid Metal Ion Source (LMIS) because of the very low melting temperature of Gallium. The principle of LMIS is to heat up to its melting temperature a reservoir of metal, pierce by a sharp needle. The liquid metal wets the needle and by application of a strong electric field a "Taylor cone " of melted metal forms at the apex of the tip which defines the ion

Fig. 10 Gold on Si reproducing a relief engraved on a Assyrian palace

source and emits metal ions. A standard ion column allows to reduce the source size on the sample surface and to scan the beam with the desired pattern. Such Gallium column are widely used, most often combined with an electron column for real time observation, for it great flexibility and the high throughput for prototyping customized devices for research and development as well as industrial applications. One of its very hot application is the slicing of thin slabs for transmission microscopy. In terms of resolution Ga FIB are usually limited to few ten of nanometers mainly because of the strong interaction and straggling of the Ga ions in matter [11]. A drawback of this technique is the propensity of Ga to implant the material leaving a disordered alloyed zone around the impact of the beam.

A newcomer in FIB system is helium ions. This a Gas Field Ion Source (GFIS) which can have an image source emitting ions almost atomic like, leading to a few nanometers resolution. Although the idea of GFIS emerged in the 50s with Müller and Tsong work [12] it is only recently that a commercial equipment is available [13]. This due to several technical hurdles that needed to be overcome in order to offer a convenient tool serviceable to a non-expert in the field of GFIS.

The principle of the GFIS is to cool down a sharp tungsten needle at about 70 K in an UHV condition (about 10^{-10} mbar). Then a small amount of He gas is introduced in the gun assembly (few 10^{-6} mBar). Helium is adsorbed on the needle and with

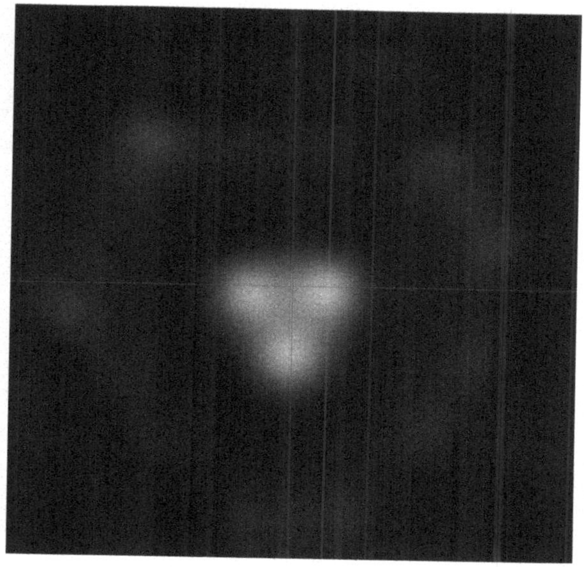

Fig. 11 Image of the trimer of W atoms emitting helium ion obtained in the mode source imaging of the ion column

the help of an extractor pupil brought to several tens of kilovolts placed at a close distance to the apex of the tip, helium is ionized and then extracted through an ion optical column. A remarkable feature of this gun is that the apex of the needle can be tailored in situ via a proprietary method such that only three atoms remain at the end of the tip. This so call trimer emits three streams of He ions. One of the metal atom of the trimer can then be selected through a mechanical tilt and shift of the overall gun assembly and aligned with the center of the ion optics column, leading to have only one of the ion streams effectively impinging the column. The actual source size is then of atomic size and the resulting beam spot on the sample after traveling through the ion column can be as small as 0.35 nm. Figure 11 shows an image of the trimer obtained by scanning the first lens of the ion column (source image mode).

6.1 Imaging with Helium Ions

Helium ions microscopy offers some advantages compare to the widely used scanning electron microscopy (SEM). First because of their heavy mass compare to electron, helium ions interact rapidly with matter. The size of the actual interaction volume between the impinging particle and the sample plays an important role to reach the ultimate resolution. Helium ions, upon reaching the surface of the sample, give rise to a strong flow of secondary electrons in a nanometer range from the point of impact (SE1 electron) and very few back-scattered electrons as SE2 far from the impinging spot. One has then a high resolution and a high surface sensitivity. Furthermore, because of the very small convergent angle of the ion column, the depth of focus is

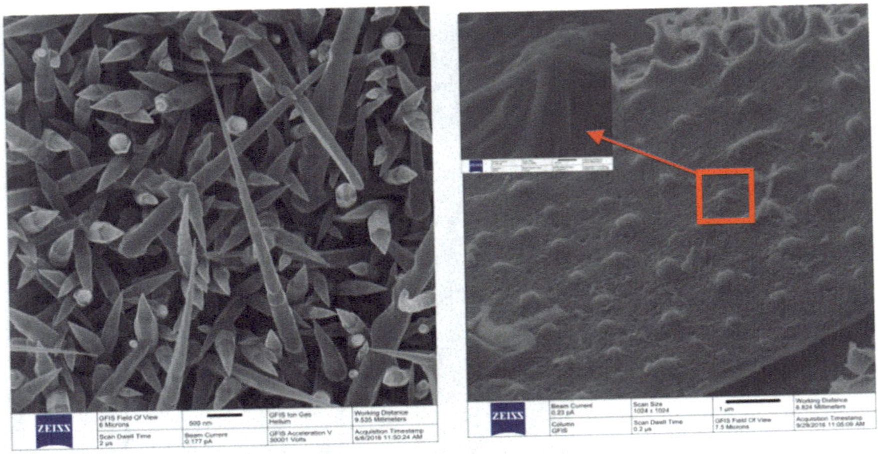

Fig. 12 Left: image of micron size Si nanowires. High depth of focus shows through the sharpness of the image on the top as well as on the bottom of the needle several micrometers long. Right: image of a diatom on the hair of the relic of Santa Magdalena, in the inset a closed view demonstrating the high resolution without coating

about 5–10 times larger than in a conventional SEM system. The high depth of focus can be appreciating on the picture Fig. 12 right of a long Si nanowire. Finally, using a low energy electron gun towards the sample one can neutralize the incoming ions and avoid any spurious charging effect even when one is dealing with insulators like biological material. Spectacular high resolution of such biological sample have been produced without any deposited conducting layer which alter strongly the resolution. An example is given on Fig. 12 left where high resolution image unveils the presence of diatom on the hair of a relic of Santa Magdalena [14].

6.2 *Helium Ion Beam Induced Deposition*

Beam Induced Deposition (BID) is a process of decomposing gaseous molecules by a focused beam leading the deposition of non-volatile fragments onto a nearby substrate. Electron or ion beam are successfully used as incident beam. The most popular precursors for deposition of elemental solids are metal carbonyls of Me(CO)x structure where Me = W, Ti, Cr, Mn, Ru, ... The presence of the CO ligand in the mixture results in the formation of an alloy Me_xC_{1-x} with a carbon content that can be elevated. The scattered electrons are responsible for the dissociation of the precursor molecule and again the extension of back scattered electron degrades the resolution. Electron BID has been widely used but the very low rate of dissociation induces an important carbon contamination. Important proximity effects also affect the resolution due to important back scattered electron spreading. Using Ga ion

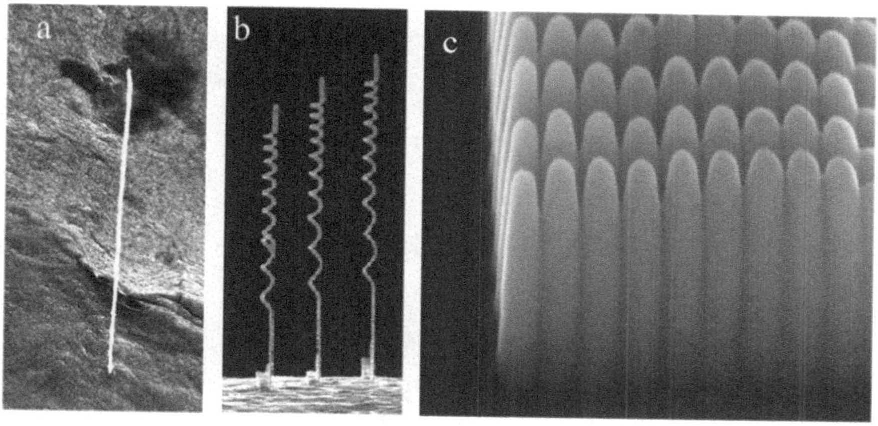

Fig. 13 **a** Single W nanowire 5 μm length and 50 nm diameter. **b** Spirals 5 μm long outer diameter 120 nm wire diameter 35 nm with variable thread **c** close packed wires 50 nm diameter and 300 nm height

greatly increased the growth rate but Ga ions introduce additional contaminations and radiation damages to the deposited structure, which can be important for applications.

Helium ions bring advantage compare to Gallium with a better resolution due to the small spot size and the absence of contamination. In the Fig. 13 we show examples of tungsten wire (in fact $W_{0.7}C_{0.3}$) vertically grown with 30 kV He ions. Because of the very high depth of focus the wire diameter is constant all over the 5 Âμm length of the wire. Some 3D structures can also be realized as shown on Fig. 13b where wires of 35 nm diameter roll on a helix of 120 nm outer diameter with a variable thread along the helix [15]. The very small straggling of the ions together with the small extension of the secondary electrons allow to form close packed wires as shown on Fig. 13c. TEM analysis of these structures found a poly-crystalline material with crystallite several tens of nanometers size. The superconducting properties of such wires show a sharp temperature transition to the superconducting state together with a high magnetic critical field[16]. It has been shown also that it was possible to bring close by within 10 nm distance two W wires without any electrical contact between them which is impossible using Ga ions due to the sidewall diffusion of Gallium ions [17].

6.3 *Helium Direct Milling*

Because of their mass, helium ions can also directly mill materials but compare to gallium beams the etch rate is much lower. There are indeed almost two orders of magnitude etch rate efficiency between the two species for gold for instance.

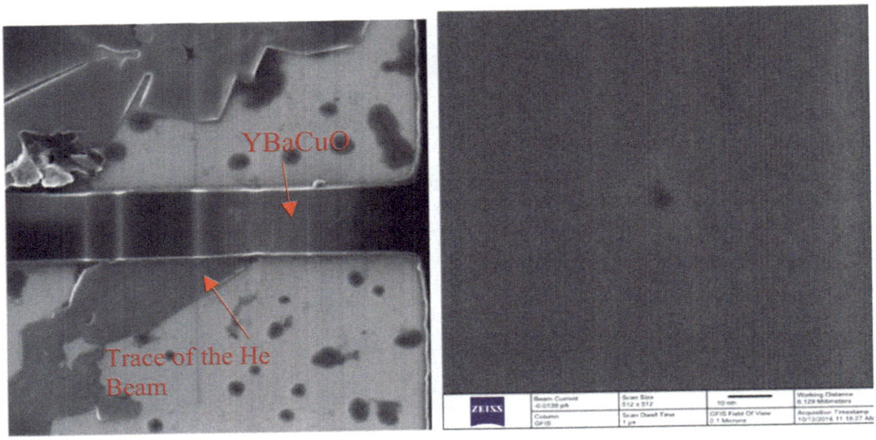

Fig. 14 Left: a Josephson junction on YBaCuO (dark ribbon). The YBaCuO layer is not cut, the ion damages induced by the beam create the insulating barrier Right: a 3 nm diameter nano-hole drilled in a suspended graphene sheet

Although the probe size can be as low as ca. 2 nm, the smallest feature size made using Ga FIB is rarely below 30 nm due to interaction with the milled sample.

The high resolution, small straggling and low damage of helium beam offer some specific applications where gallium implantation fail down: nanoscale hole drilling in membranes for DNA translocation, tailoring magnetic media at very narrow scale, fabrication of Josephson junction in HTC's, fine peeling of slab for TEM inspection defined by Ga FIB in order to remove the disordered region etc. …. In order to increase the milling efficiency, Neon ions instead of Helium gas can be introduced in the gun vacuum. One gains more than an order of magnitude in etch rate but at the drawback of spot size and lifetime of the trimer. Zeiss products propose both He and Ne gases and eventually an additional Ga column. Examples of a 3 nm hole pierced in a suspended graphene sheet and Josephson junction on HTC superconductor are shown on Fig. 14. As for magnetic materials the damages induced by the ion are enough to destroy superconductivity, it is thus no needed to etch down the material to change its properties. The small straggling of and implantation He ions compare to Ga leads to a better resolution and sharp features [18] .

6.4 *Lithography*

The very small spot size of the He GFIS pleads also in favor of lithographic applications. Another significant property of He is the almost absence of proximity effect. Proximity effect are a major drawback of electron beam lithography (see Sect. 5 of this chapter). This is due to the strong extension of the secondary electrons spreading when the high energy electron beam hit the substrate. Because of strong interaction with matter, He ions are rapidly stopped upon their arrival on the surface, the emitted secondary electrons are within few nanometers from the impact point. Figure 15 left show Monte Carlo simulation for the electron diffusion of a 100 kV electron beam on silicon covered by a layer of resist [19] (100 kV beam are commonly used in modern electron beam lithography engine) and Fig. 15 right the case for a 30 kV He beam. Back scattered electrons extend on a very large area, up to 40 μm, in the resist around the impact of the electron beam whereas for Helium there are limited to a spot about few nanometers.

The quasi absence of proximity effect gives He beams a superior patterning capability when one wants to produced nested patterns or elaborate designs at sub 100 nm scale [20]. On the one hand, the fact that He ions interact strongly with material bring a very high sensitivity to resist which counterbalances the small current intensity in the beam and, one the other hand, it enables its usage on thick resists. Actually, the main applications are limited to thin negative resists.

Figure 16 gives some examples of high resolution line on aluminum oxide based resist [21]. A single line less than 5 nm width on 50 nm height resist is shown together with a dense array of lines 20 nm wide and with a pitch of 20 nm. Note that the width of the middle line of the array does not change shape even out of the dense array showing the quasi absence of proximity effect. Sub 10 nm lithography is easily achieved with this sharp focused helium He beam equipment.

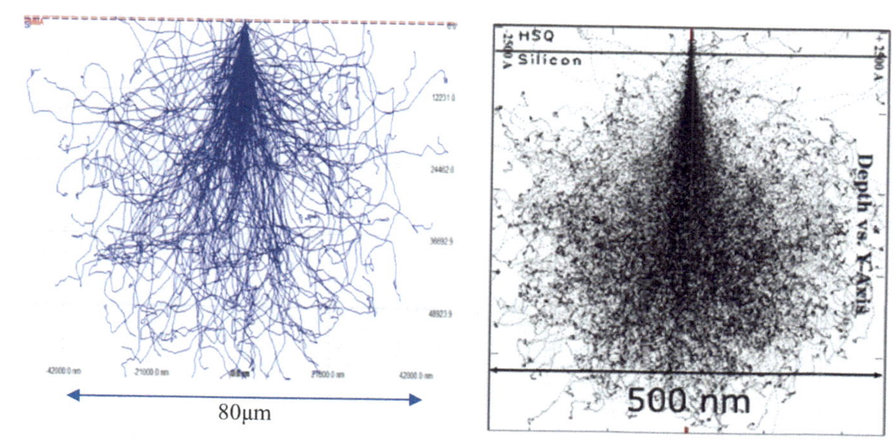

Fig. 15 Left: Monte Carlo simulation of a 100 kV electron impinging a silicon wafer coated by a fine layer of PMMA. Right: TRIM simulation for a 30 kV He beam on silicon coated by a layer of HSQ. Note the very different scale for the two figures

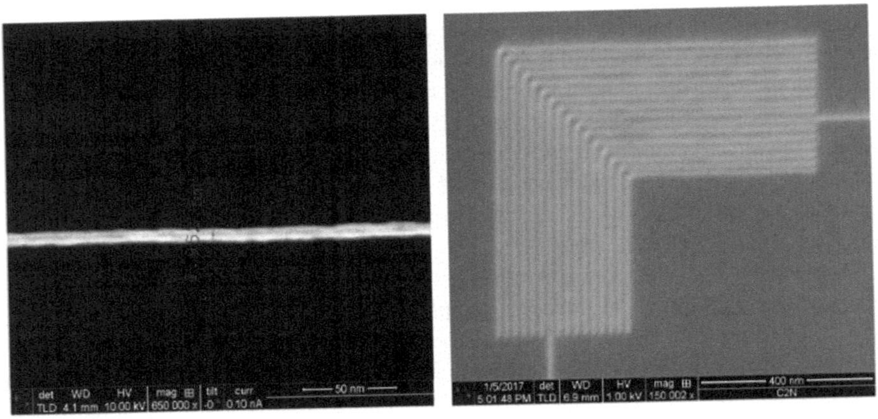

Fig. 16 Left: single 5 nm AlO_x wide line (50 nm height) Right: array of 20 nm AlO_x lines with a pitch of 20 nm

7 Nanogears Fabrication

The aluminum based negative resist has been used to define the patterns. The advantage of this resist lies in its better etching resistance and high resolution compare to the Hydrogen SilsesQuioxane (HSQ) one which is one of the best negative resist in term of resolution. A high etching resistance allows to pattern thinner resist which increases the resolution. Furthermore, the alumina resist offers a better resist profile with smaller roughness for sub 10 nm wire compare to the HSQ resist as shown Fig. 17.

On Fig. 18 we show some examples of 10 tooth nanogears on 30 nm thick alumina resist. The larger one is 180 nm outer diameter with a 50 nm hole and the smallest one 80 nm with a 26 nm hole. Typical doses are around 200 mC/cm^2. The absence

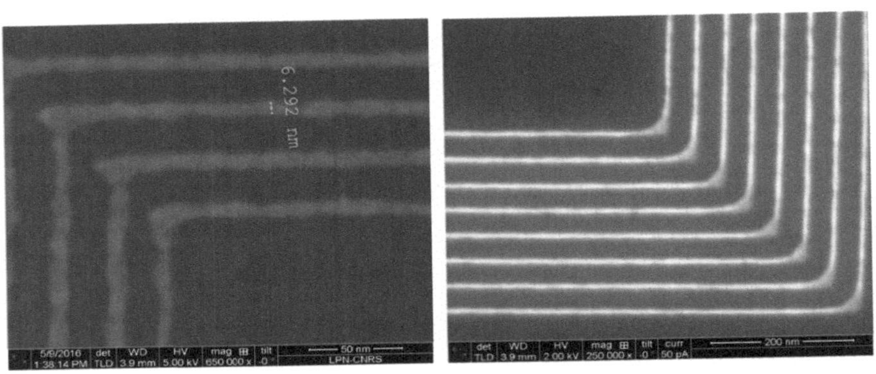

Fig. 17 Left: 6 nm lines on 30 nm thick HSQ resist. Right: 6 nm lines on 30 nm thick alumina resist

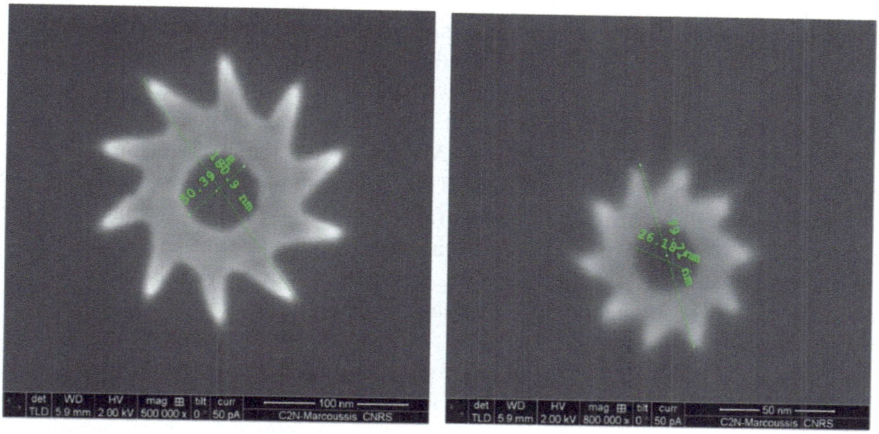

Fig. 18 Some examples of 10 tooth nanogears defined on resist. Left 180 nm outer diameter, right 80 nm outer diameter

of proximity effect shows up by the very sharp end of the tooth and by the very well defined hole in the center where an axis can stand. When trying to reach outer diameter smaller than 50 nm we have to decrease the number of teeth.

On the left side of Fig. 19 we show a 50 nm gear with 6 tooth. It is quite difficult at this scale to have an enough accurate image to determine if the hole in the center is present. This resist patterns can now serve as a mask against a fluorine based reactive ion etching (RIE) to etch the underlying silicon. This is shown on the tilted picture on the right side of the Fig. 19. The good definition of the tooth and the vertical profile of the etched edge stand for the strong resistance of the resist and the good

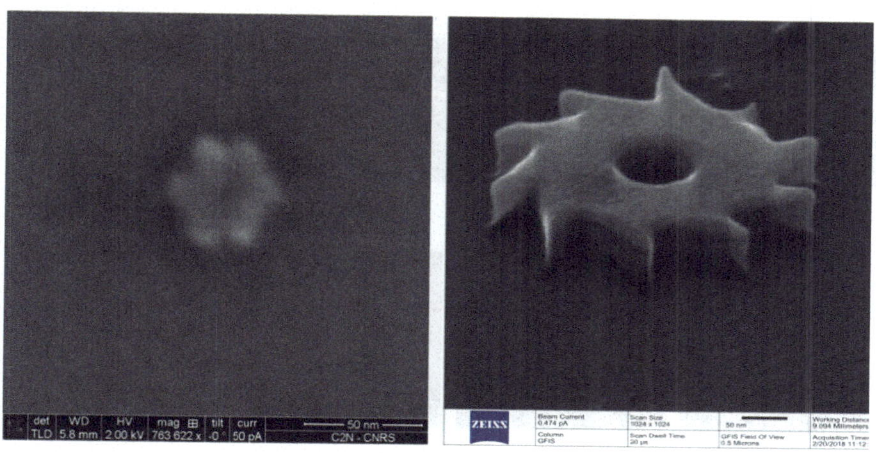

Fig. 19 Left: a 6 tooth nanogear 50 nm outer diameter. Right: a nanogear etched on silicone substrate

anisotropy of the RIE etching. 80 nm of silicon is thus etched with a remaining of 12 nm of resist out of the 30 nm used for this experiment.

In order to have a self-sustained gear we used a Si/20, SiO/25 nm Si silicon-on-insulator (SOI) substrate. The gear is first pattern on the resist providing a mask to etch the silicone down to the SiO layer. A wet HF etch removes all the SiO layer leaving the gear free from the substrate. This technique unable us to determine the exact position of the gear but will be helpful to test the possibility to rotate the gear using an AFM. A sequence of a total π rotation of a 25 nm thick silicon gear 120 nm out diameter is show as a 'cartoon' on Fig. 20. We used an AFM to both image and rotate the gear by scanning the tip in the manipulate mode acting on the longest teeth (which is acting for the carry).

Finally, the position of the gear needs to stick at a well determined position. This is necessary, firstly to rotate the gear without any translation and, secondly, to be able to couple several gears together in order to perform a complex mechanical movement. We achieve this by using the deposition ability of our tool. We first pattern the gear on the SOI substrate covered by resist. After development, we etch the top Si and the SiO layer down to the Si substrate. We then use the He beam to deposit a W wire in the hole of the gear. After dissolution of the SiO in a HF chemical wet etch, we obtain a gear free to move around its axis. The flow of the process is schematized on the Fig. 21.

Very first realizations of such process have been attempted using at first 125 nm outer diameter gears. In Fig. 22 we show an AFM image of two 25 nm thick Si gears with their central tungsten axis. We can observe very small thickness gears on the left side and in the front, being in fact the thin substrate trace of the etch process of the original gear that was sitting on the top. This is due to the fact that these Si gears disappeared during the rinse process after the SiO wet etch by sliding along the W axis. Axis with a large conical top end can be fabricated to avoid such an escape.

Fig. 20 Sequence of AFM images the manipulation of a 120 nm outer diameter 25 nm thick silicon gear using the AFM tip to manipulate the longest teeth

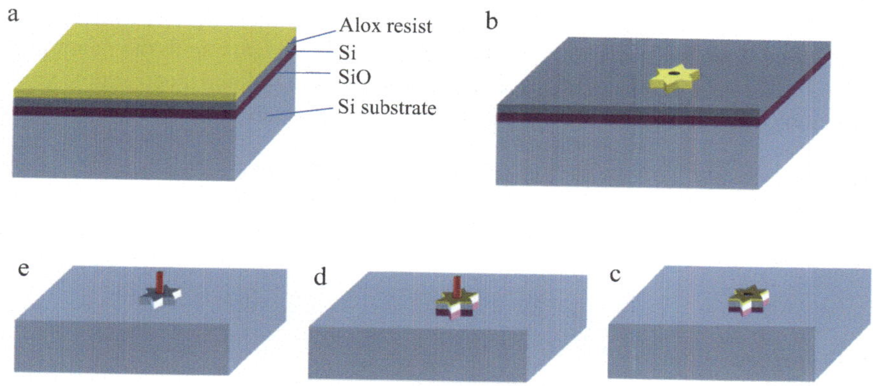

Fig. 21 Flow process for a free rotating Si gear with an axis fabricated on a Si substrate. **a** SOI substrate with resist. **b** Exposure of the gear. **c** RIE etching down to the substrate. **d** Growth of the axis. **e** Wet etching of the SiO

Fig. 22 AFM image of gears with central axis obtained using de SOI process describe above

8 Conclusion

We have listed in this chapter some of the high resolution techniques that can be used in the top-down approach to fabricate objects down to the few nanometer scale. We have also described the successful realization of solid state gears at the nanometer scale, free to move from the supporting surface, using a focused he ion beam. This top-down approach is the inverse of the bottom-up one allowing one to assemble, for instance, atom or molecule to build more complex systems up to the nanometer scale. These two approaches are used together in order to link the macroscopic word to the molecular one and this strategy is part of the purposes of the MEMO project.

Acknowledgements We thank the European Union Horizon 2020 FET open project "Mechanics with Molecule(s)" (MEMO, grant 766864) for financial support.

This work was partly supported by the French RENATECH network.

References

1. see: http://www.zyvex.com/nanotech/feynman.html
2. Atalla, M.M., Tannenbaum, E., Sheibner, E.J.: Stabilization of silicon surfaces by thermally grown oxides. Bell Syst. Tech. J. 749–783 (1959)
3. Atalla, M.M.: Semiconductor devices having dielectric coatings US Patent 3, 206, 670 (filled March 1960) 1965 and Khang, D. Electric field controlled semiconductor device US Patent 3, 102, 230 (filled Mai 1960) (1963)
4. https://www.asml.com/en/products/duv-lithography-systems
5. https://www.asml.com/en/products/euv-lithography-systems
6. https://www-03.ibm.com/press/us/en/pressrelease/659.wss#release
7. https://heidelberg-instruments.com/en/features-technologies/key-features/thermal-scanning-probe-lithography.html?gclid=CjwKCAjwp-X0BRAFEiwAheRui42iqHWFyq52v4gbOsW5T_KoAsk_SfETy_BlUVMBAxSK4zjq1DoXmhoCnuQQAvD_BwE
8. Fay, A., Thiam, N.A., Cordini, M.-L., Servin, I., Constancias, C., Lattard, L., Pain, L.: "Fast" & "Thick" e-beam resists exposed with multi-beam tool at 5 keV for implants and mature nodes: experimental and simulated model study. In: Alternative Lithographic Technologies VII, Proceedings of SPIE, 9423, 94231Q1-94231Q15 (2015)
9. Paysant, M., Ulysse, C., Faini, G.: Exposition OnLab, le musée des poussières 2009, Musée du Louvre and https://www.michelpaysant.fr/onlab/onlab2/
10. Wieland, M.J.-J., Kampherbeek, B.J., Vincent van Veen, A.H., Kruit, P.: Electron beam exposure system. US Patent 97,458 B2 (2005)
11. Gierak, J.: Focused ion beam technology and ultimate applications. Semicond. Sci. Technol. **24**(4), 043001 (2009)
12. Müller, E.W.: Das Feldionenmikroskop Zeitschrift für Phys. **131**, 136–142 (1951)
13. Ward, L.B.W., Notte, J.A., Economou, N.P.: Helium ion microscope: a new tool for nanoscale microscopy and metrology. J. Vac. Sci. Technol. **B24**, 2871 (2006)
14. Charlier, P., Weil, R., Deblock, R., Augias, A.: Deo, Helium ion microscopy (HIM): proof of the applicability on altered human remains (hairs of Holy Maria-Magdalena). Leg. Med. **24**, 84–85 (2017)
15. Córdoba, R., Mailly, D., Rezaev, R., Smirnova, E., Schmidt, O.G., Fomin, V.M., Zeitler, U., Guillamón, I., Suderow, H., De Teresa, J.M.: Three-dimensional superconducting nanohelices grown by He+focused ion beam direct writing. Nano Lett. 19(12), 8597–8604 (2019)
16. Córdoba, R., Ibarra, A., Mailly, D., De Teresa, J.M.: Vertical growth of superconducting crystalline hollow nanowires by He+ focused ion beam induced deposition. Nano Lett. 18(2), 1379–1386 (2018)
17. Basset, J., Watfa, D., Aiello, G., Féchant, M., Morvan, A., Estève, J., Gabelli, J., Aprili, M., Weil, R., Kasumov, A., Bouchiat, H., Deblock, R.: High kinetic inductance microwave resonators made by He-Beam assisted deposition of tungsten nanowires. Appl. Phys. Lett. **114**(10), 102601 (2019)

18. Couëdo, F., Amari, P., Feuillet-Palma, C., Ulysse, C., Srivastava, Y.K., Singh, R., Bergeal, N., Lesueur, J.: Scientific Reports (2020, in press)
19. Urbánek, M., Kolarik, V., Krátký, S., Matějka, M., Horacek, M., Chlumská, J.: The optical properties of photonic-crystal nanocavities containing plasmonic nanoparticles. In: JO—NANOCON 2013—Conference Proceedings, 5th International Conference
20. Fallica, R., Kirchner, R., Ekinci, Y., Mailly, D.: Comparative study of resists and lithographic tools using the Lumped Parameter Model. J. Vac. Sci. Technol. **B34**, 06K702 (2016)
21. Cattoni, A., Mailly, D., Dalstein, O., Faustini, M., Seniutinas, G., Rösner, B., David, C.: Sub-10 nm electron and helium ion beam lithography using a recently developed alumina resist. Microelectron. Eng. **193**, 18 (2018)

Prototypes of Molecular Gears with an Organometallic Piano-Stool Architecture

Seifallah Abid, Guillaume Erbland, Claire Kammerer, and Gwénaël Rapenne

Abstract In the field of Molecular Machines, molecular gears have mainly been synthesized to be studied in solution. Then, the cogwheel subunits are restricted to be only arranged in an intramolecular manner. In the last years, the possibility to arrange a train of gears at the single molecular level and observe the propagation of a rotation motion from one molecule to its neighbor using Scanning Tunneling Microscopy (STM) opened new perspectives with the opportunity to have intermolecular arrangements on surfaces. In this chapter, we describe the research background of single molecular gears and our strategy using organometallic piano-stool complexes to anchor such gears on surfaces. Our molecules incorporate two subunits linked together through a ruthenium center acting as a ball bearing. The lower part is the anchoring tripodal ligand and the upper part the cogwheel. Various functionalities have been explored to behave as teeth, ranging from mono-dimensional phenyl rings to bi-dimensional porphyrin fragments.

Keywords Molecular machines · Molecular gear · Ruthenium complexes · Single molecule · Porphyrins · STM

1 Introduction

1.1 Molecular Machines

A revolution in the way chemists consider molecular systems and their associated motions took place over the last decades with the rise of a new field dedicated to Molecular Machines. Macromolecular biological systems acting as machines have

S. Abid · G. Erbland · C. Kammerer · G. Rapenne (✉)
CEMES, Université de Toulouse, CNRS, 29, Rue Jeanne Marvig, 31055 Toulouse, France
e-mail: gwenael.rapenne@cemes.fr

G. Rapenne
Division of Materials Science, Nara Institute of Science and Technology, Takayama, Ikoma, Nara 8916-5, Japan

C. Joachim (ed.), *Building and Probing Small for Mechanics*, Advances in Atom and Single Molecule Machines, https://doi.org/10.1007/978-3-030-56777-4_5

been long optimized by Nature and are key elements to many functions in living organisms. However, in artificial molecular systems, the design and synthesis of molecular machines for which motion and function can be controlled remains a highly challenging task. In this context, Jean-Pierre Sauvage, Sir Fraser Stoddart and Ben Feringa were awarded the 2016 Nobel Prize of Chemistry for their seminal and fundamental contributions to this field [1–3].

A large variety of artificial molecular machines and motors have been developed in a monumentalization approach, starting with small molecular building blocks and building step by step a complexified architecture until the molecule becomes a machine. In particular, daily life macroscopic objects have been miniaturized according to a technomimetic approach [4] leading to nanometer scale syringes [5], wheels [6–9], wheelbarrows [10–12], vehicles [13–17], scissors [18], elevators [19], gyroscopes [20], motors [21–32] among others.

In spite of the tremendous developments witnessed in this field, applications of ultimately miniaturized machines in the macroscopic world are still rare. One of the strategies to exploit such nanoscale units is to deposit them on surface and use them as self-assembled monolayers or single molecules thanks to the amazing capabilities offered by Scanning Tunneling Microscopy (STM). Beyond the high-resolution imaging possibilities, the use of the STM tip can allow to induce and control precise molecular motions from the single molecular scale [33] to the switch of large domains [34].

In this context, one fascinating target is a train of molecular gears rotating synchronously on a surface to be able to transfer energy at the nanoscale from a molecule to another one over long distances.

1.2 Gears in the Insects' World

The oldest known machine involving gearing mechanisms is the Antikythera machine, built around 200 BC. However, in contradiction with the common idea that gears are a human invention, micrometric gears were recently discovered in the nymphs of the planthopper *Issus coleoptratus* [35]. Indeed, the exoskeleton of their hind legs incorporates a curved strip with up to 12 micrometric gear teeth, as observed by Scanning Electron Microscopy (Fig. 1). Both legs are thus mechanically coupled through this gearing mechanism, leading to a synchronized motion from the preparation to the propulsion of the jump, that can reach 2 m. Interestingly, it must be noted that the dissymmetric character of the teeth allows only one direction of powered rotation.

According to this recent discovery, synthesis of molecular gears can be defined not only as technomimetic [4] but also as biomimetic [36].

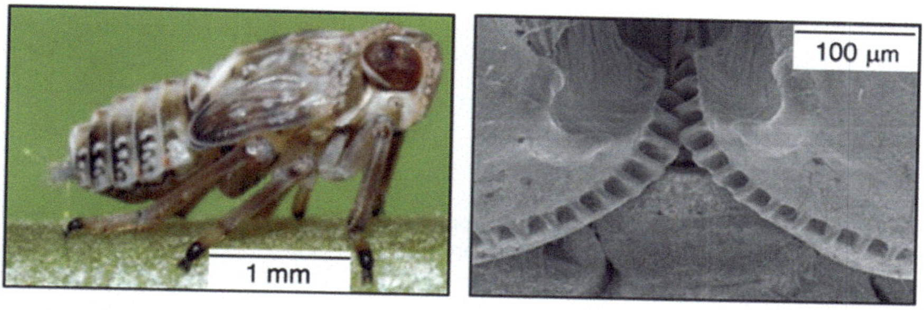

Fig. 1 Side view of a planthopper *Issus coleoptratus* nymph (left) and scanning electron microscopy image of meshed micrometric gears located on the hind legs (right). From Ref. [35]. Adapted with permission from AAAS

2 Molecular Gears from Solution to Surface

2.1 Gears in Solution: Drawbacks and Limitations

The first examples of synthetic molecular gears appeared forty years ago as intramolecular systems involving two intertwined subunits undergoing correlated motion in solution. The two pioneers in this field, Kawada and Iwamura [37] and Hounshell et al. [38], first reported compounds exhibiting bevel-gear mechanisms, with two triptycene moieties linked by a very short spacer such as a methylene group in the case of Mislow (Fig. 2, left) or an oxygen atom in the case of Iwamura (Fig. 2, center). The first molecular spur gears with two triptycene units having parallel axes of rotation (Fig. 2, right) were designed and synthesized in 2012 by J. Siegel, a former Ph.D. student of K. Mislow, with a particular attention paid to the inter-axle distance as key parameter in molecular gearing mechanisms [39].

In the last years, M. Shionoya reported a new series of molecular gears which can be controlled at will by an external stimulus. These complexes are based on platinum(II) with two azaphosphatriptycene ligands acting as cogwheels. Under thermal

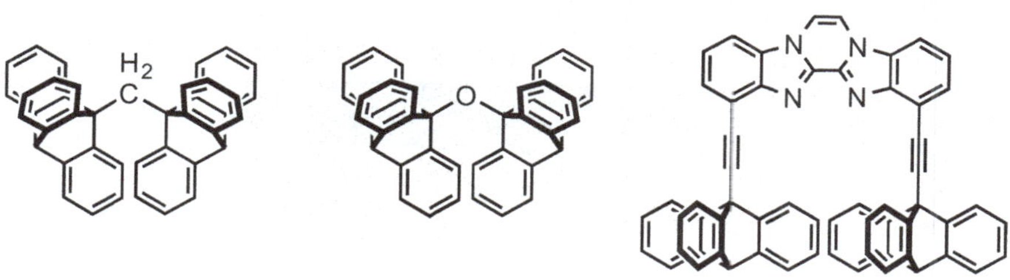

Fig. 2 Molecular gears proposed by Hounshell et al. [38] (left), Kawada and Iwamura [37] (center) and Frantz et al. [39] (right), all are based on triptycene cogwheels

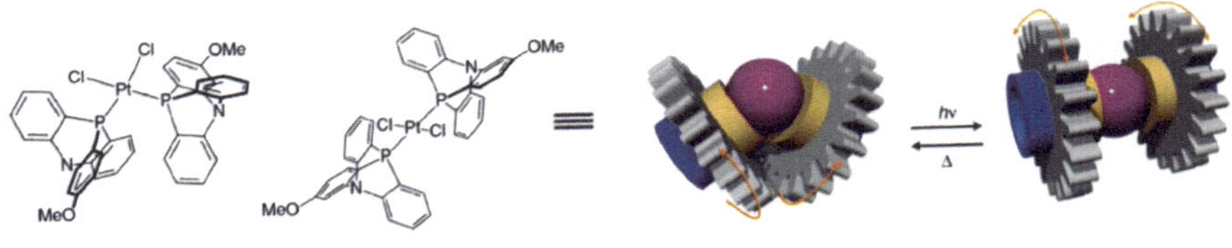

Fig. 3 Chemical structure (left) and schematic representation (right) of the engaged and disengaged molecular gears proposed by Shionoya et al. (adapted from Ref. [40])

or photo-activation, *cis-trans* isomerization of the complex leads very elegantly to the engagement or disengagement of the cogwheels (Fig. 3) [40].

A vast majority of the studies on correlated motion, such as the ones presented above, have been performed in solution, at the intramolecular level. The observed behavior is thus the average behavior of an assembly of molecules and not of a single molecule. Moreover, the lack of orientation in solution prevents transfer of motions (i.e. energy) over long distances, which is a limitation for applications. As an alternative, intermolecular gearing becomes possible on a supporting surface, which enables to maintain the distance between cogwheels and the relative orientation of the axles.

2.2 *Gears on Surface: Single Molecules and Atomic Resolution*

Intermolecular gearing on a surface was evidenced for the first time in 2007 by Moresco et al. in a rack and pinion device made of penta(*p*-*tert*-butylphenyl)-(*p*-*tert*-butylpyrimido)benzene molecules on a Cu(111) substrate (Fig. 4) [41]. Indeed,

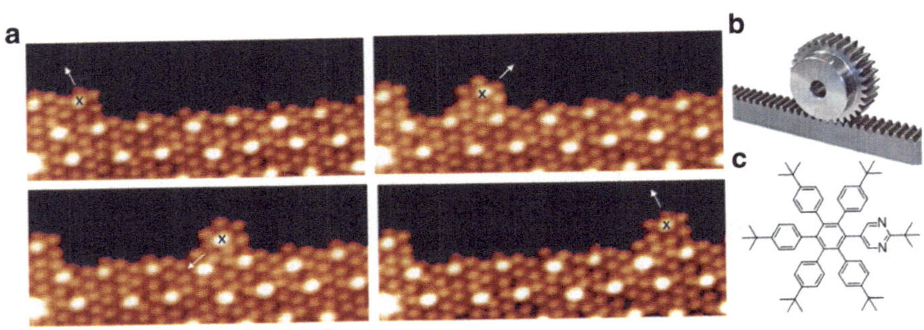

Fig. 4 STM images (**a**) showing the displacement of a pinion (single molecule) along a rack (self-assembled monolayer of the same molecule) on Cu(111). The white arrow gives the orientation of the molecule according to the pyrimidine tagged tooth. A macroscopic rack and pinion (**b**) and the chemical structure of the molecule (**c**). From Ref. [41], adapted with permission from Springer Nature

STM lateral manipulation of one single molecule along the edge of the close-packed island obtained after on-surface deposition resulted in a rotary motion of this per-arylated benzene pinion, as a consequence of the teeth intertwining. This rotation was demonstrated by comparing successive STM images and following the trajectory of the chemically-tagged pyrimidine tooth.

The same gear prototype was further investigated on Au(111) as a single molecule. When mounted on an atomic defect, the controlled and stepwise rotation of this per-arylated benzene molecule was successfully achieved [42], thus opening the way for intermolecular gearing on surface at the single molecular scale.

Unfortunately, when a second molecule was added in close vicinity to form a train of two gears, no correlated motion could be observed due to diffusion of the cogwheels. This underlines the prime importance of rotation axles to anchor cogwheels on surface at optimized intermolecular distances, as a prerequisite to favor transmission of rotary motion over diffusion.

A promising strategy to build intermolecular gearing systems consists in mounting molecular cogwheels on top of single atoms deposited on surface. This has been recently demonstrated by Joachim et al., who successfully mounted by STM lateral manipulation the per-arylated benzene cogwheel depicted in Fig. 4 on top of a Cu single atom on a Pb(111) surface. In the presence of this centered rotational axle, controlled rotation of single cogwheels and correlated motion within a train of two gears was achieved [43].

2.3 *Our Strategy*

As an alternative, we envisioned to pre-assemble the cogwheel on top of a single atom axle in the frame of a coordination complex, synthesized in solution and subsequently deposited on surface. The architecture of our previously-reported ruthenium-based motor [29] appeared particularly promising, since it has been shown that such structures can be displaced and arranged in a very precise assembly through STM lateral manipulation at 77 K, while diffusion is frozen at 5 K. This would thus enable the specific arrangement of trains of organometallic gears with finely tuned intermolecular distances at 77 K and would prevent gears disengagement due to diffusion by lowering the temperature to 5 K. Indeed, at that temperature, the motion of such ruthenium complexes is efficiently restricted by the strong interaction between the surface and the tripodal scorpionate ligand [44], which not only prevents translation but also gives a very efficient control on the geometry of the whole architecture thanks to the presence of three anchoring points.

This rigid tripodal ligand also holds a second key function as it supports the ruthenium ion, acting as single atom axle for the star-shaped cogwheel. The structure of the cogwheel should be as rigid as possible with few degrees of freedom in order to obtain a maximum control over the mechanically-driven motion. Indeed, in these experiments, the STM tip will be used as a source of mechanical energy to induce the rotary motion of the first cogwheel, that will be mechanically transmitted to the

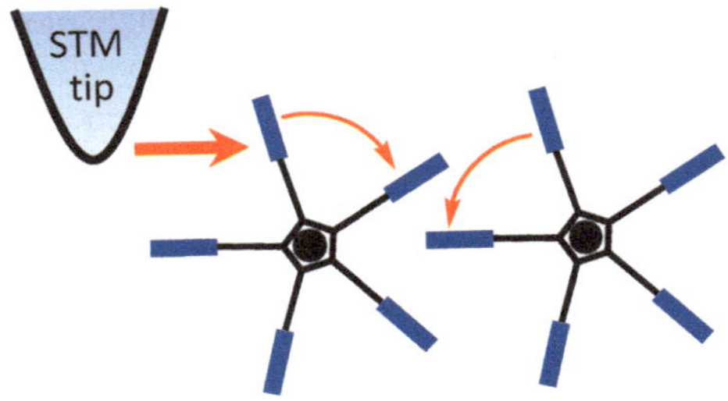

Fig. 5 Expected transmission of rotation from one molecule to its neighbor using a STM tip as source of mechanical energy

next one if gearing mechanism is effective (Fig. 5). The propagation of rotary motion within the train of gears will then be evidenced by comparing the STM images of the initial and final states. As we will see later, it will be helpful to have one of the teeth chemically or sterically tagged.

In this chapter we will focus on the design and molecular synthesis of cogwheel prototypes to build a train of gears based on star-shaped organometallic ruthenium complexes.

3 Molecular Gears Based on Organometallic Ruthenium Complexes

3.1 *Design*

As mentioned above, our design for molecular gears including a rotational axle is based on the previously-reported ruthenium-based motor [29] shown Fig. 6 (left). The lower part of gears is a tripodal thioether-functionalized hydrotris(indazolyl)borate ligand to anchor molecules on metallic surfaces, but also to alleviate the ruthenium rotational axle and thus the upper cogwheel. A pentaphenylcyclopentadienyl moiety is used in this case as a central platform to modularly attach various peripheral groups acting as teeth. We explored two types of architectures, which are schematically

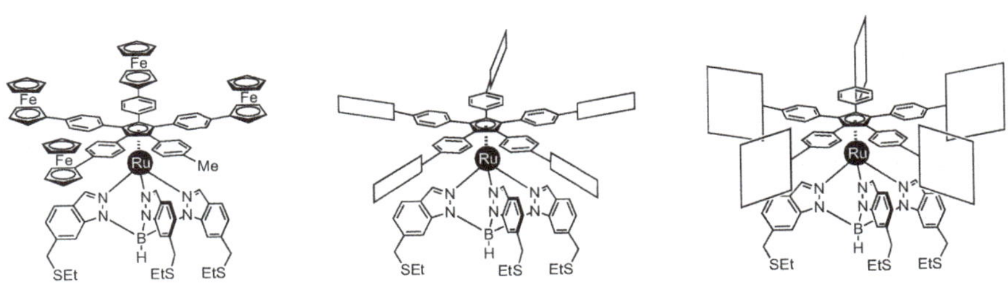

Fig. 6 Structure of the parent compound, a molecular motor [29] (left), and of the cogwheels with five mono-dimensional arms (center) and with five bi-dimensional paddles (right)

presented in Fig. 6 (center and right): the first architecture integrates five mono-dimensional arms and the second one five bi-dimensional paddles.

3.2 Efficient Synthesis of the Ruthenium Complex Precursor

From a synthetic point of view, both architectures of molecular gears featuring five identical mono- or bi-dimensional teeth stem from the same precursor via five-fold cross-coupling reactions. The key intermediate in this strategy is thus a ruthenium(II) complex bearing the tripodal scorpionate ligand on one side and a penta(*p*-bromophenyl)cyclopentadienyl ligand on the other side (Scheme 1). Owing to the importance of this building block for the synthesis of ruthenium-based rotary molecular machines, its synthesis was recently optimized [45]. It is now prepared in a reproducible manner in 5 steps and 31% overall yield on the longest linear sequence, compared to 8 steps formerly with a maximum yield of 4%.

3-Amino-4-methylbenzoic acid was first protected as an ethyl ester under standard conditions, and the o-toluidine moiety was next converted into an indazole scaffold using the modified one-pot Jacobson procedure with isoamylnitrite as nitrosating agent. Step economy was achieved at that stage in this optimized route thanks to a highly efficient one-pot "*N*-deprotection/ester reductive sulfidation" sequence mediated by indium(III), allowing the direct conversion of *N*-acetyl indazolyl ester into the corresponding *N*-deprotected indazolyl thioether in a single step (instead of three). The reaction proceeds in the presence of indium(III) iodide as Lewis acid, an excess of 1,1,3,3-tetramethyldisiloxane as reducing agent and ethanethiol as nucleophile. Importantly, this one-pot sequence not only shortens the synthetic route towards the key ruthenium complex but also avoids the troublesome isolation of intermediates.

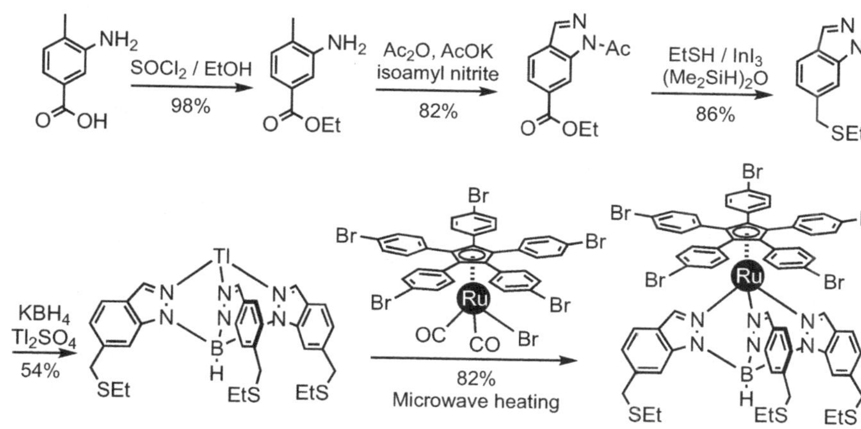

Scheme 1 Optimized synthetic route yielding the ruthenium key precursor coordinated to the tripodal ligand (lower side) and a pentabrominated functionalized cyclopentadiene (upper side)

The next major improvement in the synthetic sequence relies on the replacement of potassium by thallium during the formation of the tripodal ligand. Indeed, the thallium salt of the tris(indazolyl)borate ligand was obtained via a thermally-controlled reaction of indazolyl thioether with potassium borohydride in the presence of thallium sulfate, leading to a highly pure crystalline product. Finally, the ruthenium(II) key precursor resulted from a ligand exchange on the bromidodicarbonyl-1,2,3,4,5-penta(*p*-bromophenyl)cyclopentadienyl ruthenium(II) complex, with a positive effect of microwave irradiation on this reaction.

3.3 Molecular Gears Based on Cogwheels with Mono-Dimensional Arms

With an optimized and shortened access to this ruthenium complex precursor, the synthesis of molecular gear prototypes has been pursued and cogwheels with mono-dimensional teeth were first targeted. In this family, the cyclopentadienyl ring is thus surrounded with five 4,4'-biphenyl moieties, each carrying one or two methyl or *tert*-butyl substituents on the outer positions (Fig. 7). The synthesis of these three prototypes relies on a five-fold Suzuki-Miyaura cross-coupling reaction of the penta(*p*-bromophenyl)cyclopentadienyl ruthenium(II) key precursor in the presence of an excess of the appropriate boronic acid derivative [46]. Using $Pd(OAc)_2$/SPhos as catalytic system and potassium phosphate as base in anhydrous toluene, the penta-substituted gear prototypes were obtained in 52–82% yield. These optimized conditions thus allow an efficiency as high as 88–96% for each single C–C bond formation. Full characterization of each complex confirmed the C_5-symmetric character of the upper cogwheel.

According to preliminary studies, such gear prototypes seem to be ineffective in propagating rotary motion to their neighbors. Theoretical calculations performed on

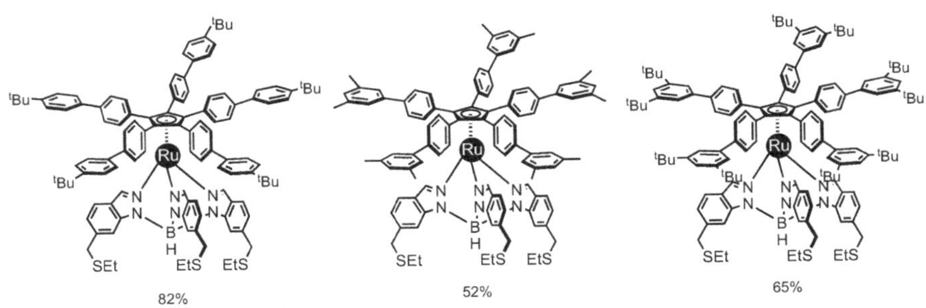

Fig. 7 Chemical structure of the three gear prototypes with mono-dimensional arms and respective yields of the quintuple cross-coupling reaction to form them

intricated gears revealed that upon rotation of the first cogwheel, a vertical deformation of the teeth can occur, thus preventing the transfer of rotation to the next molecule.

To shut down this deformation pathway, the design of gear prototypes was adapted with 2D-paddles (Fig. 6, right) instead of 1D-teeth (Fig. 6, center). The very large steric hindrance of the groups surrounding the cyclopentadienyl core in this second generation of cogwheels is expected to prevent gear slippage. Indeed, according to the size and rigidity of these paddles, it seems impossible to rotate the first gear without rotating its neighbor provided the intermolecular distance is appropriate, and mechanical transmission of motion should thus be favored with this new design.

3.4 Molecular Gears Based on Cogwheels with Bi-dimensional Paddles

Three different chemical fragments exhibiting flat structures and large areas were selected as bi-dimensional paddles for this second generation of cogwheels. Four gear prototypes incorporating carbazoles, BODIPYs and porphyrins as paddles and exhibiting increasing diameters up to 5.2 nm were thus synthesized [46]. Their chemical structures are shown in Fig. 8, and one can notice that bulky alkyl groups have systematically been added to paddles to increase the solubility during the synthesis and above all, to avoid π-stacking between neighboring molecules on surface, as this could hamper the gears motion.

The four gear prototypes bearing 2D-paddles were prepared via a five-fold cross-coupling reactions applied to the appropriate brominated or iodinated penta(*p*-halogenophenyl)cyclopentadienyl ruthenium(II) key precursor. By way of example, a five-fold copper-catalyzed *N*-arylation reaction in the presence of 3,6-di-*tert*-butyl-*9H*-carbazole afforded the pentacarbazole gear (Fig. 8, top right). Starting from penta(*p*-bromophenyl)cyclopentadienyl ruthenium(II) precursor, the target compound was obtained in 19% yield under optimized conditions involving copper iodide and (±)-*trans*-cyclohexane-1,2-diamine as catalytic system in combination with potassium phosphate as base in refluxing 1,4-dioxane. Interestingly, when the same reaction conditions were applied to the more reactive iodinated precursor (same ruthenium complex with 5 iodine atoms instead of five bromines), the yield could be increased to 28%, corresponding to 78% per C–N coupling.

BODIPY fragments were next selected to vary the nature of the paddles and enlarge the cogwheel diameter. The standard Suzuki-Miyaura conditions used previously for the preparation of gears with 1D-arms failed to efficiently deliver the desired penta-substituted product, leading to mixture of compounds resulting from one to five C–C couplings. After optimization, it appeared that higher coupling efficiency could be reached in a ternary solvent system combining toluene, ethanol and water (2:2:1), and the penta-BODIPY gear prototype was obtained in 36% yield.

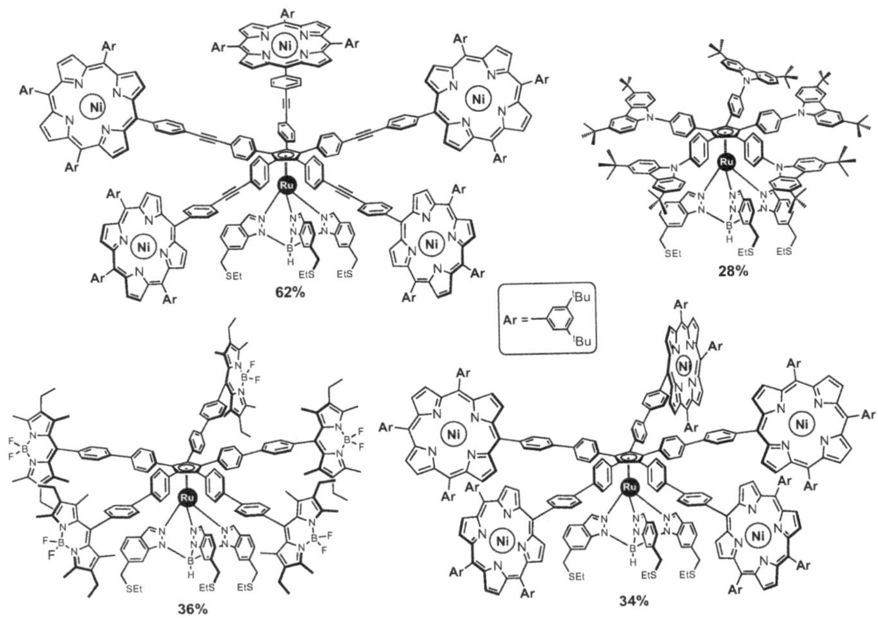

Fig. 8 Chemical structure of the four gears with bi-dimensional paddles and respective yields of the quintuple cross-coupling reaction to form them

Finally, the largest paddle scaffold relied on a nickel porphyrin macrocycle featuring 3,5-di-*tert*-butylphenyl groups on three *meso* positions to enhance stability as well as solubility and limit π-stacking. Metallation of the porphyrin moieties with nickel(II) was meant to prevent any undesired metallation on surface during STM studies, owing to the robustness of nickel porphyrins.

As a strategy to modulate the diameter of the cogwheel, it was envisioned to vary the length of the spacer between the cyclopentadienyl and tetrapyrrole cores. Starting from penta(*p*-halogenophenyl)cyclopentadienyl ruthenium(II) key precursor, a Suzuki-Miyaura coupling would deliver a biphenyl spacer whereas a Sonogashira reaction would provide a longer phenylethynylphenyl linker. The required A_3B-type porphyrin precursors incorporating a phenylboronic ester moiety or a phenylethynyl group on the distinct meso position were synthesized under statistical conditions according to Lindsey's method [47].

The pentabrominated key precursor and its porphyrinic boronic acid pinacol ester coupling partner were submitted to Suzuki-Miyaura reaction conditions in the presence of $Pd(dppf)Cl_2$ as catalyst and cesium carbonate as base at 100 °C in a DMF/water system (99:1). The desired gear carrying five porphyrin paddles and 4,4'-biphenyl spacers was obtained in 34% yield.

Extension of the cogwheel diameter was next attempted by reacting the pentaiodinated precursor with the alkynylporphyrin under milder conditions involving $Pd(PPh_3)_2Cl_2$ and CuI as catalysts in triethylamine at 45 °C. The pentaporphyrin gear resulting from a five-fold Sonogashira reaction and exhibiting a diameter over 5 nm was obtained in 62% yield.

All these gear prototypes were fully characterized by standard techniques involving mass spectrometry and NMR spectroscopy. The latter unambiguously showed that free rotation of the upper platform is maintained at room temperature, even in the most sterically hindered architectures.

In view of on-surface studies, it was anticipated that a dissymmetry element in the cogwheels, acting as a tag, would be desirable to monitor by STM possible gearing effect. The design and synthesis of ruthenium complexes carrying a dissymmetrized cogwheel was thus tackled.

3.5 Dissymmetrized Cogwheels

3.5.1 How Can We Probe the Rotation Using STM?

In the absence of time-resolution in STM imaging, molecular motion is probed by comparing images taken before and after the rotation event. When deposited on surface, the gear prototypes synthesized above will appear as pseudo C_5-symmetric objects, in line with the cogwheel structures. Tracking and quantifying motion might thus be challenging. The presence of a tag, i.e. a chemical or steric variation lowering the object symmetry, appears highly desirable for in-depth mechanical studies at the molecular scale. As depicted on Fig. 9, dissymmetrized cogwheels incorporating one distinct arm (in red) would allow for a straightforward differentiation between the initial configuration and the final state, resulting from disrotatory motion of neighboring molecules.

It was thus envisioned to synthesize a family of star-shaped ruthenium complexes incorporating one tagged porphyrinic paddle. To this aim, the possibility to tune the length of the spacer between cyclopentadienyl and porphyrinic cores was exploited to incorporate one longer arm in the cogwheel structure. A single Sonogashira coupling on one halogenophenyl position of the ruthenium(II) key precursor would

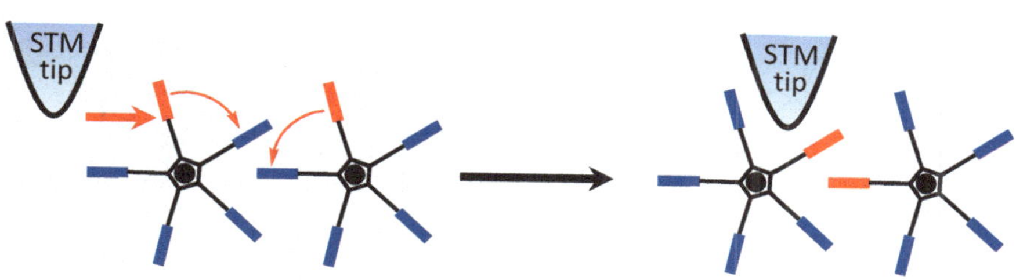

Fig. 9 Principle of the motion transmission through a molecular gear train actuated by a STM tip. Initial configuration (left) and final configuration (right) after a fifth of a turn (72°). The first gear rotates clockwise pushed by the STM tip and induces the counterclockwise rotation of the next molecule

generate the longer phenylethynylphenyl spacer and subsequent four-fold Suzuki-Miyaura cross-couplings on the remaining positions would lead to the four shorter 4,4'-biphenyl linkers.

In view of the functionalization of one single arm via Sonogashira coupling, it is important to note that the five halogenated positions in the key building block are equivalent and thus react independently when submitted to statistical reactions conditions, leading to mixture of mono- and polysubstituted products. To overcome this issue, a chemoselective single Sonogashira reaction relying on the higher reactivity of aryl iodides compared to aryl bromides was targeted, and the corresponding penta(*p*-halogenophenyl)cyclopentadienyl ruthenium(II) precursor carrying one iodine and four bromine atoms was prepared [48].

3.5.2 Synthesis of a Dissymmetric Ruthenium Precursor

The dissymmetric ruthenium key precursor bearing one pre-activated position in view of a chemoselective single cross-coupling was obtained in four steps from tetra(*p*-bromophenyl)cyclopentadienone (Scheme 2). Introduction of the iodinated arm was performed by 1,2-addition of 4-iodophenyllithium to give the corresponding cyclopentadienol in 94% yield. Bromination using a mixture of hydrobromic and acetic acids almost quantitatively afforded bromocyclopentadiene as a mixture of three regioisomers. Subsequent reaction with ruthenium(0) cluster $Ru_3(CO)_{12}$ led to a selective insertion into the central C-Br bond to yield a single ruthenium(II) complex carrying the dissymmetrized penta(*p*-halogenophenyl)cyclopentadienyl

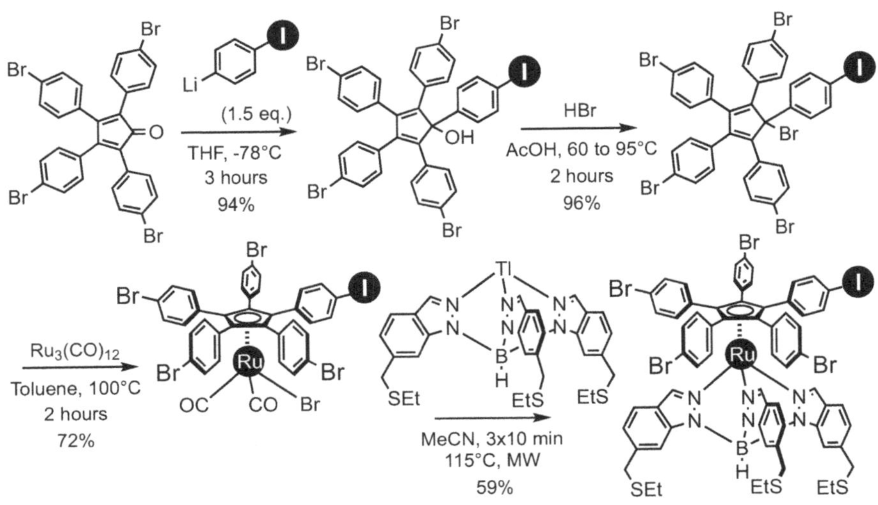

Scheme 2 Synthesis of 1-(*p*-iodophenyl)-2,3,4,5-tetra(*p*-bromophenyl)cyclopentadienyl hydrotris(indazolyl)borate ruthenium(II)

ligand. Finally, ligand exchange in the presence of the thallium salt of thioether-functionalized tripod afforded the desired dissymmetric precursor in 38% overall yield.

This dissymmetric penta(*p*-halogenophenyl)cyclopentadienyl ruthenium(II) key building block was then used for the selective synthesis of a gear prototype with a sterically-tagged cogwheel.

3.5.3 Synthesis of a Dissymmetric Pentaporphyrinic Cogwheel

As shown on Scheme 3, the dissymmetrized key precursor was engaged in sequential cross-coupling reactions, to selectively install the longer tooth in a first step and subsequently insert the four remaining shorter teeth. It is important to note that the nickel porphyrin structure remains identical in the five paddles and dissymmetry only arises from the length of the linkers.

The Sonogashira coupling was carried out under mild conditions in the presence of the A_3B-type porphyrin **A** bearing a terminal alkyne to chemoselectively functionalize the iodinated position of the precursor. Using $PdCl_2(PPh_3)_2$ and CuI as

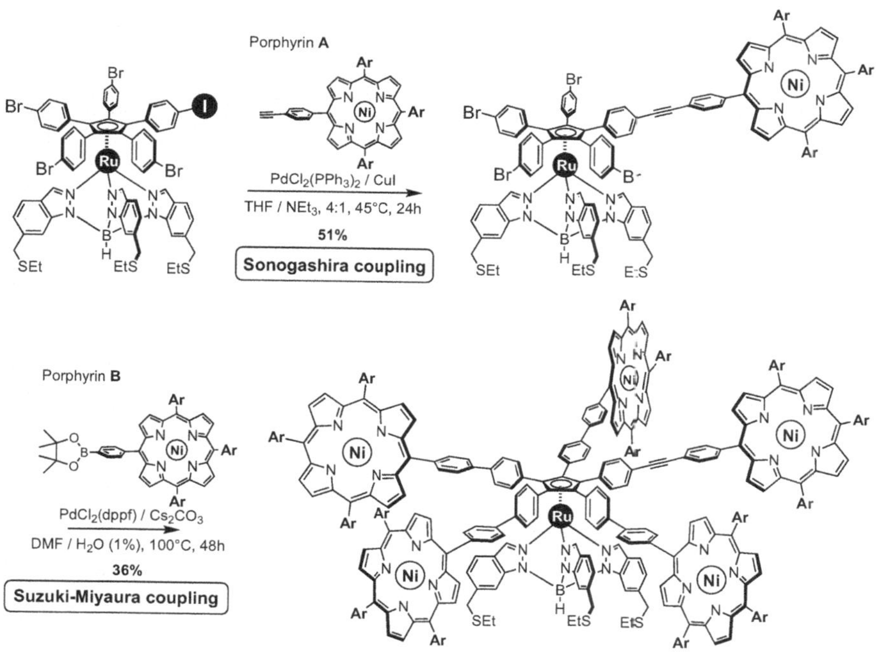

Scheme 3 Synthetic route to the dissymmetrized gear prototype. One of the spacers between the cyclopentadienyl ligand and the porphyrins contains an additional ethynyl fragment. Ar = 3,5 di-*tert*-butylphenyl

catalysts in a mixture of tetrahydrofuran and triethylamine (4:1), the desired mono-porphyrinic ruthenium complex was isolated in 51% yield after 24 h at 45 °C. This monofunctionalized intermediate was next reacted with a large excess of A_3B-type porphyrin **B** bearing a boronic ester in the presence of cesium carbonate as base and $PdCl_2$(dppf) as catalyst. After 48 h at 100 °C in a DMF/water (99:1) mixture, all the remaining *p*-bromophenylene groups were functionalized and the gear prototype carrying five identical porphyrin fragments but different spacers was obtained in 36% yield [48].

This first example of a dissymmetric ruthenium-based molecular gear was thus obtained in two steps from 1-(*p*-iodophenyl)-2,3,4,5-tetra(*p*-bromophenyl)cyclopentadienyl hydrotris(indazolyl)borate ruthenium(II) *via* sequential chemoselective Sonogashira/four-fold Suzuki-Miyaura cross-couplings.

4 Conclusion and Perspectives

Alternative strategies are now developed to obtain dissymmetric gears with spacers of identical length, using various porphyrin structures. For that purpose, chemoselective conditions of Suzuki-Miyaura couplings must be identified to discriminate iodoaryl versus bromoaryl groups. Work is also underway to study such dissymmetric molecular gears on surface.

Acknowledgements This work was supported by the University Paul Sabatier (Toulouse, France) and the CNRS. It has received funding from the Agence Nationale de la Recherche (ANR) (ACTION project ANR-15-CE29-0005) and from the European Union's Horizon 2020 research and innovation program under the project MEMO, grant agreement No 766864. This research was also partly supported by the JSPS KAKENHI grant in aid for Scientific Research on Innovative Areas "Molecular Engine (No. 8006)" 18H05419. Dr. Colin Martin is warmly acknowledged for his careful reading and improving of our manuscript.

References

1. Sauvage, J.-P.: From chemical topology to molecular machines (Nobel Lecture). Angew. Chem. Int. Ed. **56**, 11080 (2017)
2. Stoddart, J.F.: Mechanically interlocked molecules (MIMs)-molecular shuttles, switches, and machines (Nobel Lecture). Angew. Chem. Int. Ed. **56**, 11094 (2017)
3. Feringa, B.L.: The art of building small: from molecular switches to motors (Nobel Lecture) Angew. Chem. Int. Ed. **56**, 11060 (2017)
4. Kammerer, C., Erbland, G., Gisbert, Y., Nishino, T., Yasuhara, K., Rapenne, G.: Biomimetic and technomimetic single molecular machines. Chem. Lett. **48**, 299 (2019)
5. Ikeda, A., Tsudera, T., Shinkai, S.: Molecular design of a "molecular syringe" mimic for metal cations using a 1,3-alternate calix[4]arene cavity. J. Org. Chem. **62**, 3568 (1997)
6. Joachim, C., Rapenne, G.: Molecule concept nanocars: chassis, wheels, and motors? ACS Nano **7**, 11 (2013)

7. Chu, P.-L.E., Wang, L.-Y., Khatua, S., Kolomeisky, A.B., Link, S., Tour, J.M.: Synthesis and single-molecule imaging of highly mobile adamantane-wheeled nanocars. ACS Nano **7**, 35 (2013)
8. Jacquot de Rouville, H.-P., Garbage, R., Ample, F., Nickel, A., Meyer, J., Moresco, F., Joachim, C., Rapenne, G.: Synthesis and STM imaging of symmetric and dissymmetric ethynyl-bridged dimers of boron-subphthalocyanine bowl-shaped nano-wheels. Chem. Eur. J. **18**, 8925 (2012)
9. Nickel, A., Meyer, J., Ohmann, R., Jacquot de Rouville, H.-P., Rapenne, G., Joachim, C., Cuniberti, G., Moresco, F.: STM Manipulation of boron-subphthalocyanine nano-wheel dimers on Au(111). J. Phys. Condens. Matter **24**, 404001 (2012)
10. Jimenez-Bueno, G., Rapenne, G.: Technomimetic molecules: synthesis of a molecular wheelbarrow. Tetrahedron Lett. **44**, 6261 (2003)
11. Rapenne, G., Jimenez-Bueno, G.: Molecular machines: synthesis and characterization of two prototypes of molecular wheelbarrows. Tetrahedron **63**, 7018 (2007)
12. Grill, L., Rieder, K.-H., Moresco, F., Jimenez-Bueno, G., Wang, C., Rapenne, G., Joachim, C.: Imaging of a molecular wheelbarrow by scanning tunneling microscopy. Surf. Science **584**, 153 (2005)
13. Shirai, Y., Osgood, A., Zhao, Y., Kelly, K.F., Tour, J.M.: Directional control in thermally driven single-molecule nanocars. Nano Lett. **5**, 2330 (2005)
14. Vives, G., Tour, J.M.: Synthesis of single-molecule nanocars. Acc. Chem. Res. **42**, 473 (2009)
15. Jacquot de Rouville, H.-P., Garbage, R., Cook, R.E., Pujol, A.R., Sirven, A.M., Rapenne, G.: Synthesis of polycyclic aromatic hydrocarbon-based nanovehicles equipped with triptycene wheels. Chem. Eur. J. **18**, 3023 (2012)
16. Kudernac, T., Ruangsupapichat, N., Parschau, M., Maciá, B., Katsonis, N., Harutyunyan, S.R., Ernst, K.-H., Feringa, B.L.: Electrically driven directional motion of a four-wheeled molecule on a metal surface. Nature **479**, 208 (2011)
17. Rapenne, G., Joachim, C.: World's first nanocar race: a single molecule piloted per team. Nature Rev. Mater. **2**, 17040 (2017)
18. Muraoka, T., Kinbara, T., Kobayashi, Y., Aida, T.: Light-driven open−close motion of chiral molecular scissors. J. Am. Chem. Soc. **125**, 5612 (2003)
19. Badjić, J.D., Balzani, V., Credi, A., Silvi, S., Stoddart, J.F.: A molecular elevator. Science **303**, 1845 (2004)
20. Khuong, T.-A.V., Dang, H., Jarowski, P.D., Maverick, E.F., Garcia-Garibay, M.A.: Rotational dynamics in a crystalline molecular gyroscope by variable-temperature ^{13}C NMR, ^{2}H NMR, X-ray diffraction, and force field calculations. J. Am. Chem. Soc. **129**, 839 (2007)
21. Kelly, T.R., de Silva, H., Silva, R.A.: Unidirectional rotary motion in a molecular system. Nature **401**, 150 (1999)
22. Koumura, N., Zijlstra, R.W.J., van Delden, R.A., Harada, N., Feringa, B.L.: Light-driven monodirectional molecular motor. Nature **401**, 152 (1999)
23. Ruangsupapichat, N., Pollard, M.M., Harutyunyan, S.R., Feringa, B.L.: Reversing the direction in a light-driven rotary molecular motor. Nature Chem. **3**, 53 (2011)
24. Hernandez, J.V., Kay, E.R., Leigh, D.A.: A reversible synthetic rotary molecular motor. Science **306**, 1532 (2004)
25. Kassem, S., van Leeuwen, T., Lubbe, A.S., Wilson, M.R., Feringa, B.L., Leigh, D.A.: Artificial molecular motors. Chem. Soc. Rev. **46**, 2592 (2017)
26. Jian, H., Tour, J.M.: En route to surface-bound electric field-driven molecular motors. J. Org. Chem. **68**, 5091 (2003)
27. Garcia-Lopez, V., Liu, D., Tour, J.M.: Light-activated organic molecular motors and their applications. Chem. Rev. **120**, 79 (2020)
28. Vives, G., Jacquot de Rouville, H.-P., Carella, A., Launay, J.-P., Rapenne, G.: Prototypes of molecular motors based on star-shaped organometallic ruthenium complexes. Chem. Soc. Rev. **38**, 1551 (2009)
29. Perera, U.G.E., Ample, F., Echeverria, J., Kersell, H., Zhang, Y., Vives, G., Rapenne, G., Joachim, C., Hla, S.-W.: Controlled clockwise and anticlockwise rotational switching of a molecular motor. Nature Nanotechnol. **8**, 46 (2013)

30. Tierney, H.L., Murphy, C.J., Jewell, A.D., Baber, A.E., Iski, E.V., Khodaverdian, H.Y., McGuire, A.F., Klebanov, N., Sykes, E.C.H.: Experimental demonstration of a single-molecule electric motor. Nature Nanotechnol. **6**, 625 (2011)
31. Baroncini, M., Silvi, S., Credi, A.: Photo-and redox-driven artificial molecular motors. Chem. Rev. **120**, 200 (2020)
32. Zhang, Y., Calupitan, J.P., Rojas, T., Tumbleson, R., Erbland, G., Kammerer, C., Ajayi, T.M., Wang, S., Curtiss, L.C., Ngo, A.T., Ulloa, S.E., Rapenne, G., Hla, S.W.: A chiral molecular propeller designed for unidirectional rotations on a surface. Nature Commun. **10**, 3742 (2019)
33. Grill, L., Rieder, K.-H., Moresco, F., Rapenne, G., Stojkovic, S., Bouju, X., Joachim, C.: Rolling a single molecular wheel at the atomic scale. Nature Nanotechnol. **2**, 95 (2007)
34. Zhang, Y., Kersell, H., Stefak, R., Echeverria, J., Iancu, V., Perera, U.G.E., Li, Y., Deshpande, A., Braun, K.-F., Joachim, C., Rapenne, G., Hla, S.W.: Simultaneous and coordinated rotational switching of all molecular rotors in a network. Nature Nanotechnol. **11**, 706 (2016)
35. Burrows, M., Sutton, G.: Interacting gears synchronize propulsive leg movements in a jumping insect. Science **341**, 1254 (2013)
36. Breslow, R.: Biomimetic chemistry: biology as an inspiration. J. Biol. Chem. **284**, 1337 (2008)
37. Kawada, Y., Iwamura, H.: Unconventional synthesis and conformational flexibility of bis(1-triptycyl)ether. J. Org. Chem. **45**, 2547 (1980)
38. Hounshell, W.D., Johnson, C.A., Guenzi, A., Cozzi, F., Mislow, K.: Stereochemical consequences of dynamic gearing in substituted bis(9-triptycyl)methanes and related molecules. Proc. Natl. Acad. Sci. U.S.A. **77**, 6961 (1980)
39. Frantz, D.K., Linden, A., Baldridge, K.K., Siegel, J.: Molecular spur gears comprising triptycene rotators and bibenzimidazole-based stators. J. Am. Chem. Soc. **134**, 1528 (2012)
40. Ube, H., Yasuda, Y., Sato, H., Shionoya, M.: Metal-centred azaphosphatriptycene gear with a photo- and thermally driven mechanical switching function based on coordination isomerism. Nature Commun. **8**, 14296 (2017)
41. Chiaravalloti, F., Gross, L., Rieder, K.-H., Stojkovic, S.M., Gourdon, A., Joachim, C., Moresco, F.: A rack-and-pinion device at the molecular scale. Nature Mater. **6**, 30 (2007)
42. Manzano, C., Soe, W.-H., Wong, H.S., Ample, F., Gourdon, A., Chandrasekhar, N., Joachim, C.: Step-by-step rotation of a molecule-gear mounted on an atomic-scale axis. Nature Mater. **8**, 576 (2009)
43. Soe, W.-H., Srivastava, S., Joachim, C.: Train of single molecule-gears. J. Phys. Chem. Lett. **10**, 6462 (2019)
44. Kammerer, C., Rapenne, G.: Scorpionate hydrotris(indazolyl)borate ligands as tripodal platform for surface-mounted molecular motors and gears. Eur. J. Inorg. Chem. 2214 (2016)
45. Erbland, G., Gisbert, Y., Rapenne, G., Kammerer, C.: Expedient synthesis of thioether-functionalized hydrotris(indazolyl) borate as an anchoring platform for rotary molecular machines. Eur. J. Org. Chem. 4731 (2018)
46. Erbland, G., Abid, S., Gisbert, Y., Saffon-Merceron, N., Hashimoto, Y., Andreoni, L., Guérin, T., Kammerer, C., Rapenne, G.: Star-shaped ruthenium complexes as prototypes of molecular gears. Chem. Eur. J. **25**, 16328 (2019)
47. Lindsey, J.S., Hsu, H.C., Schreiman, I.C.: Synthesis of tetraphenylporphyrin under very mild conditions. Tetrahedron Lett. **27**, 4969 (1986)
48. Gisbert, Y., Abid, S., Bertrand, G., Saffon-Merceron, N., Kammerer, C., Rapenne, G.: Modular synthesis of pentaarylcyclopentadienyl Ru-based molecular machines via sequential Pd-catalysed cross couplings. Chem. Commun. **55**, 14689 (2019)

Design and Synthesis of a Nano-winch

Yohan Gisbert, Agnès M. Sirven, Gwénaël Rapenne, and Claire Kammerer

Abstract Technical progress in the field of Scanning Probe Microscopy (SPM) has opened the way for the development of new surface-mounted artificial molecular machines, which can be addressed at the single molecule scale. In this context, a ruthenium-based molecular motor has been shown to undergo controlled unidirectional and reversible rotation when fueled with electrons delivered by the tip of a Scanning Tunneling Microscope. In this chapter, we report our efforts towards a deeper understanding of the mechanical properties of this molecular motor. In view of complementary force measurements to be performed at the single molecule scale using SPM techniques, the organometallic structure of the motor has been derivatized to append a long chain terminated by a hook. We detail here the design of this nano-winch architecture and the modular synthesis of a first prototype dedicated to Atomic Force Microscopy-based Single Molecule Force Spectroscopy experiments.

Keywords Molecular motor · Nano-winch · Ruthenium complex · Single molecule · Scanning Probe Microscopy

1 Introduction

In the field of molecular machines [1], molecular motors represent a subclass of systems able to convert various sources of energy into unidirectional mechanical motion. In Nature, biological molecular motors are ubiquitous and commonly exploit chemical energy recovered from ATP hydrolysis or electrochemical potential gradients as energy input [2]. Well-known examples of biomolecular motors inducing linear motion include myosins and kinesins, responsible in particular for muscle contraction and intracellular cargo transport. In contrast, rotary motion is generated

Y. Gisbert · A. M. Sirven · G. Rapenne · C. Kammerer (✉)
CEMES, Université de Toulouse, CNRS, 29, rue Jeanne Marvig, 31055 Toulouse, France
e-mail: claire.kammerer@cemes.fr

G. Rapenne
Division of Materials Science, Nara Institute of Science and Technology, 8916-5, Takayama, Ikoma, Nara, Japan

C. Joachim (ed.), *Building and Probing Small for Mechanics*, Advances in Atom and Single Molecule Machines, https://doi.org/10.1007/978-3-030-56777-4_6

in bacterial flagellar motors and in ATP synthases, for which reversibility of the rotation allows mutual conversion of chemical energy and electrochemical potential. These motors thus accomplish a variety of complex chemical and mechanical tasks, vital to biological systems. The operating conditions at the nanoscale involve high viscosity and constant Brownian motion, which has to be harnessed in order to ensure directionality in the movements. Strategies to reduce degrees of freedom and allow out-of-equilibrium operating conditions have been implemented by evolution, such as tracks for linear motors or compartmentalization by lipid barriers for rotary motors. This organization is critical for the collective emergence of useful work.

Far from the complexity of biological systems, chemists have designed and synthesized artificial molecular machines over the past forty years, with the aim of triggering controlled large-amplitude motions in response to a broad range of stimuli (light, electrochemistry, pH, bond formation/cleavage, ion binding, etc.…) [3–5]. The tremendous efforts in this new field were recognized by the 2016 Nobel Prize in Chemistry, awarded to Jean-Pierre Sauvage [6], Sir Fraser Stoddart [7] and Ben Feringa [8] for their major contributions. In particular, control of directionality in translational and rotational systems was tackled in order to ultimately obtain molecular motors, able to perform work by progressive and repetitive operations. Since the late 1990s, many types of chemically-fueled and light-fueled motors have been reported, based on diverse architectures and mechanisms [9–12]. Seminal work by Kelly et al. on a ratchet-type compound based on triptycene and helicene units led to an important milestone, with a 120° unidirectional rotation around a C–C single bond driven by successive chemical steps [13]. Light driven directional rotary motion was initially developed by Feringa et al. in chiral overcrowded alkenes, in which successive photoisomerization and thermal relaxation steps induced repetitive unidirectional 360° rotation around C=C double bonds [14]. Chemically- and light-fueled translational motors were also conceived, as exemplified by the molecular walkers [15] reported by Leigh et al., and by various mechanically-interlocked molecules such as catenanes [16] and rotaxanes [17, 18].

These systems have been mostly studied in solution, which hampers the recovery of work due to random orientation of the population of molecular motors. Collective behaviors were however obtained after suitable organization of the assemblies of motors, leading to the emergence of controlled motion up to the macroscale [19]. In this regard, the light-fueled overcrowded alkenes developed by Feringa et al. have been intensively studied, among others, and led upon irradiation to the rotation of a micrometric glass rod deposited over a liquid crystal phase doped with the motor [20], or to the macroscopic contraction of a polymeric gel incorporating the motor as reticulation unit [21]. Organization on surface also promoted collective behaviors of rotaxane-based machines, leading to reversible bending of cantilever beams in response to chemical stimuli [22], or to directional transport of a droplet under irradiation [23]. The latter examples highlight the frequent adaptation to interfacial conditions of systems initially designed to be operated in solution.

Alternatively, several research groups adopted the opposite strategy and considered adsorption on surface as key requirement for the molecular design, in view of controlling the relative orientation of the motors. On-surface studies were also considerably facilitated by the technical progress made in the field of Scanning Probe Microscopy. In particular, thanks to the high-resolution imaging capabilities of Scanning Tunneling Microscopy (STM), it has been possible to precisely study motions on surface of entire molecules and within molecules, at the single molecule scale and in self-assemblies. STM is not only a tool to probe organization and motion on surface, but it can also be used for the mechanical manipulation of atoms and molecules, which is crucial to gain knowledge in fundamental processes occurring at the nanoscale [24, 25]. Moreover, the tip of the STM can act as a localized atomic scale electrode, particularly appealing for the development of electron-fueled molecular motors. Since the report of a surface-bound butyl methyl sulfide by Sykes in 2011 [26], several electrically-driven systems have been developed, undergoing on metallic surfaces controlled unidirectional translational [27] or rotational motion. In the latter case, rotation of the whole molecule around an anchoring point can occur (the surface acts as stator) [26, 28, 29] or motion can be restricted to one submolecular fragment rotating around an internal degree of freedom (the stator subunit is then anchored onto the surface) [30, 31]. It is important to note here that such molecular motors have been studied at low temperature, thereby decreasing random thermal fluctuations and increasing control over targeted electrically-driven motion. In these conditions, our group demonstrated the reversible unidirectional rotation of a piano-stool ruthenium complex studied at the single molecule scale [30].

From macro- to nano-scale motors, central questions concern their efficiency and the useful work they are able to perform [32]. On these aspects, biological systems outcompete artificial molecular motors, as their evolution was driven by functionality, and as they achieve energy conversion efficiencies over 40%, along with a force-per-mass ratio comparable to that of modern macroscale engines [33, 34]. To get further insight on the intrinsic performances of our surface-bound electron-fueled ruthenium-based molecular motor, incorporation of the organometallic scaffold into a winch structure was envisioned, allowing for various force measurements at the single molecule scale.

2 A Reversible Electron-Fueled Molecular Motor

Since the early 2000s, our group has designed and synthesized a series of electrically-driven azimuthal rotors [35–37], to be studied on surface as single molecules [30, 31] or as self-assemblies [38] by means of STM. Electric energy delivered by the STM tip is converted into a rotation movement, which was shown to be unidirectional and reversible in the case of a piano-stool ruthenium complex incorporating a dissymmetric rotor subunit (Fig. 1) [30].

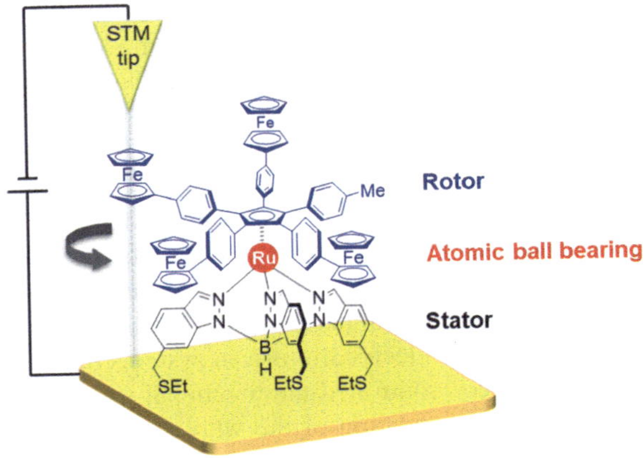

Fig. 1 Structure of the ruthenium-based molecular motor adsorbed on a gold surface and related electron-fueled unidirectional rotation

2.1 Architecture of the Organometallic Motor

The architecture of the unidirectional and reversible electron-fueled molecular motor is based on a piano-stool ruthenium(II) complex, involving a hydrotris(indazolyl)borate ligand as tripodal stator and a pentaarylcyclopentadienyl platform as rotor. The central ruthenium atom thus acts as a linker between both subunits, but also as ball bearing to favour azimuthal rotation over other degrees of freedom (Fig. 1).

The rigid hydrotris(indazolyl)borate scaffold exhibits a dual function: the facial coordination of the nitrogens to the ruthenium center lifts the upper rotor subunit away from the surface, while an appropriate functionalization on the opposite face allows tight anchoring on surface [39]. Indeed, the three ethylthioether groups appended to the indazole moieties provide high stability on metallic surfaces by preventing translation, rotation and rocking motions of the stator.

The rotating subunit is a rigid aromatic platform based on a central cyclopentadienyl moiety carrying four identical ferrocenylphenyl arms and one shorter tolyl arm. This structural dissymmetry finds two origins. First, in the absence of time-resolution in STM, molecular motion is probed by comparing static images and following a chemically or sterically "tagged" part of the molecule. Here, the shorter tolyl arm acts as a tag, which is easily traceable on STM images. Second, with the aim of achieving unidirectional rotation, the symmetry of the rotor had to be lowered as compared to the parent pseudo C_5 penta(ferrocenylphenyl)cyclopentadienyl ligand, so as to generate a dissymmetric energy landscape for rotary motion.

2.2 Synthesis and On-surface Characterization

Molecular motor **6** was obtained in five steps starting from cyclopenta-2,4-dienone **1**, incorporating four identical *p*-bromophenylene arms in view of further functionalization with ferrocene moieties in the final step (Scheme 1) [40]. Introduction of the fifth arm was achieved via 1,2-addition of *p*-tolylmagnesium bromide to yield the corresponding cyclopentadienol, which was subsequently converted into bromocyclopentadiene **2** under acidic conditions. Bromocyclopentadiene **2** was obtained in 55% yield over two steps, as a 2:2:1 mixture of regioisomers. Oxidative addition on ruthenium(0) cluster $Ru_3(CO)_{12}$ afforded half-sandwich ruthenium complex **3**, which underwent ligand exchange in the presence of the thioether-functionalized potassium hydrotris(indazolyl)borate **4.K** to yield the precursor of the dissymmetrized motor **5**. Finally, a four-fold Pd-catalyzed Suzuki-Miyaura coupling with ferroceneboronic acid led to molecular motor **6** in 52% yield.

After full characterization, the solid ruthenium complex **6** was deposited by sublimation on a Au(111) surface and subsequently studied by STM under ultra-high vacuum (UHV) at low temperature at the single molecule scale. STM images confirmed that some complexes remain intact after deposition and are adsorbed on the surface via the three thioether anchoring groups, as expected. At 80 K, the molecule undergoes rotation of the upper cyclopentadienyl ligand around the B-Ru axis, thereby confirming that this is the privileged degree of freedom. However, this rotation is purely related to Brownian motion and is thus a random oscillation with no directionality. Interestingly, when the temperature is lowered to 5 K, the

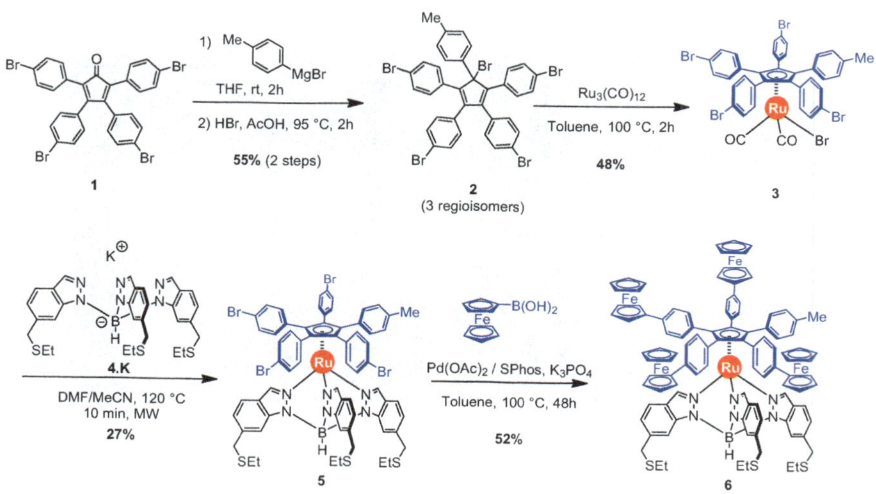

Scheme 1 Synthesis of the ruthenium-based molecular motor in five steps, starting from 2,3,4,5-tetra(*p*-bromophenyl)cyclopenta-2,4-dienone

motion is stopped and a contrast between the two different types of arms is observed on STM images, as expected from the incorporation of a tag on the rotor subunit [30].

2.3 Electron-Fueled Unidirectional and Reversible Rotation

At 5 K, excitation of the motionless complex by a voltage pulse delivered on one arm by the STM tip induces a unidirectional rotation of the upper rotor subunit. In this set-up, motion is thus triggered by an inelastic electron tunnelling process leading to stepwise rotation with angles being multiples of 24°, in line with the predicted 15 potential wells in a full 360° rotation [30]. Directionality originates from the asymmetric character of the energy landscape in the excited state, itself related to the dissymmetric architecture of the complex.

Moreover, it was observed that the direction of rotation directly depends on the nature of the excited arm, i.e. on the nature of the excited state that is addressed. The direction can thus be reversed at will by locating the STM tip apex above one of the ferrocenyl arms or above the truncated arm (Fig. 2).

This organometallic structure, deposited on surface and studied as a single molecule at low temperature under ultra-high vacuum, was thus shown to convert electric energy provided by the STM tip into reversible directed motion. Following this breakthrough, new questions arose about the forces developed by this rotary motor and its ability to provide useful work, and thorough mechanical studies were thus envisioned.

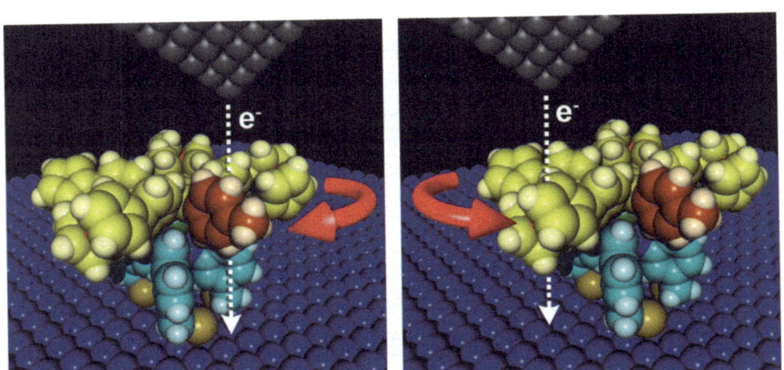

Fig. 2 Electrically-driven unidirectional and reversible rotation of the ruthenium-based molecular motor, on Au(111) surface at low temperature. The direction of rotation can be reversed at will by locating the STM tip apex above one of the ferrocenyl arms (right image, counterclockwise rotation) or above the truncated tolyl arm (left image, clockwise rotation)

3 Towards Complementary Mechanical Studies of the Molecular Motor: The Nano-winch

To gain insight in the mechanical properties of the ruthenium-based rotary motor, complementary direct and indirect studies were devised. Both approaches allow to probe forces exerted by single molecules, but they rely on different operating conditions and scanning probe microscopy techniques.

In a direct approach, force spectroscopy experiments will be performed by means of an atomic force microscope (AFM) at the solid-liquid interface in ambient conditions to determine the stall force of the rotary motor. In a second approach, the surface-bound motor will be loaded with various molecular fragments adsorbed on surface, and the electrically-driven mechanical motion of the motor at low temperature in UHV will be studied by STM as a function of the loading.

In both cases, it appears necessary to derivatize the motor structure and append a long chain, crucial for Single Molecule Force Spectroscopy (SMFS) experiments and for the connection of molecular loads. Following technomimetic principles of molecular machine design [41, 42], this derivatized rotary motor displaying a long chain can be viewed as a nano-winch.

3.1 *Design of a Nano-winch Architecture*

As in a macroscopic winch, the nanometer-sized counterpart will include three main parts: the motor, a long rope and a terminal hook to attach loads. The structure of the ruthenium-based rotary motor will thus be integrated in the nano-winch, in a way to maintain directionality and reversibility in the electron-fueled motion. This implies to operate minimal changes both in the architecture of the complex and in its electronic structure. It was thus envisioned to connect the long chain to the truncated arm of the cyclopentadienyl rotor, and use a chemical spacer to minimize steric and electronic interactions (Fig. 3).

Polyethylene glycol (PEG) was selected as polymeric chain, since it exhibits the required adsorption properties for AFM-based force spectroscopy experiments.

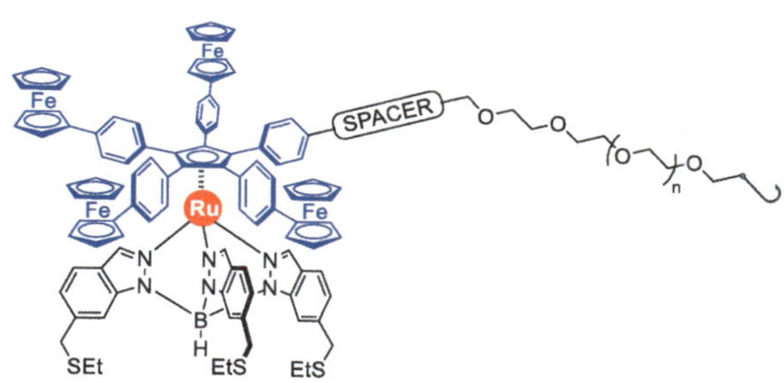

Fig. 3 Architecture of the nano-winch, including the rotary motor subunit linked via a chemical spacer to a long PEG chain terminated by a hook

Additionally, the availability of monodisperse PEG chains with a broad range of polymerisation degrees is an asset to fine tune mechanical studies, and adjust the length of the chain depending on the operating conditions in the direct and indirect approaches.

Finally, a chemical hook located at the PEG chain extremity will allow the connection of various molecular loads.

3.2 Direct Strategy: AFM-Based Single Molecule Force Spectroscopy on an Unloaded Nano-winch

AFM-based Single Molecule Force Spectroscopy allows to probe mechanical forces in individual molecules and has been used to investigate various biological and artificial systems [43], including small molecular machines such as rotaxanes [44] and catenanes [45]. AFM-based force spectroscopy is performed at the solid-liquid interface on surface-bound single molecules, i.e. the nano-winch will be adsorbed on a gold surface in a dilute distribution and immersed in a high boiling-point solvent. Bringing the AFM tip into contact with the surface will allow the PEG chain of a single nano-winch to adsorb on the tip, and the latter will be progressively moved away from the surface, leading to a stretching of the PEG chain (Fig. 4). It is important to note that these experiments will be carried out at room temperature, meaning that the ruthenium-based motor will undergo random thermal motion and particularly random oscillation of the cyclopentadienyl platform. This rotation will be stopped by

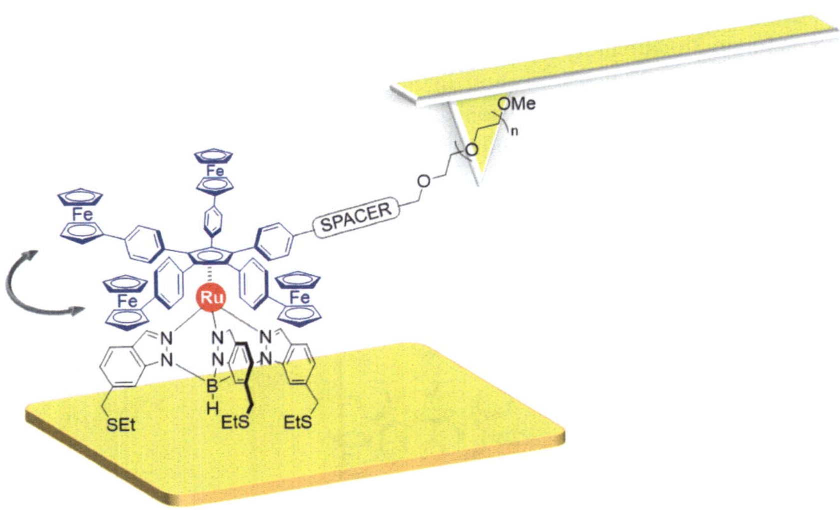

Fig. 4 Principle of the AFM-based Single Molecule Force Spectroscopy experiments, performed at the solid-liquid interface under ambient conditions

moving the tip further away from the surface, and the force applied on the nano-winch at that stage will give access to the stall force of the motor.

From a structural point of view, the nano-winch used in such force spectroscopy experiments will require a long PEG chain of ca. 8 nm, with a non-functionalized end. As starting point, a methoxy-terminated PEG-24 chain could thus be connected to the motor, and further fine tuning of the tether length will be possible provided a modular synthetic route towards nano-winches is devised.

3.3 *Indirect Strategy: STM Study of a Loaded Nano-winch*

In this alternative strategy, our aim is to study by UHV-STM at low temperature the electrically-driven rotary motion of the loaded motor, i.e. of the nano-winch connected to various molecular fragments (Fig. 5) [46]. After deposition on Au(111) and subsequent adsorption of the motor subunit linked to its load, lateral mechanical manipulation of the load will be performed with the STM tip at 77 K to extend the PEG chain. This layout will then be frozen by lowering the temperature to 5 K and the rotation of the motor will be triggered by a voltage pulse on one of the motor arms. Directional rotation of the rotor subunit is expected to result in a translation of the load if the force output of the motor is sufficient. It is important to note that gravity is irrelevant at the nanoscale, and the term "load" is thus not related here

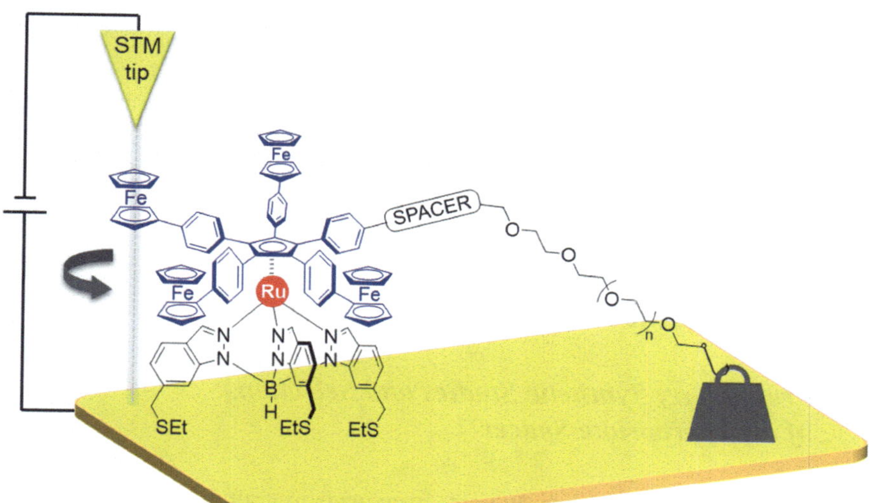

Fig. 5 Principle of the LT-UHV-STM experiments on the nano-winch connected to various molecular loads, adsorbed on a gold surface. Rotary motion of the motor is triggered by a voltage pulse and subsequent translational motion of the load occurs if the maximal output force of the motor is sufficient

Fig. 6 Structures of the three molecular loads already available, displaying a terminal alkyne function and incremental numbers of triptycenyl units. The yields over the complete synthetic sequence are given below

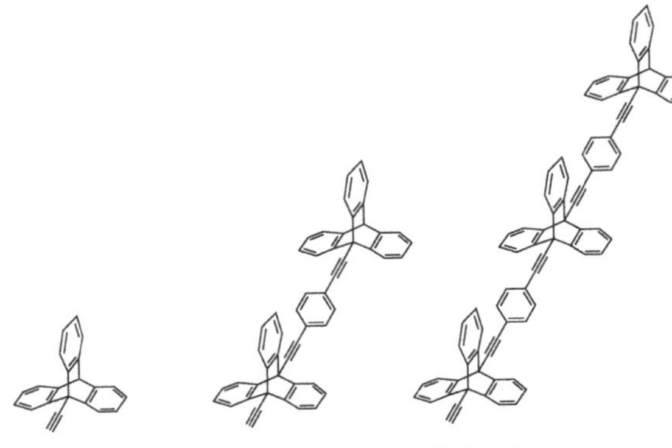

to weight but refers to the force exerted by the molecular fragment on the motor. Using fragments with increasing diffusion energies upon adsorption on surface, the maximal output force of the motor will be surpassed at some point, as witnessed by the absence of translation of the load. Comparison of the diffusion energies of the last translated fragment and of the first motionless one will thus give an estimation of the maximal work delivered by the single-molecule motor, and by extension of its maximal output force.

From a structural point of view, the length of the PEG chain should be much shorter than for force spectroscopy experiments, and a PEG-8 building block is envisioned as starting point. The loads should be covalently bound to the chain under reliable, mild and high-yielding conditions, which is reminiscent of click reaction requirements [47]. An azide function will thus act as a hook to allow for azide-alkyne cycloaddition, yielding a 1,2,3-triazole as connector between the chain and the load.

Three different molecular loads displaying a terminal alkyne function have been synthesized, with incremental numbers of triptycenyl units to induce increasing diffusion energies on surface (Fig. 6) [48].

4 Synthesis of the Nano-winch

4.1 Preliminary Synthetic Studies and Selection of an Appropriate Spacer

In preliminary synthetic studies, attempts were made to graft the PEG chain of the nano-winch directly on the truncated arm of the motor by replacing the tolyl moiety with a PEGylated phenyl ether group [49]. However, this approach failed to afford the desired compound and introduction of a spacer between the rotor phenylene unit and the PEG chain was envisioned.

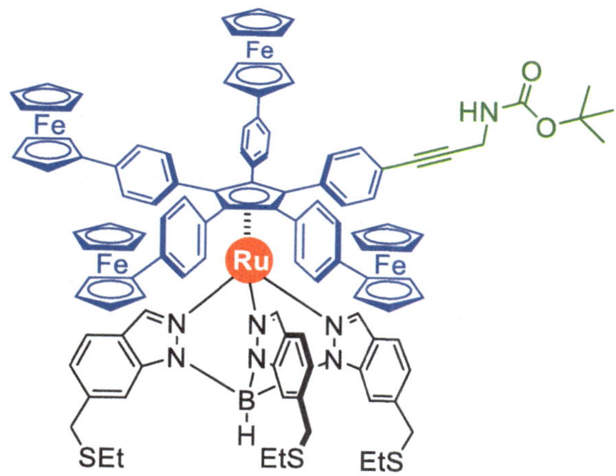

7

Fig. 7 Structure of the key precursor to various nano-winches, featuring a *tert*-butyloxycarbonyl-protected propargyl amine fragment

This spacer should exhibit two functions: it should allow spacial and electronic decoupling of the motor from the chain, and favour modular syntheses of nano-winches via late-stage functionalization of the motor scaffold with variable PEG derivatives. Following these guidelines, the *N*-propargyl amide fragment appeared particularly attractive due to the high availability of monodisperse carboxylic acid-terminated PEG-chains.

In our modular synthetic strategy, the common key precursor thus incorporates a dissymmetrized cyclopentadienyl ligand, with four phenylferrocenyl arms and a single phenylpropargyl amine arm. Final stage condensation with the appropriate PEGylated carboxylic acid derivative delivers the targeted nano-winch structure. However, due to the limited stability of the free propargyl amine derivative, it was decided to use the *tert*-butyloxycarbonyl-protected (BOC-protected) parent compound **7** as common precursor (Fig. 7).

4.2 Modular Synthetic Strategy Towards a Common Nano-winch Precursor

With the *N*-propargyl amide fragment selected as spacer for the diverse nano-winch prototypes, the synthetic route to the BOC-protected propargyl amine common precursor **7** was devised.

As for the parent molecular motor, the presence of a dissymmetrized rotating subunit requires the discrimination of one arm along the synthesis. Introduction of the distinct phenylpropargyl amine moiety may be performed at an early stage via a nucleophilic addition on 2,3,4,5-tetra(*p*-bromophenyl)cyclopenta-2,4-dienone **1**, similarly to the molecular motor synthetic route described above

(Scheme 1). However, this linear strategy inherently limits the structural diversity of the winch spacers, and issues for the coordination to ruthenium may also be encountered with a pre-functionalized cyclopentadiene derivative. A modular strategy was thus devised, involving a late-stage functionalization of a 1,2,3,4,5-penta(*p*-halogenophenyl)cyclopentadienyl hydrotris(indazolyl)borate ruthenium(II) precursor, carrying one single activated *p*-iodophenyl group and four *p*-bromophenyl moieties [50]. This strategy exploits the enhanced reactivity of aryl iodides compared to aryl bromides in palladium-catalyzed cross-couplings, and more particularly under Sonogashira conditions. The dissymmetrized precursor **11**, with a single *p*-iodophenyl arm, would then allow for a chemoselective Sonogashira cross-coupling to install the *N*-protected propargyl amine function, followed by a four-fold cross-coupling to incorporate the ferrocenyl moieties on the remaining brominated positions. More generally, this modular strategy was also applied to the synthesis of dissymmetric molecular gear prototypes via sequential cross-couplings (see chapter "Prototypes of Molecular Gears with an Organometallic Piano-Stool Architecture" of this volume).

The synthesis of the dissymmetrized 1-(*p*-iodophenyl)-2,3,4,5-tetra(*p*-bromophenyl)cyclopentadienyl hydrotris(indazolyl)borate ruthenium(II) complex **11** was carried out in four steps starting from 2,3,4,5-tetra(*p*-bromophenyl)cyclopenta-2,4-dienone **1** (Scheme 2). Introduction of the iodinated arm was performed by 1,2-addition of 4-iodophenyllithium to give the corresponding alcohol **8**, which was smoothly converted to 5-bromocyclopenta-1,3-diene **9**, obtained as a mixture of three regioisomers. Subsequent reaction

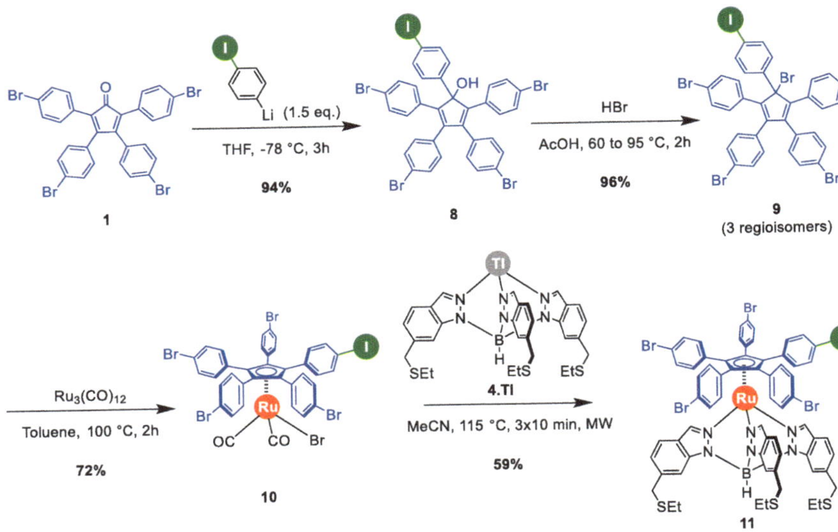

Scheme 2 Synthesis of 1-(*p*-iodophenyl)-2,3,4,5-tetra(*p*-bromophenyl)cyclopentadienyl hydrotris(indazolyl)borate ruthenium(II) key intermediate

with ruthenium(0) cluster $Ru_3(CO)_{12}$ led to a selective insertion into the central C-Br bond to yield ruthenium(II) complex **10**, carrying a dissymmetrized penta(*p*-halogenophenyl)cyclopentadienyl ligand. Finally, ligand exchange in the presence of the thallium salt of thioether-functionalized hydrotris(indazolyl)borate **4.Tl** afforded the desired precursor **11** in 38% overall yield.

The key pentahalogenated building block, carrying one pre-activated *p*-iodophenyl arm, was next engaged in sequential chemoselective cross-couplings to introduce successively the *N*-protected propargyl amine moiety followed by the four ferrocenyl groups (Scheme 3). The Sonogashira reaction was carried out under mild conditions using *tert*-butyloxycarbonyl-protected propargyl amine as coupling partner in the presence of Pd(0) and Cu(I) co-catalysts in a THF/triethylamine mixture at 40 °C. Chemoselective functionalization of the *p*-iodophenyl group was achieved to afford ruthenium complex **12** in 75% yield. In the next step, the four remaining *p*-bromophenyl moieties were engaged in Suzuki-Miyaura cross-couplings with ferrocene boronic acid pinacol ester as a coupling partner, in the presence of $PdCl_2(dppf)$ as a catalyst and a large excess of cesium carbonate as a base in a DMF/water system. The desired tetraferrocenyl product **7** was obtained in 16% yield, which corresponds to a yield of 63% per C-C bond newly formed.

The ruthenium complex **7**, carrying four phenylferrocenyl arms and a single BOC-protected phenylpropargyl amine arm, was obtained in six steps from 2,3,4,5-tetra(*p*-bromophenyl)cyclopenta-2,4-dienone **1** in 4.6% overall yield. This compound was used as key building block in the synthesis of various nano-winch prototypes, since it proved stable enough to be stored for several months, as opposed to the corresponding deprotected amine. The preparation of each nano-winch prototype, dedicated to targeted mechanical studies, was thus further achieved in two steps involving propargyl amine deprotection followed by a condensation with the appropriate PEG chain.

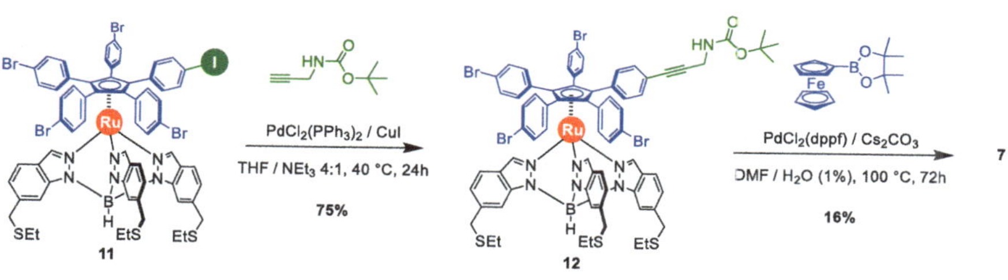

Scheme 3 Two-step synthesis of nano-winch key precursor **7** via chemoselective sequential cross-coupling reactions

4.3 Application to the Synthesis of an Unloaded Nano-winch for Single Molecule Force Spectroscopy Experiments

In view of Single Molecule Force Spectroscopy experiments, the first prototype of unloaded nano-winch was subsequently synthesized in two steps from the BOC-protected propargyl amine intermediate **7** (Scheme 4). For this first prototype, a long polyethylene glycol (PEG-24) chain was selected to ensure an efficient adsorption of the chain on the AFM tip.

Deprotection of the propargyl amine moiety in **7** was conducted under mild conditions, using trimethylsilyl triflate in the presence of lutidine to give the corresponding free amine in 81% yield. Such unusual conditions for BOC cleavage were required to maintain a buffered medium, as the ruthenium(II) complex appeared unstable under standard acidic deprotection conditions. The free propargyl amine **13** then underwent condensation with the carboxylic acid-functionalized PEG chain to generate the desired *N*-propargyl amide spacer and thereby covalently link the long chain to the motor scaffold. To maximize the efficiency of this final synthetic step, monodisperse methoxy-PEG24-propionic acid activated as an *N*-hydroxysuccinimidyl ester was selected as coupling partner. Under very mild conditions involving triethylamine

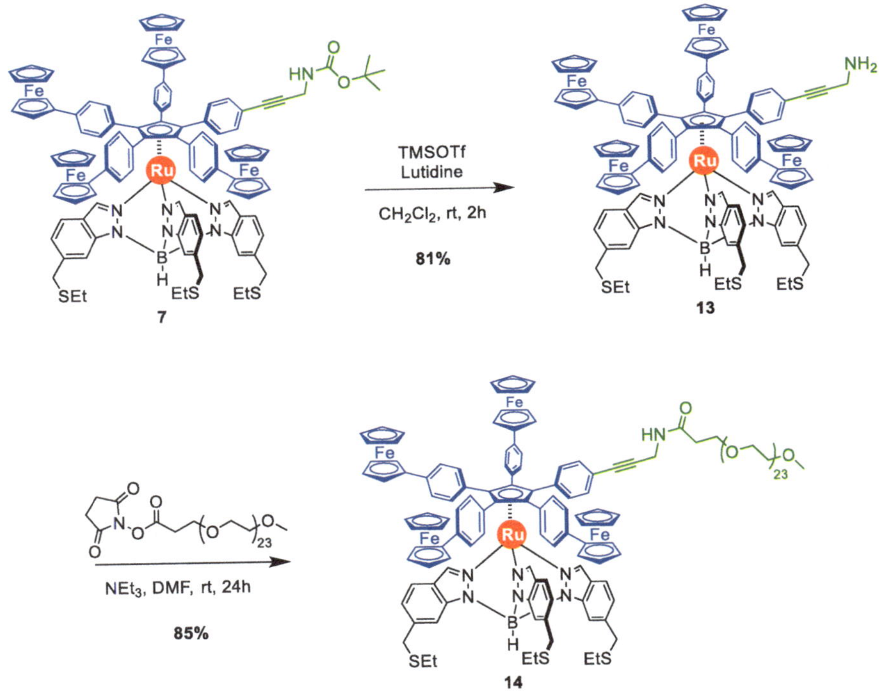

Scheme 4 Synthesis of the first nano-winch prototype dedicated to Single Molecule Force Spectroscopy experiments, in two steps from BOC-protected propargyl amine derivative **7**

in DMF at room temperature, the target compound **14** was obtained in 85% yield. Full characterization unambiguously confirmed the connection of both the *N*-propargyl amide spacer and the PEG-24 chain to the molecular motor structure, leading to a first nano-winch prototype. AFM-based Single Molecule Force Spectroscopy studies on nano-winch **14** are currently underway.

5 Conclusion and Perspectives

A nano-winch architecture has been designed to get further insight in the mechanical properties of the electron-fueled molecular motor developed in our group. In view of complementary force measurements at the single molecule scale, derivatization of the organometallic motor structure was performed by appending a long polyethylene glycol chain, possibly terminated by a hook. This general nano-winch architecture has to be adjusted according to the investigation technique, such as AFM-based Single Molecule Force Spectroscopy on the oscillating unloaded motor or as low-temperature STM studies on the loaded motor.

The first nano-winch prototype incorporating a methoxy-terminated PEG24-chain has been successfully synthesized following a highly modular approach, and force spectroscopy measurements are currently underway to determine the stall force of the ruthenium-based motor.

Synthetic efforts are now dedicated to the preparation of a new nano-winch prototype displaying a terminal chemical hook, such as an azide moiety, to covalently link molecular loads (Fig. 8). Electron-fueled motion of the resulting loaded nano-winch will then indirectly give access to the maximal work delivered by this rotary motor.

Acknowledgements This work was supported by the University Paul Sabatier (Toulouse, France) and CNRS. It has received funding from the Agence Nationale de la Recherche (ACTION project ANR-15-CE29-0005) and from the European Union's Horizon 2020 research and innovation program under the project MEMO, grant agreement No 766864. This research was also partly

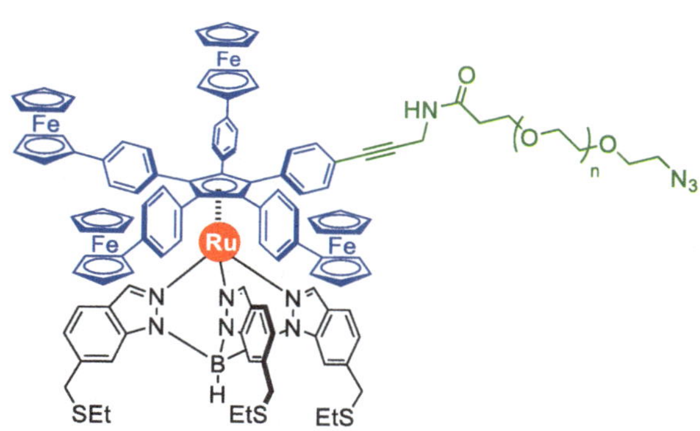

Fig. 8 Structure of a future nano-winch prototype carrying an azide hook, in view of subsequent loading with alkynyl-derived molecular fragments

supported by the JSPS KAKENHI grant in aid for Scientific Research on Innovative Areas "Molecular Engine (No. 8006)" 18H05419. Y.G. thanks the MESR for a Ph.D. Fellowship. A.M.S. thanks the Physics Institute of CNRS and the Région Midi-Pyrénées for a Ph.D. Fellowship.

References

1. Balzani, V., Credi, A., Venturi, M.: Molecular Devices and Machines. Concepts and Perspectives for the Nanoworld, 2nd edn. Wiley-VCH, Weinheim (2008)
2. Saper, G., Hess, H.: Synthetic systems powered by biological molecular motors. Chem. Rev. **120**, 288 (2020)
3. Kay, E.R., Leigh, D.A., Zerbetto, F.: Synthetic molecular motors and mechanical machines. Angew. Chem. Int. Ed. **46**, 72 (2007)
4. Erbas-Cakmak, S., Leigh, D.A., McTernan, C.T., Nussbaumer, A.L.: Artificial molecular machines. Chem. Rev. **115**, 10081 (2015)
5. Baroncini, M., Casimiro, L., de Vet, C., Groppi, J., Silvi, S., Credi, A.: Making and operating molecular machines: a multidisciplinary challenge. ChemistryOpen **7**, 169 (2018)
6. Sauvage, J.-P.: From chemical topology to molecular machines (Nobel Lecture). Angew. Chem. Int. Ed. **56**, 11080 (2017)
7. Stoddart, J.F.: Mechanically interlocked molecules (MIMs)-molecular shuttles, switches, and machines (Nobel Lecture). Angew. Chem. Int. Ed. **56**, 11094 (2017)
8. Feringa, B.L.: The art of building small: from molecular switches to motors (Nobel Lecture). Angew. Chem. Int. Ed. **56**, 11060 (2017)
9. Kassem, S., van Leeuwen, T., Lubbe, A.S., Wilson, M.R., Feringa, B.L., Leigh, D.A.: Artificial molecular motors. Chem. Soc. Rev. **46**, 2592 (2017)
10. García-López, V., Liu, D., Tour, J.M.: Light-activated organic molecular motors and their applications. Chem. Rev. **120**, 79 (2020)
11. Baroncini, M., Silvi, S., Credi, A.: Photo- and redox-driven artificial molecular motors. Chem. Rev. **120**, 200 (2020)
12. Goswami, A., Saha, S., Biswas, P.K., Schmittel, M.: (Nano)mechanical motion triggered by metal coordination: from functional devices to networked multicomponent catalytic machinery. Chem. Rev. **120**, 125 (2020)
13. Kelly, T.R., Silva, H.D., Silva, R.A.: Unidirectional rotary motion in a molecular system. Nature **401**, 150 (1999)
14. Koumura, N., Zijlstra, R.W.J., Delden, R.A., van Harada, N., Feringa, B.L.: Light-driven monodirectional molecular rotor. Nature **401**, 152 (1999)
15. von Delius, M., Geertsema, E.M., Leigh, D.A.: A synthetic small molecule that can walk down a track. Nat. Chem. **2**, 96 (2010)
16. Leigh, D.A., Wong, J.K.Y., Dehez, F., Zerbetto, F.: Unidirectional rotation in a mechanically interlocked molecular rotor. Nature **424**, 174 (2003)
17. Chatterjee, M.N., Kay, E.R., Leigh, D.A.: Beyond switches: ratcheting a particle energetically uphill with a compartmentalized molecular machine. J. Am. Chem. Soc. **128**, 4058 (2006)
18. Serreli, V., Lee, C.-F., Kay, E.R., Leigh, D.A.: A molecular information ratchet. Nature **445**, 523 (2007)
19. Dattler, D., Fuks, G., Heiser, J., Moulin, E., Perrot, A., Yao, X., Giuseppone, N.: Design of collective motions from synthetic molecular switches, rotors, and motors. Chem. Rev. **120**, 310 (2020)
20. Eelkema, R., Pollard, M.M., Vicario, J., Katsonis, N., Ramon, B.S., Bastiaansen, C.W.M., Broer, D.J., Feringa, B.L.: Nanomotor rotates microscale objects. Nature **440**, 163 (2006)
21. Li, Q., Fuks, G., Moulin, E., Maaloum, M., Rawiso, M., Kulic, I., Foy, J.T., Giuseppone, N.: Macroscopic contraction of a gel induced by the integrated motion of light-driven molecular motors. Nat. Nanotechnol. **10**, 161 (2015)

22. Huang, T.J., Brough, B., Ho, C.-M., Liu, Y., Flood, A.H., Bonvallet, P.A., Tseng, H.-R., Stoddart, J.F., Baller, M., Magonov, S.: A nanomechanical device based on linear molecular motors. Appl. Phys. Lett. **85**, 5391 (2004)
23. Berná, J., Leigh, D.A., Lubomska, M., Mendoza, S.M., Pérez, E.M., Rudolf, P., Teobaldi, G., Zerbetto, F.: Macroscopic transport by synthetic molecular machines. Nat. Mater. **4**, 704 (2005)
24. Grill, L., Rieder, K.-H., Moresco, F., Rapenne, G., Stojkovic, S., Bouju, X., Joachim, C.: Rolling a single molecular wheel at the atomic scale. Nat. Nanotechnol. **2**, 95 (2007)
25. Manzano, C., Soe, W.-H., Wong, H.S., Ample, F., Gourdon, A., Chandrasekhar, N., Joachim, C.: Step-by-step rotation of a molecule-gear mounted on an atomic-scale axis. Nat. Mater. **8**, 576 (2009)
26. Tierney, H.L., Murphy, C.J., Jewell, A.D., Baber, A.E., Iski, E.V., Khodaverdian, H.Y., McGuire, A.F., Klebanov, N., Sykes, E.C.H.: Experimental demonstration of a single-molecule electric motor. Nat. Nanotechnol. **6**, 625 (2011)
27. Kudernac, T., Ruangsupapichat, N., Parschau, M., Maciá, B., Katsonis, N., Harutyunyan, S.R., Ernst, K.-H., Feringa, B.L.: Electrically driven directional motion of a four-wheeled molecule on a metal surface. Nature **479**, 208 (2011)
28. Eisenhut, F., Meyer, J., Krüger, J., Ohmann, R., Cuniberti, G., Moresco, F.: Inducing the controlled rotation of single o-MeO-DMBI molecules anchored on Au(111). Surf. Sci. **678**, 177 (2018)
29. Simpson, G.J., García-López, V., Boese, A.D., Tour, J.M., Grill, L.: How to control single-molecule rotation. Nat. Commun. **10**, 4631 (2019)
30. Perera, U.G.E., Ample, F., Echeverria, J., Kersell, H., Zhang, Y., Vives, G., Rapenne, G., Joachim, C., Hla, S.-W.: Controlled clockwise and anticlockwise rotational switching of a molecular motor. Nat. Nanotechnol. **8**, 46 (2013)
31. Zhang, Y., Calupitan, J.P., Rojas, T., Tumbleson, R., Erbland, G., Kammerer, C., Ajayi, T.M., Wang, S., Curtiss, L.C., Ngo, A.T., Ulloa, S.E., Rapenne, G., Hla, S.-W.: A chiral molecular propeller designed for unidirectional rotations on a surface. Nat. Commun. **10**, 3742 (2019)
32. Browne, W.R., Feringa, B.L.: Making molecular machines work. Nat. Nanotechnol. **1**, 25 (2006)
33. Marden, J.H.: Scaling of maximum net force output by motors used for locomotion. J. Exp. Biol. **208**, 1653 (2005)
34. Hess, H., Saper, G.: Engineering with biomolecular motors. Acc. Chem. Res. **51**, 3015 (2018)
35. Vives, G., Jacquot de Rouville, H.-P., Carella, A., Launay, J.-P., Rapenne, G.: Prototypes of molecular motors based on star-shaped organometallic ruthenium complexes. Chem. Soc. Rev. **38**, 1551 (2009)
36. Stefak, R., Sirven, A.M., Fukumoto, S., Nakagawa, H., Rapenne, G.: Desymmetrization on both ligands of pentaphenylcyclopentadienylhydrotris(indazolyl)borate ruthenium(II) complexes: prototypes of organometallic molecular gears and motors. Coord. Chem. Rev. **287**, 79 (2015)
37. Kammerer, C., Erbland, G., Gisbert, Y., Nishino, T., Yasuhara, K., Rapenne, G.: Biomimetic and technomimetic single molecular machines. Chem. Lett. **48**, 299 (2019)
38. Zhang, Y., Kersell, H., Stefak, R., Echeverria, J., Iancu, V., Perera, U.G.E., Li, Y., Deshpande, A., Braun, K.-F., Joachim, C., Rapenne, G., Hla, S.-W.: Simultaneous and coordinated rotational switching of all molecular rotors in a network. Nat. Nanotechnol. **11**, 706 (2016)
39. Kammerer, C., Rapenne, G.: Scorpionate hydrotris(indazolyl)borate ligands as tripodal platform for surface-mounted molecular motors and gears. Eur. J. Inorg. Chem. 2214 (2016)
40. Vives, G., Rapenne, G.: Directed synthesis of symmetric and dissymmetric molecular motors built around a ruthenium cyclopentadienyl tris(indazolyl)borate complex. Tetrahedron **64**, 11462 (2008)
41. Rapenne, G.: Synthesis of technomimetic molecules: towards rotation control in single-molecular machines and motors. Org. Biomol. Chem. **3**, 1165 (2005)
42. Gakh, A.A.: Molecular Devices. An introduction to Technomimetics and Biological Applications. Wiley, Hoboken (2018)
43. Willet, N., Duwez, A.-S.: Molecular Manipulation with Atomic Force Microscopy. CRC Press, Boca Raton (2011)

44. Lussis, P., Svaldo-Lanero, T., Bertocco, A., Fustin, C.-A., Leigh, D.A., Duwez, A.-S.: A single synthetic small molecule that generates force against a load. Nat. Nanotechnol. **6**, 553 (2011)
45. Quaethem, A.V., Lussis, P., Leigh, D.A., Duwez, A.-S., Fustin, C.-A.: Probing the mobility of catenane rings in single molecules. Chem. Sci. **5**, 1449 (2014)
46. Ohmann, R., Meyer, J., Nickel, A., Echeverria, J., Grisolia, M., Joachim, C., Moresco, F., Cuniberti, G.: Supramolecular rotor and translator at work: on-surface movement of single atoms. ACS Nano **9**, 8394 (2015)
47. Kolb, H.C., Finn, M.G., Sharpless, K.B.: Click chemistry: diverse chemical function from a few good reactions. Angew. Chem. Int. Ed. **40**, 2004 (2001)
48. Sirven, A.M., Garbage, R., Qiao, Y., Kammerer, C., Rapenne, G.: Synthesis of functionalized mono-, bis-, and trisethynyltriptycenes for one-dimensional self-assembly on surfaces. Chem. Eur. J. **21**, 15013 (2015)
49. Unpublished results
50. Gisbert, Y., Abid, S., Bertrand, G., Saffon-Merceron, N., Kammerer, C., Rapenne, G.: Modular synthesis of pentaarylcyclopentadienyl Ru-based molecular machines via sequential Pd-catalysed cross couplings. Chem. Commun. **55**, 14689 (2019)

Chemical Anchoring of Molecular Rotors

Oumaima Aiboudi and Franziska Lissel

Abstract A reliable anchoring on the substrate is the fundamental prerequisite to investigate surface-bound molecular rotors. The choice of the anchor group is dependent on the used substrate, and the surface-molecule bond must be sufficiently strong to endure under electrical operation. Here, we give an overview of anchor groups suitable to immobilize molecules on gold and other coinage metals via chemisorption. Sulfur-, nitrogen- and oxygen-based anchors are reviewed, *N*-heterocyclic carbenes as well as selected examples of other carbon-based anchors are considered, and examples of anchor groups reported for surface-bound molecular rotors are given. Anchoring is discussed in terms of the surface-molecule binding mode, i.e. radical adsorption and lone pair interaction. Green's ligand classification, Pearson's *hard/soft- acid/base* (HSAB) principle as well as the concepts denticity and podality are considered. Emphasis is placed on chemical aspects, e.g. the need to protect and controllably deprotect reactive anchors such as thiols and acetylenes.

Keywords Chemical anchoring · Molecular rotor · Surface immobilization · Binding mode

1 Introduction

Synthetic organic chemistry forms the basis for creating functional molecular systems able to convert external chemical, electrical, or thermal energy into controlled mechanical motion, i.e. molecular machinery such as molecular motors [1, 2], gears

O. Aiboudi · F. Lissel (✉)
Institute of Macromolecular Chemistry (IMC), Leibniz Institute for Polymer Research Dresden (IPF), Hohe Str. 6, Dresden 01069, Germany
e-mail: lissel@ipfdd.de

Institute of Chemistry, Technical University Dresden, Dresden 01062, Germany

F. Lissel
Cfaed Center for Advancing Electronics Dresden, Technical University Dresden, Dresden 01062, Germany

C. Joachim (ed.), *Building and Probing Small for Mechanics*, Advances in Atom and Single Molecule Machines, https://doi.org/10.1007/978-3-030-56777-4_7

[3, 4] and rotors [5–7]. Molecular rotors are systems in which a part of the molecule rotates against another part of the molecule [8], and can be classified as solution phase, solid phase, and surface-anchored molecular rotors [9].

Looking at the structural layout, a molecular rotor consists of three parts: The rotator, which is the part of the smallest moment of inertia, a stator with a higher inertia and an atomic-scale axle connecting these two components. While for rotors in liquid and solid phases it can be difficult to distinguish between rotator and stator, for surface-mounted rotors, the stator is always connected to the surface, and in some cases, the surface itself plays the role of a stator. The strong molecule-substrate interactions provide further advantages: Fixing the rotor's position in two dimensions provides easy accessibility and enhanced control by external fields, as well as the option to couple the rotor to other nanomechanical devices.

Using the surface as focal point, two types of orientation are distinguished (Fig. 1) [8]: The azimuthal rotor has a rotation axis perpendicular to the solid surface [10, 11], whereas an altitudinal rotor spins around an axle perpendicular to the surface [8]. Like for other types of energy-to-movement conversions on the molecular level, the rotational motion can be induced by external stimuli such as thermal energy, light, electrons or chemical conversions [1, 6, 12, 13].

For surface-mounted rotors, the connection between molecule and surface is a key descriptor. The atom or atom group binding to the solid substrate is denoted as anchor group, and for rotors, the anchor group(s) will compulsory form part of the stator. Molecules must be designed in a fashion that allows the anchor group to bind reliably and reproducibly to a given surface, and the surface-affinity of the anchor group must be higher than for any other part of the molecule, e.g. the one of a rotating blade. The design of the anchor group is invariably guided by the surface: An oxide layer requires a different anchor group than Au (111) or a graphene flake. In this chapter we will focus on anchor groups able to bind to coinage metal surfaces, give an overview of the physicochemical properties of the resulting bond, and detail

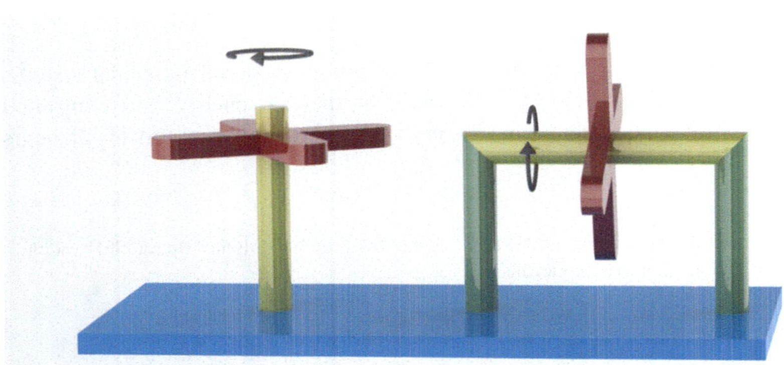

Fig. 1 Schematic illustration of an altitudinal rotor (right) having two stands as stator (green) an axle (yellow) and a rotator in red. On the left an azimuthal rotor using the surface as stator

different classes of chemical anchors and their binding modes. The discussion will concentrate on experiments and theory using Au (111) as substrate, with selected examples of other coinage metals, e.g. Ag (111) [14] and Cu (111) [15].

2 Physiochemical Properties

When an atom or molecule is adsorbed on a surface, new electronic states are formed. Two types of adsorption are distinguished depending on the strength of the interaction between the adsorbate and the substrate: Physisorption describes all weak electrostatic interactions including *van der Waals* interactions, dipole-dipole and London forces. The bond energies of these weak interactions between adsorbate and substrate typically range from 0.2 to 4 kJ/ mol [16]. In the case of chemisorption, the interaction energy is substantially higher (e.g. the thiolate-Au bond is in the range of 150–200 kJ/mol [17]): Substrate and adsorbed molecule interact by sharing or transferring electrons, and rearrangements of the valence levels of the molecule occur.

The anchor group defines the electronic coupling between the metal and the molecule, which essentially describes the degree of the hybridization with metal states. The bonding between a metal and a molecule is weaker than the typical organic bonds forming molecules. Compared with covalent bonding like e.g. a carbon–hydrogen bond with a bond energy of around 450 kJ/mol, a metal–molecule bond is usually weaker, in the order of 50–200 kJ/mol [17]. The relative strength of the formed bond determines if a given chemical anchor group is good for immobilizing a rotor on a specific surface. For instance, investigating thiolate self-assembled monolayers (SAMs) on the (111) surfaces of coinage metals and catalytically active Ni revealed that the SAM stability and the strength of the sulfur-metal bond depend on both the nature of the surface and the type of interactions between the sulfur terminated anchor group and the metal substrate [18].

The ideal anchor group must provide a well-defined and reproducible binding, which is sufficiently strong and robust under electrical operation. Chemical anchor groups can be classified according to several aspects, namely the number and character of surface-binding atoms, the charge of the anchor group and the type of bond formed between anchor group and surface.

Denticity and podality are both descriptors of the number of surface-binding atoms: While denticity refers to the number of binding atoms in the anchor group, podality gives the number of anchor groups. Monodentate anchor groups bind through one atom only (e.g. thiols, amine, pyridines) while bidentate anchor groups bind through two atoms as e.g. carboxylates, which adsorb via both oxygen atoms after the loss of the acidic proton. Multidentate anchor groups include larger π systems such as fullerene and pyrene, which can bind through partial or whole facets of the group.

Utilizing multipodal platforms is a good strategy to reinforce the immobilization of a single molecule on metal surfaces while simultaneously controlling the spatial

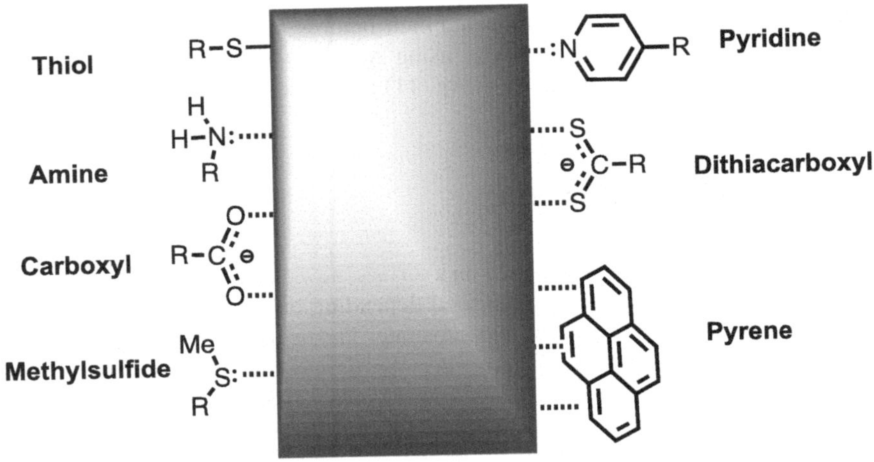

Fig. 2 Chemical structure of typical anchor groups for metal surfaces. Solid substrate-anchor bonds indicate a high covalent character

arrangement. These platforms are characterized by the presence of several—usually chemically equivalent—interaction sites that are not in a line [19]. The multisite cooperative interaction between anchor groups and substrate increases the binding stability of molecular rotors bearing a sterically demanding rotating unit: If one site of a multipodal anchor group is detached from the surface, the molecule is still attached by the other contact points. A basic example of these platforms are tripods with three conventional anchors groups working together to achieve a better surface binding, similar to the chelate effect observed in classic metal-ligand complexes [20]. The anchor groups are often at the tip of rigid aromatic units functioning as legs, which are attached to a branching point, e.g. a sp^3 hybridized carbon.

Some examples of standard anchors are shown in Fig. 2, different multipodal systems are described in Sect. 3. In all cases, the choice of the surface determines the choice of the anchor groups.

3 Binding Modes

There are two possible scenarios describing chemical anchoring on metal surfaces: Radical adsorption and lone pair interaction. As described in detail by Green [21], the elementary types of metal-ligand interactions can be classified according to the number of electrons that each atom contributes to the surface-anchor bond (Fig. 3).

An X-type ligand interacts with a metal surface via a normal two-electron covalent bond (e.g. Au-S) to which each atom or molecule contributes a single electron. For L and Z-types ligands, one atom supplies both electrons, the interaction corresponds to a dative (coordinative) bond with varying degrees of covalent character [22]. When

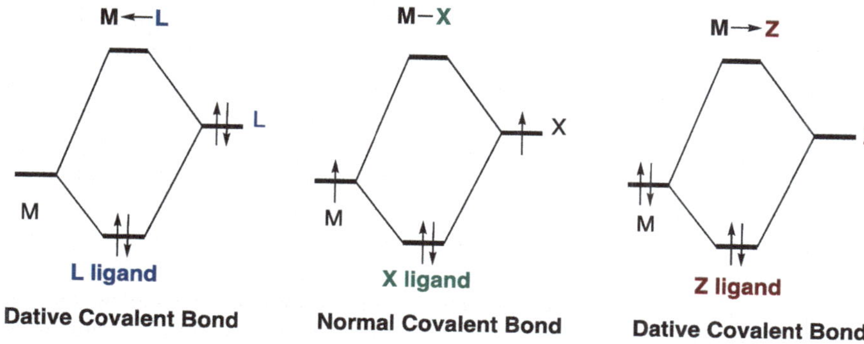

Fig. 3 The covalent bond classification (CBC), modified after Green [21]

the anchor group is a donor, L-type ligand bonding occurs via the lone pair provided by the anchor group (e.g. Au-NR_2) whereas Z-type ligands accept electron pairs from the surface. Another concept used to classify and predict the preference of metals for specific ligands is the *hard/soft—acid/base* (HSAB) principle by Pearson [23]. Hard-hard or soft-soft combinations contribute to the stabilization and strength of the bonds between donor and acceptor [24]. For instance, while S-based anchors are ubiquitously used for metal surfaces, gold (a soft acid) exhibits a higher affinity towards the softer selenium, and Se-based anchor groups bind more stable to gold than the corresponding sulfur derivatives [25, 26].

In the case of radical adsorption (e.g. of thiyls), the anchor group atoms have unpaired electrons available for covalent interaction with unsaturated electronic states on the metal surface, and the open-shell electrons on the anchor (X Ligand) atom form electron pairs with the metal atoms at the surface. In most cases, these radicals are generated through the spontaneous dissociation of the molecule on the surface [27].

In the case of lone pair adsorption, a donor-acceptor bond (dative bond) forms through the delocalization of the lone pair of the L ligand to an undercoordinated metal atom. Apart from sulfide, this category also includes several groups containing N atoms such as amine, pyridine and nitrile.

4 Chemical Anchor Groups

4.1 Sulfur-Gold Linkage

To date, the most widely used anchors to adsorb molecules on a gold surface are organosulfur compounds [28, 29]. Due to the strong affinity between sulfur and coinage metals, these ligands are chemisorbed onto the surface with high binding energies of around 0.45—1.72 eV in the case of gold [14, 30]. In the field of molecular electronics, thiols (R-SH) are ubiquitously used chemical anchor groups for various metal surfaces [18, 31]. The Au-thiolate bond is generated by the dissociation of the S-H bond, which spontaneously occurs on all clean metal surfaces, resulting in a thiyl radical (RS.) [27, 32]. The subsequent metal-sulfur bond shows a varying degree of covalency for all coinage metals [33], which is strongest for the Au-thiolate. However, thiols, and in particular aromatic thiols, are oxidation labile, hindering the ability of the molecule to successfully chemisorb. To overcome this, different protecting groups have been used as alternative thiol-like anchoring agents [34, 35]. Thioacetate groups for instance are well known to allow the formation of chemisorbed Au-thiolates. Acetyl-protected thiols can be converted to free thiols either by an in situ acidic [36] or base-mediated [31] deprotection or by spontaneous cleavage of the acetate group upon exposure to gold [37, 38].

Another type of organosulfur anchor group used to immobilize molecules onto metal substrates while forming similar surface species to those observed for thiols are dialkylsulfides [27, 39]. The binding of this ligand involves a spontaneous S–S bond scission upon contact with transition metals such as Au and Ag during the adsorption process, leading to the chemisorption of the two generated thiolates [40].

In the case of self-assembled thiolate monolayers on Au (111), several binding modes have been proposed (Fig. 4), from an early standard model describing the binding at atop, bridge or hollow sites on an unreconstructed Au (111) surface, to more complex motifs including a disulfide bonding, a thiolate complex with an Au adatom, and a chain structure with thiolates bridging Au adatoms [33, 41]. Depending on the organic part, thiolates show different degrees of mobility on the surface, leading to time-dependent segregation of mixed thioalkyl monolayers.

Unlike thiols and dialkyldisulfides, thioethers and methylsulfides belong to the class of L-ligands and chemisorb via a lone pair. During the adsorption process, these anchor groups form donor-acceptor bonds with undercoordinated gold atoms on the surface [42]. Besides the organosulfur compounds introduced above, the utilization of bidentate sulfur based anchors such as dithiocarboxylic acids and dithiocarbamates has also been reported [43, 44].

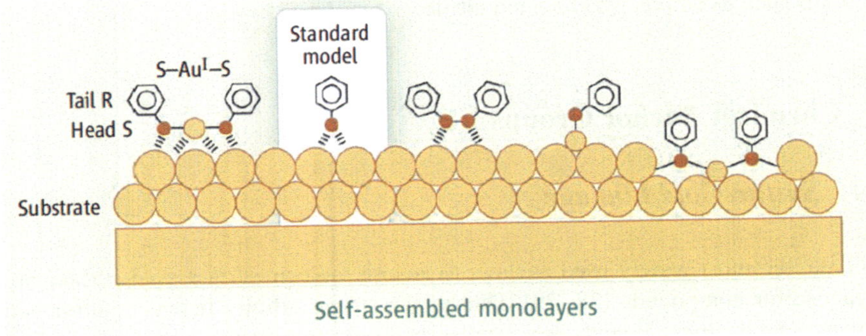

Fig. 4 Suggested binding mode for thiols to Au surface. Reprinted from Ref. [33] with permission

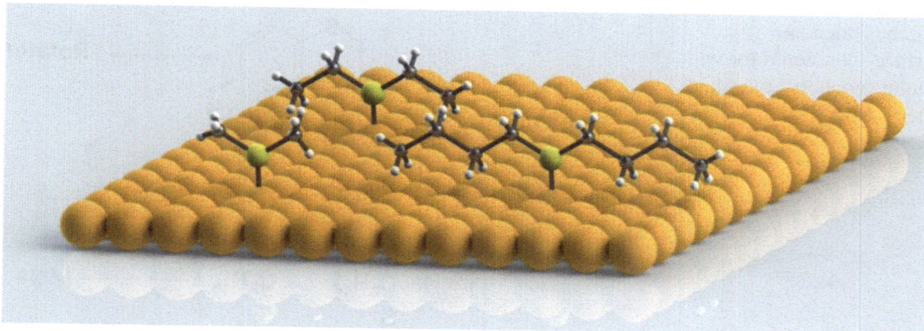

Fig. 5 Schematic illustration of mondendate thioether rotors on gold modified after Ref. [46]

5 Monopodal Sulfur Anchors for Molecular Rotors

Thioethers (RSR') represent a simple monodentate rotor system, in which molecular rotation can be induced both electrically and thermally [45]. In 2008, Sykes et al. reported an LT-STM single-molecule study of the motion of a set of thioethers with various chain length on a Au (111) surface [46].

The thioether rotors were chemisorbed on a Au (111) surface (stator) through a coordinative S-Au bond (the axle) as shown in Fig. 5, the alkyl tails interact weakly with the surface and play the role of the rotator.

6 Bipodal Sulfur Anchors for Molecular Rotors

Feringa and coworkers developed a light driven rotary motor able of unidirectional rotation in solution controlled by a single stereocenter [47], and later demonstrated that this motor can be mounted on the surface of gold nanoparticles [48]. The rotary motor design is based on a chiral helical alkene with an upper half serving as a propeller. The propelling unit undergoes a repetitive 360° rotation in four distinct steps upon irradiation with light of the appropriate wavelength. The rotation axis consists of a carbon–carbon double bond which is connected to the propeller and to a lower half of the molecule serving as a stator. The stator carries two thiol-functionalized 'legs', resulting in a bipodal anchoring scheme immobilizing the molecule on the surface (Fig. 6).

7 Multipodal Sulfur Anchors for Molecular Rotors

Different sulfur-based tri- and tetrapodal anchors structures were reported (see Fig. 7 for selected examples). Perera et al. used a tripodal stator to mount a multi-component rotary motor on the Au (111) surface [10]. The motor consists of a one toluene and

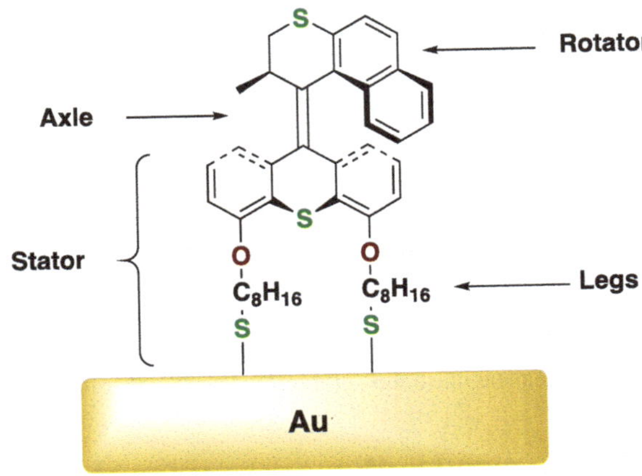

Fig. 6 Bidendate surface-mounted rotor with thiolate anchors modified after [48]

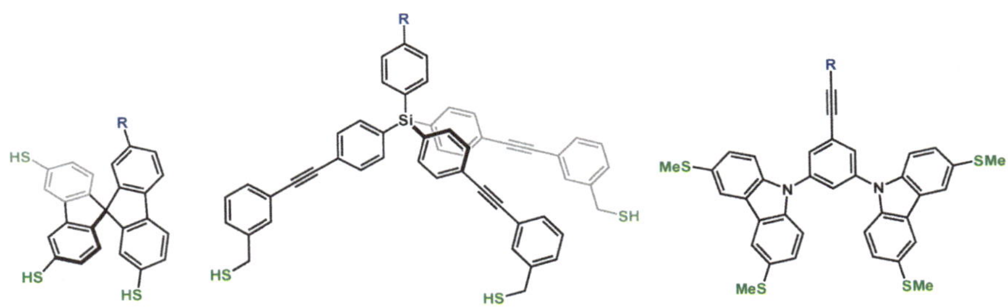

Fig. 7 Chemical structures of different sulfur-based multipodal platforms modified after [80–82]

four ferrocene arms as rotator, allowing controlled a rotation and a better monitoring of the rotation steps due to the structural dissymmetry, a ruthenium center connecting the rotator and stator units, and a thioether-terminated hydro-*tris* (indazolyl) borate tripodal stator, which enables stable anchoring through its three Au-S bonds. Once chemisorbed on Au (111), clockwise or anti-clockwise rotation of the motor was observed by selective inelastic electron tunneling through different subunits of the motor.

The branching point of a multipodal anchor determines the angle between surface and functional, e.g. rotating, group. Müllen et al. recently reported a tetrapodal anchor enforcing a near upright orientation of the functional head group (Fig. 8) [49].

7.1 Nitrogen-Metal Bond

Nitrogen-based anchor groups are an often used alternative to organosulfur anchor groups. This category encompasses various functional groups such as amines, pyridine, and nitrile [50–53]. In contrast to sulfur-containing anchor groups which form

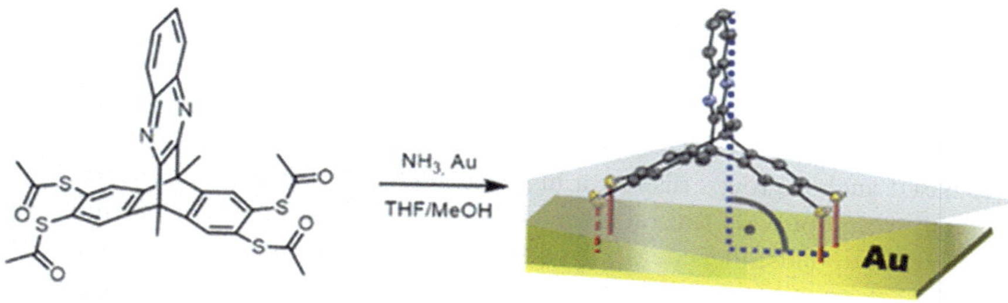

Fig. 8 Schematic illustration of a sulfur-based tetrapod on a Au surface. Reprinted from Ref. [49] with permission of the American Chemical Society

a robust linkage with metal surfaces either via radical or lone pair adsorption, N-based ligands connect exclusively via coordinative interactions. These L-type ligand interactions involve electron donation from their lone pair to an undercoordinated (Lewis acidic) gold atom [50, 54].

As discussed earlier, one strategy for achieving a more robust surface attachment and better control of molecular orientation is to use multiple anchoring atoms (Fig. 9). In 2011, Aso and working group developed a set of pyridine based tripods able to anchor on gold surfaces [55]. According to CV measurements of their monolayers on gold, the tripodal structure shows enhanced adsorption ability and robustness compared to a single pyridine anchor even under biased conditions due to the multiplier effect of the tripodal pyridine anchor. In addition, XPS measurements showed that the π orbital of pyridine contributes to the physical adsorption of the developed tripodal anchor on gold.

Gao et al. reported the rotation of tetra-*tert*-butyl zinc phthalocyanine ((*t*-Bu)$_4$-ZnPc) molecules on a Au (111) surface [56]: A coordinative bond is formed between a Au adatom and one of the outer isoindole nitrogens, leading to an off-center rotation axis and allowing the tilted macrocycle to rotate.

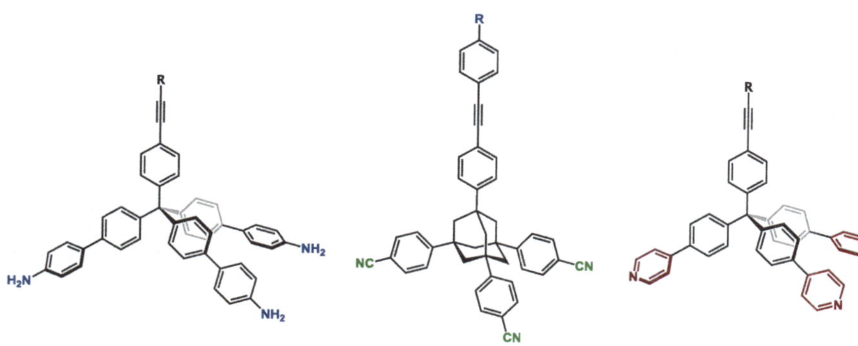

Fig. 9 Chemical structures of various N-based tripodal anchor groups modified after [55, 83, 84]

7.2 Carben-Metal Bond

N-heterocyclic carbenes [57] (NHCs) have demonstrated high potential as anchor groups for metal surfaces [58]: Compared to traditional anchor groups such as thiols, NHC-based anchor groups display a higher chemical, oxidative and electrochemical stability and provide several advantages due to their capability to form strong carbene–metal bonds, their delocalized bonding and structural diversity [58, 59]. NHCs display a combination of strong σ-donating and moderately π-accepting properties [60], which makes them especially suitable for anchoring to late-transition metals such as gold [61]. The surface deposition of NHCs can be achieved via several methods. A main approach is wet-deposition in which deprotonation of the corresponding imidazolium salts is facilitated by a strong base such as KO^tBu and KHMDS to form an active carbene able to anchor on metal substrates [62, 63]. This process must be carried out under inert atmosphere due to the high sensitivity towards air and moisture. In addition, depending on the base, residual salt and solvent impurities may be formed, leading to metal surface contamination and consequently a low-quality monitoring and imaging via STM.

An alternative approach is vapor deposition [64, 65], in which an NHC salt, e.g. an NHC carboxylate, binds onto the surface under ultrahigh vacuum (UHV) conditions (Fig. 10). Nguyen [59] et al. reported the first successful fabrication of well-defined patterns of NHCs on gold surfaces by Microcontact Printing (μCP) monitored by XPS.

Recent studies demonstrated that the substituents at the *N,N*-positions of the NHC are key in determining the binding behavior of NHCs on metal surfaces [65, 66]. In general, as shown in Fig. 11, four binding mode can be distinguished: The work of Fuchs and coworkers [65] revealed that methyl substituted carbenes (IMe) bind to a gold adatom perpendicular to the surface in an Up_{ad} configuration, while NHCs bearing butyl groups (IBu) lay flat in an $Down_{surf}$ configuration, leading to the formation of dimeric NHC-Au-NHC complexes. The observed change in binding mode upon elongating the alkyl chain can be explained by the increased *van der Waals* interaction between the saturated chains and the gold surface. The same group further examined the mobility of a variety of NHC anchor groups on planar gold surfaces using STM. Experimental images revealed a circular shape of diisopropilphenyl substituted NHC (IPr) at temperatures around 77 K on Au (111), indicating the ability of the free part of the molecule to rotate around a single carbene–gold bond

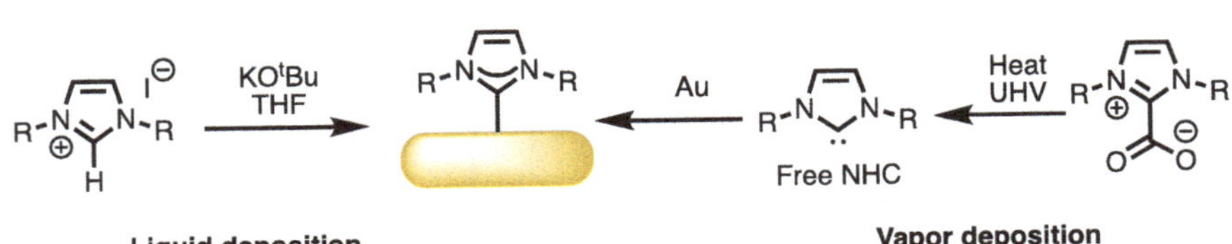

Fig. 10 Schematic illustration of wet and vapour deposition of NHCs

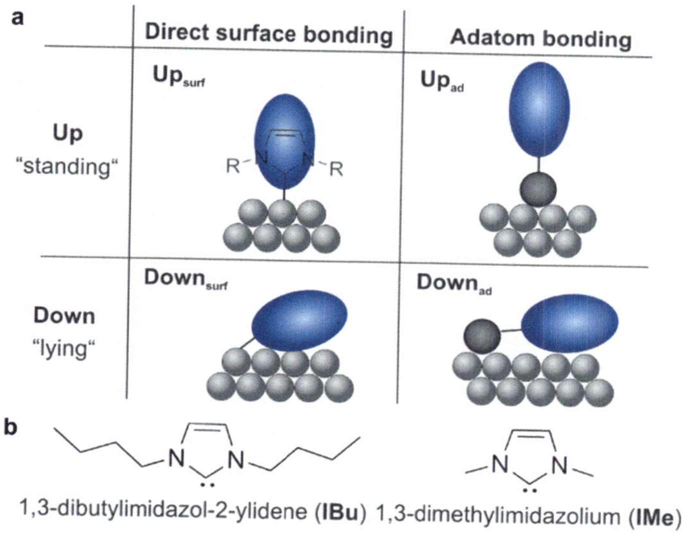

Fig. 11 **a** Schematic illustration of the different binding modes of NHCs on metal surfaces. **b** Chemical structure of IBu and IMe. Reprinted from Ref. [65] with permission of Nature publishing

on the surface. The IPr molecule pulls an Au atom out of the surface inducing adatom formation, which enables a molecular "ballbot-like" motion of NHCs sticking to a single adatom [67].

7.3 Oxygen-Metal Bond

Another way of anchoring molecules on metal substrates by chemisorption is the use of oxygen-based ligands as anchor group. This category includes several functional groups, e.g. carboxylic acids (-COOH), carboxylic esters (-COOR), nitro groups (-NO_2), and alkoxy groups (-O-R) [68–71]. Of these, carboxylic acid is the most commonly used anchor group for standard immobilization on metal substrates (e.g. Au, Ag and Cu) [70, 72, 73]. Carboxylic acids connect via coordinative interactions: The acidic hydrogen of the -COOH group is cleaved, leading to a carboxylate (COO-) able to chemisorb onto the surface with the two oxygen atoms [74, 75]. Several monodentate and bidentate binding modes have been observed (Fig. 12). Of these, the bridging bidentate configuration is the preferred mode [76]. Nitro groups connect to the metallic surface with both oxygen atoms in a bidentate chelating configuration, while esters adopt the monodentate one. Analogous to thioether, alkoxyalkanes (ROR) are L-type ligands forming dative interactions with the metal substrate via an active lone pair. However, the bond strength and stability of ether-based anchor (hard base) on Au (soft acid) is weaker than the corresponding thioether [23].

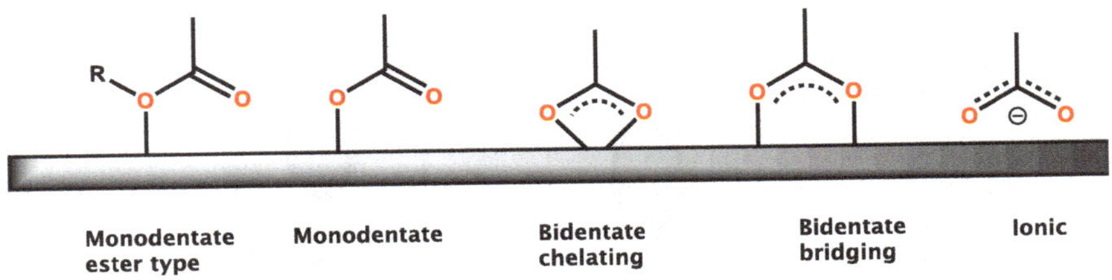

Fig. 12 Suggested binding modes for carboxylic acid- and ester-based anchors

7.4 Other Anchor Groups

Besides the anchor classes reviewed above, various other ligands demonstrate anchoring ability on metal surfaces. For instance, a stable and conductive covalent carbon-Au bond can be generated using alkynes [29]. As terminal alkynes are reactive and prone to oxidative coupling, these anchor groups are usually capped with a protective group for easier handling. Trimethylsilyl (TMS) protected alkynes are made reactive by in situ cleavage of the TMS group using e.g. tetrabutylammonium fluoride (TBAF) [77]. An anchoring pathway requiring no active deprotecting agent is the utilization of trimethylstannyl caps for the alkyne [26].

As discussed in Sect. 2, selenium has a high affinity to gold, and the formed Se-Au bonder is stronger than the corresponding S-Au bond. Both isothiocyanate (R-NCS) and isoselenocyanate (R-NCSe) are L-type ligands for coinage metals, but the binding of the latter is more stable [26]. In term of achieving strong covalent binding, selenols (R-SeH) function as X-type ligands. Yet selenols oxidize rapidly in air to produce diselenide at a faster rate than thiols. Acetyl moieties are a commonly used protecting group, which can be cleaved by external reagents to obtain the active selenium species [78].

In the group of pnigtogens, also phosphorous can be used in anchor groups, and, like nitrogen-based ligands, phosporous-containing anchor groups bind as L-type via a lone pair. Phosphines (R_3P) are excellent ligands for defined transition metal complexes, and similarly can be used to form donor-acceptor bond with metal substrates [42].

Another appealing anchoring strategy is the use of larger groups such as C_{60}, which have a large contact area to metallic surfaces. The binding of C_{60} is strongly dependant on its orientation and can be achieved in several ways, among others through a single carbon atom η^1, via the centre of a hexagonal ring η^6, or the centre of pentagonal ring η^5 [79].

8 Summary

To investigate surface-bound molecular rotors, a reliable bond between molecule and substrate is a fundamental prerequisite. In this chapter we presented several approaches to understand the interactions between molecules and surface of coinage metals, discussed different binding modes and the nature of the resulting coupling between chemical anchor group and surface. Concepts such as Green's ligand classification and Pearson's *hard/soft—acid/base* (HSAB) principle can be helpful to consciously choose or design the best anchor group for a given problem set. In addition, structural features such as multisite binding (multipodal platforms) can increase the chemisorption of the adsorbed molecules.

While the library of chemical groups enabling the immobilization of molecular units on metallic surfaces is obviously large, only a limited number of anchor groups are frequently used–often despite significant drawbacks in the handling of the free anchor group as well as the stability of the resulting binding to the surface.

Organosulfur compounds with terminations such as -SH, RSSR, SR are ubiquitously used anchor groups in single-molecule electronics as well as to obtain monolayers on metal surfaces as they provide strong bonds with coinage metals such as gold and copper. Nevertheless, there is still a need for optimization with regards to their limited oxidative and thermal stability. Carbon-based anchors, such as alkynes and especially NHCs, have emerged as attractive alternative offering enhanced stability and robustness of surface attachment on metallic substrates.

Acknowledgements O.A. thanks the Helmholtz International Research School for Nanoelectronic Networks (NanoNet) for a PhD fellowship, F.L. the Fonds der Chemischen Industrie (FCI) for a Liebig fellowship.

References

1. Fletcher, S.P., Dumur, F., Pollard, M.M., Feringa, B.L.: Chemistry: a reversible, unidirectional molecular rotary motor driven by chemical energy. Science 310(5745), 80–82 (2005)
2. Kulago, A.A., Mes, E.M., Klok, M., Meetsma, A., Brouwer, A.M., Feringa, B.L.: Ultrafast light-driven nanomotors based on an acridane stator. J. Org. Chem. **75**(3), 666–679 (2010)
3. Manzano, C., et al.: Step-by-step rotation of a molecule-gear mounted on an atomic-scale axis. Nat. Mater. **8**(7), 576–579 (2009)
4. Molecular gears: wheels in motion. NPG Asia Mater. **1**(1), 13 (2009)
5. Stipe, B.C., Rezaei, M.A., Ho, W.: Inducing and viewing the rotational motion of a single molecule. Science **279**(5358), 1907–1909 (1998)
6. Gimzewski, J.K., Joachim, C., Schlittler, R.R., Langlais, V., Tang, H., Johannsen, I.: Rotation of a single molecule within a supramolecular bearing. Science **281**(5376), 531–533 (1998)
7. Zheng, X., Mulcahy, M.E., Horinek, D., Galeotti, F., Magnera, T.F., Michl, J.: Dipolar and nonpolar altitudinal molecular rotors mounted on an Au(111) surface. J. Am. Chem. Soc. **126**(14), 4540–4542 (2004)
8. Kottas, G.S., Clarke, L.I., Horinek, D., Michl, J.: Artificial molecular rotors. Chem. Rev **105**(4), 1281–1376 (2005)
9. Xue, M., Wang, K.L.: Molecular rotors as switches. Sensors (Switzerland) **12**(9), 11612–11637 (2012)
10. Perera, U.G.E., et al.: Controlled clockwise and anticlockwise rotational switching of a molecular motor. Nat. Nanotechnol. **8**(1), 46–51 (2013)
11. Carroll, G.T., Pollard, M.M., van Delden, R., Feringa, B.L.: Controlled rotary motion of light-driven molecular motors assembled on a gold film. Chem. Sci. **1**(1), 97–101 (2010)

12. Carroll, G.T., London, G., Landaluce, T.F., Rudolf, P., Feringa, B.L.: Adhesion of photon-driven molecular motors to surfaces via 1, 3-dipolar cycloadditions: effect of interfacial interactions on molecular motion. ACS Nano **5**(1), 622–630 (2011)
13. Stipe, B.C., Rezaei, M.A., Ho, W. Inducing and viewing the rotational motion of a single molecule. Science **279**(5358), 1907–1909 (1998)
14. Ulman, A.: Formation and structure of self-assembled monolayers. Chem. Rev. **96**(4), 1533–1554 (1996)
15. Rao, B.V., Kwon, K.Y., Liu, A., Bartels, L.: 2, 5-dichlorothiophenol on Cu(111): initial adsorption site and scanning tunnel microscope-based abstraction of hydrogen at high intramolecular selectivity. J. Chem. Phys. **119**(20), 10879–10884 (2003)
16. Lippert, E.: The strengths of chemical bonds, von T.L. Cottrell. Butterworths Publications Ltd., London 1958. 2. Aufl., X, 317 S., geb.t—/32/—. Angew. Chemie **72**(16), 602–602 (1960)
17. Pensa, E., et al.: The chemistry of the sulfur-gold interface: in search of a unified model. Acc. Chem. Res. **45**(8), 1183–1192 (2012)
18. Vericat, C., et al.: Self-assembled monolayers of thiolates on metals: a review article on sulfur-metal chemistry and surface structures. RSC Advances **4**(53), 27730–27754 (2014). Royal Society of Chemistry
19. Valášek, M., Lindner, M., Mayor, M.: Rigid multipodal platforms for metal surfaces. Beilstein J. Nanotechnol. **7**(1), 374–405 (2016). Beilstein-Institut Zur Forderung der Chemischen Wissenschaften
20. Kitagawa, T., et al.: Rigid molecular tripod with an adamantane framework and thiol legs. Synthesis and observation of an ordered monolayer on Au(111). J. Org. Chem. **71**(4), 1362–1369 (2006)
21. Green, M.L.H.: A new approach to the formal classification of covalent compounds of the elements. J. Organomet. Chem. (1995)
22. Haaland, A.: Covalent versus dative bonds to main group metals, a useful distinction. Angew. Chemie Int. Ed. English **28**(8), 992–1007 (1989)
23. Pearson, R.G.: Hard and soft Acids and bases. J. Am. Chem. Soc. **85**(22), 3533–3539 (1963)
24. Chemical Hardness : Ralph, G. Pearson : 9783527294824. https://www.bookdepository.com/Chemical-Hardness-Ralph-G-Pearson/9783527294824. Accessed: 22 Apr 2020
25. Patrone, L., Palacin, S., Bourgoin, J.P., Lagoute, J., Zambelli, T., Gauthier, S.: Direct comparison of the electronic coupling efficiency of sulfur and selenium anchoring groups for molecules adsorbed onto gold electrodes. Chem. Phys. **281**(2–3), 325–332 (2002)
26. Lissel, F., et al.: Organometallic single-molecule electronics: tuning electron transport through X(diphosphine)2FeC4Fe(diphosphine)2X building blocks by varying the Fe-X-Au anchoring scheme from coordinative to covalent. J. Am. Chem. Soc. **136**(41), 14560–14569 (2014)
27. Grönbeck, H., Curioni, A., Andreoni, W.: Thiols and disulfides on the Au(111) surface: the headgroup-gold interaction. J. Am. Chem. Soc. **122**(16), 3839–3842 (2000)
28. Reed, M.A., Zhou, C., Muller, C.J., Burgin, T.P., Tour, J. M.: Conductance of a molecular junction. Science **278**(5336), 252–254 (1997)
29. Schwarz, F., et al.: High-conductive organometallic molecular wires with delocalized electron systems strongly coupled to metal electrodes. (2014)
30. Yaliraki, S.N., Kemp, M., Ratner, M.A.: Conductance of molecular wires: influence of molecule-electrode binding. J. Am. Chem. Soc. **121**(14), 3428–3434 (1999)
31. Valkenier, H., Huisman, E.H., van Hal, P.A., de Leeuw, D.M., Chiechi, R.C., Hummelen, J.C.: Formation of high-quality self-assembled monolayers of conjugated dithiols on gold: base matters. J. Am. Chem. Soc. **133**(13), 4930–4939 (2011)
32. Tielens, F., Santos, E.: AuS and SH bond formation/breaking during the formation of alkanethiol SAMs on Au(111): a theoretical study. J. Phys. Chem. C **114**(20), 9444–9452 (2010)
33. Whetten, R.L., Price, R.C.: Chemistry: nano-golden order. Science **318**(5849), 407–408 (2007). American Association for the Advancement of Science
34. Lukkari, J., et al.: Organic thiosulfates (bunte salts): novel surface-active sulfur compounds for the preparation of self-assembled monolayers on gold. Langmuir **15**(10), 3529–3537 (1999)

35. Niebel, C., Calard, F., Jarrosson, T., Lère-Porte, J.P., Breton, T., Serein-Spirau, F.: Spontaneous assembly of silylethane-thiol derivatives on Au(111): a chemically robust thiol protecting group as the precursor for the direct formation of aromatic gold thiolate monolayers. Chem. Commun. **51**(36), 7622–7625 (2015)
36. Cai, L., Yao, Y., Yang, J., Price, D.W., Tour, J.M.: Chemical and potential-assisted assembly of thiolacetyl-terminated oligo(phenylene ethynylene)s on gold surfaces. Chem. Mater. **14**(7), 2905–2909 (2002)
37. Béthencourt, M.I., Srisombat, L.O., Chinwangso, P., Lee, T.R.: SAMs on gold derived from the direct adsorption of alkanethioacetates are inferior to those derived from the direct adsorption of alkanethiols. Langmuir **25**(3), 1265–1271 (2009)
38. Tour, J.M., et al.: Self-assembled monolayers and multilayers of conjugated thiols, α, ω-dithiols, and thioacetyl-containing adsorbates. Understanding attachments between potential molecular wires and gold surfaces. J. Am. Chem. Soc. **117**(37), 9529–9534 (1995)
39. Biebuyck, H.A., Bain, C.D., Whitesides, G.M.: Comparison of organic monolayers on polycrystalline gold spontaneously assembled from solutions containing dialkyl disulfides or alkanethiols. Langmuir **10**(6), 1825–1831 (1994)
40. Hagenhoff, B., Benninghoven, A., Spinke, J., Liley, M., Knoll, W.: Time-of-flight secondary ion mass spectrometry investigations of self-assembled monolayers of organic thiols, sulfides, and disulfides on gold surfaces. Langmuir **9**(7), 1622–1624 (1993). American Chemical Society
41. Häkkinen, H.: The gold-sulfur interface at the nanoscale. Nat. Chem. **4**(6), 443–455 (2012). Nature Publishing Group
42. Park, Y.S., et al.: Contact chemistry and single-molecule conductance: A comparison of phosphines, methyl sulfides, and amines. J. Am. Chem. Soc. **129**(51), 15768–15769 (2007)
43. Morf, P., et al.: Dithiocarbamates: functional and versatile linkers for the formation of self-assembled monolayers. Langmuir **22**(2), 658–663 (2006)
44. Colorado, R., Villazana, R.J., Lee, T.R.: Self-assembled monolayers on gold generated from aliphatic dithiocarboxylic acids. Langmuir **14**(22), 6337–6340 (1998)
45. Tierney, H.L., et al.: Experimental demonstration of a single-molecule electric motor. Nat. Nanotechnol. **6**(10), 625–629 (2011)
46. Baber, A.E., Tierney, H.L., Sykes, E.C.H.: A quantitative single-molecule study of thioether molecular rotors. ACS Nano **2**(11), 2385–2391 (2008)
47. Koumura, N., Geertsema, E.M., Meetsma, A., Feringa, B.L.: Light-driven molecular rotor: unidirectional rotation controlled by a single stereogenic center [6]. J. Am. Chem. Soc. **122**(48), 12005–12006 (2000). American Chemical Society
48. Van Delden, R.A., Ter Wiel, M.K.J., Pollard, M.M., Vicario, J., Koumura, N., Feringa, B.L.: Unidirectional molecular motor on a gold surface. Nature **437**(7063), 1337–1340 (2005)
49. Benneckendorf, F.S., et al.: Tetrapodal diazatriptycene enforces orthogonal orientation in self-assembled monolayers. ACS Appl. Mater. Interfaces. **12**(5), 6565–6572 (2020)
50. Arroyo, C.R., Leary, E., Castellanos-Gómez, A., Rubio-Bollinger, G., González, M.T., Agraït, N.: Influence of binding groups on molecular junction formation. J. Am. Chem. Soc. **133**(36), 14313–14319 (2011)
51. Venkataraman, L., Klare, J.E., Tam, I.W., Nuckolls, C., Hybertsen, M.S., Steigerwald, M.L.: Single-molecule circuits with well-defined molecular conductance (2006)
52. Zhang, L., Cole, J.M.: Anchoring groups for dye-sensitized solar cells. ACS Appl. Mater. Interfaces **7**(6), 3427–3455 (2015). American Chemical Society
53. Mishchenko, A., et al.: Single-molecule junctions based on nitrile-terminated biphenyls: a promising new anchoring group. J. Am. Chem. Soc. **133**(2), 184–187 (2010)
54. Venkataraman, L., Klare, J.E., Tam, I.W., Nuckolls, C., Hybertsen, M.S., Steigerwald, M.L.: Single-molecule circuits with well-defined molecular conductance. Nano Lett. **6**(3), 458–462 (2006)
55. Ie, Y., et al.: Nature of electron transport by pyridine-based tripodal anchors: Potential for robust and conductive single-molecule junctions with gold electrodes. J. Am. Chem. Soc. **133**(9), 3014–3022 (2011)

56. Gao, L., et al.: Constructing an array of anchored single-molecule rotors on gold surfaces. Phys. Rev. Lett. **101**(19), 197209 (2008)
57. Kirmse, W.: The beginnings of N-Heterocyclic carbenes. Angew. Chemie Int. Ed. **49**(47), 8798–8801 (2010)
58. Zhukhovitskiy, A.V., MacLeod, M.J., Johnson, J.A.: Carbene ligands in surface chemistry: from stabilization of discrete elemental allotropes to modification of nanoscale and bulk substrates. Chem. Rev. **115**(20), 11503–11532 (2015)
59. Nguyen, D.T., et al.: Versatile micropatterns of N-Heterocyclic carbenes on gold surfaces: increased thermal and pattern stability with enhanced conductivity. Angew. Chemie Int. Ed. **57**(35), 11465–11469 (2018)
60. H. Jacobsen, A. Correa, A. Poater, C. Costabile, and L. Cavallo, "Understanding the M–(NHC) (NHC = N-heterocyclic carbene) bond. Coord. Chem. Rev. **253**(5–6), 687–703. Elsevier (2009)
61. Crudden, C.M., et al.: Simple direct formation of self-assembled N-heterocyclic carbene monolayers on gold and their application in biosensing. Nat. Commun. **7**(1), 1–7 (2016)
62. Zhukhovitskiy, A.V., Mavros, M.G., Van Voorhis, T., Johnson, J.A.: Addressable Carbene Anchors for Gold Surfaces. J. Am. Chem. Soc. **135**(20), 7418–7421 (2013)
63. Crudden, C.M., et al.: Ultra stable self-assembled monolayers of N-heterocyclic carbenes on gold. Nat. Chem. **6**(5), 409–414 (2014)
64. Dery, S. et al.: Flexible NO_2 -functionalized N-Heterocyclic carbene monolayers on Au (111) surface. Chem.A Eur. J. **25**(66), 15067–15072 (2019)
65. Bakker, A., et al.: Elucidating the binding modes of N-Heterocyclic carbenes on a gold surface. J. Am. Chem. Soc. **140**(38), 11889–11892 (2018)
66. Larrea, C.R., Baddeley, C.J., Narouz, M.R., Mosey, N.J., Horton, J.H., Crudden, C.M.: *N* -Heterocyclic carbene self-assembled monolayers on copper and gold: dramatic effect of wingtip groups on binding, orientation and assembly. ChemPhysChem **18**(24), 3536–3539 (2017)
67. Wang, G., et al.: Ballbot-type motion of N-heterocyclic carbenes on gold surfaces. Nat. Chem. **9**(2), 152–156 (2017)
68. Zhang, Z., Imae, T.: Hydrogen-bonding stabilized self-assembled monolayer film of a functionalized diacid, protoporphyrin IX Zinc(II), onto a gold surface. Nano Lett. **1**(5), 241–243 (2001)
69. Rochford, J., Chu, D., Hagfeldt, A., Galoppini, E.: Tetrachelate porphyrin chromophores for metal oxide semiconductor sensitization: Effect of the spacer length and anchoring group position. J. Am. Chem. Soc. **129**(15), 4655–4665 (2007)
70. Zotti, L.A., et al.: Revealing the role of anchoring groups in the electrical conduction through single-molecule junctions. Small **6**(14), 1529–1535 (2010)
71. Eisenhut, F., Meyer, J., Krüger, J., Ohmann, R., Cuniberti, G., Moresco, F.: Inducing the controlled rotation of single o-MeO-DMBI molecules anchored on Au(111). Surf. Sci. **678**, 177–182 (2018)
72. Tao, Y.T.: Structural comparison of self-assembled monolayers of n-Alkanoic acids on the surfaces of silver, copper, and aluminum. J. Am. Chem. Soc. **115**(10), 4350–4358 (1993)
73. Krzykawska, A., Ossowski, J., Zaba, T., Cyganik, P.: Binding groups for highly ordered SAM formation: carboxylic: versus thiol. Chem. Commun. **53**(42), 5748–5751 (2017)
74. Han, S.W., Ha, T.H., Kim, C.H., Kim, K.: Self-assembly of anthraquinone-2-carboxylic acid on silver: fourier transform infrared spectroscopy, ellipsometry, quartz crystal microbalance, and atomic force microscopy study. Langmuir **14**(21), 6113–6120 (1998)
75. Ahn, S., et al.: Electronic transport and mechanical stability of carboxyl linked single-molecule junctions. Phys. Chem. Chem. Phys. **14**(40), 13841–13845 (2012)
76. Matsuo, Y., Kanaizuka, K., Matsuo, K., Zhong, Y.W., Nakae, T., Nakamura, E.: Photocurrent-generating properties of organometallic fullerene molecules on an electrode. J. Am. Chem. Soc. **130**(15), 5016–5017 (2008)
77. Olavarria-Contreras, I.J., Perrin, M.L., Chen, Z., Klyatskaya, S., Ruben, M., Van Der Zant, H.S.J.: C-Au covalently bonded molecular junctions using nonprotected Alkynyl anchoring groups. J. Am. Chem. Soc. **138**(27), 8465–8469 (2016)

78. Liu, J., et al.: Triptycene-terminated thiolate and selenolate monolayers on Au(111). Beilstein J. Nanotechnol. **8**(1), 892–905 (2017)
79. Schull, G., Néel, N., Becker, M., Kröger, J., Berndt, R.: Spatially resolved conductance of oriented C_{60}. New J. Phys. **10**(6), 065012 (2008)
80. Valášek, M., Edelmann, K., Gerhard, L., Fuhr, O., Lukas, M., Mayor, M.: Synthesis of molecular tripods based on a rigid 9, 9'-spirobifluorene scaffold. J. Org. Chem. **79**(16), 7342–7357 (2014)
81. Shirai, Y., Cheng, L., Chen, B., Tour, J.M.: Characterization of self-assembled monolayers of fullerene derivatives on gold surfaces: Implications for device evaluations. J. Am. Chem. Soc. **128**(41), 13479–13489 (2006)
82. O'Driscoll, L.J., et al.: Carbazole-based tetrapodal anchor groups for gold surfaces: synthesis and conductance properties. Angew. Chemie **132**(2), 892–899 (2020)
83. Hirose, T., Ie, Y., Aso, Y.: Synthesis of tripodal-anchor units having pyridine or amine functional groups and their adsorption behavior on metal electrodes. Chem. Lett. **40**(2), 204–205 (2011)
84. Zarwell, S., Dietrich, S., Schulz, C., Dietrich, P., Michalik, F., Rück-Braun, K.: Preparation of an Indolylfulgimide-Adamantane Linker Conjugate with Nitrile Anchoring Groups through Palladium-Catalyzed transformations. European J. Org. Chem. **2009**(13), 2088–2095 (2009)

Anchoring Molecular Rotors by On-Surface Synthesis

Kwan Ho Au Yeung, Tim Kühne, Frank Eisenhut, and Francesca Moresco

Abstract Single molecular rotors are important components for constructing bottom-up molecular mechanical machines and a window for shedding light on complex physical and chemical questions about motions of organic molecules on surfaces. Stability of each component in such a molecular construction site is a crucial prerequisite. To realize a stable stepwise rotation of a molecule by a low temperature scanning tunneling microscope (LT-STM), atomic scale axles is particularly important. An ideal atomic scale axle is expected to balance between anchoring and mobility of rotating a single molecule on a metal surface under external excitations. In this Chapter, several chemical anchoring strategies on how to pin a molecular rotor are tested and discussed. Tip-induced manipulation and motion analysis are used as tools to investigate the properties and functionality of the proposed strategies.

Keywords Scanning tunneling microscope · STM manipulation · Molecular rotor · Anchoring strategy · Metal-organic complex · On-surface synthesis

1 Introduction

For building a molecular mechanical machine, the balance between anchoring and mobility is crucial for a single molecular rotor (schematically shown in Fig. 1). The first demonstration of a stepwise rotation of a molecular gear on an atomic scale axle was reported from Manzano and colleagues in 2009 [1]. This interesting strategy proposed to push a HB-NBP molecule onto an atomic defect by STM manipulation, which is natively presented on a herringbone elbow from the reconstruction of the Au(111) surface. Step-by-step and concentric rotations of individual molecules were presented. However, due to the long distance between such defects, no gears could be interlocked following this strategy. Another possibility was to produce a metal adatom by gently crashing the STM tip on the surface and to push the gear onto the adatom. Promising results were shown recently by Soe and colleagues, demonstrating

K. H. Au Yeung · T. Kühne · F. Eisenhut · F. Moresco (✉)
Center for Advancing Electronics Dresden, TU Dresden, 01069 Dresden, Germany
e-mail: francesca.moresco@tu-dresden.de

C. Joachim (ed.), *Building and Probing Small for Mechanics*, Advances in Atom and Single Molecule Machines, https://doi.org/10.1007/978-3-030-56777-4_8

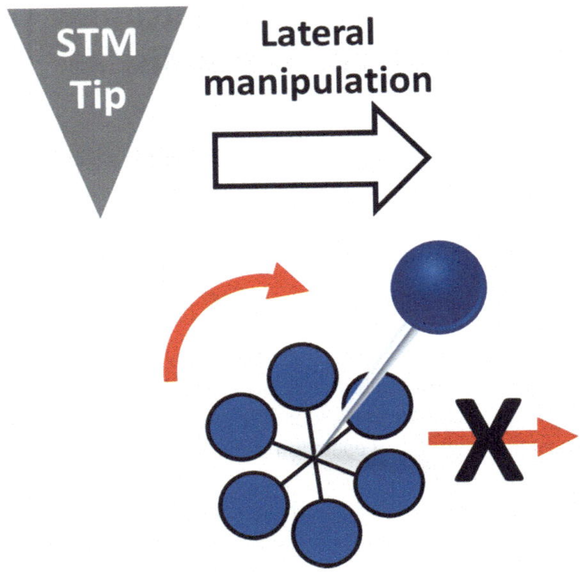

Fig. 1 Model representing an anchored/pinned molecular rotor rotated by a STM tip. An ideal anchoring strategy allows the rotational motion around the axle (blue pin) but hinders the lateral displacement

rotations between a train of gears on Cu adatoms [2]. In short, the above strategies aim at constructing axles actively and manually after molecule deposition on surface. The possibility to anchor the molecules during deposition could be an alternative strategy.

Investigations on alternative strategies were attempted with the help of advanced synthetic chemistry. A rotating molecular rotor composed of a tripodal stator for vertical positioning, a star-shaped molecule for rotations, and a ruthenium atomic ball working as an axle was reported [3]. Taking this idea further from the view of the chemical bonding between the star-shaped molecule and ruthenium atomic ball, one can imagine that forming a new bond between an organic molecule and a metal surface could be a new direction by taking the advantage of on-surface synthesis.

Table 1 summarizes different anchoring strategies recently investigated by STM experiments that allows rotations of molecular gears on metal surfaces.

In this Chapter, we discuss several chemical anchoring strategies on the Au(111) surface by employing a STM at low temperature (5 K) in ultra-high vacuum condition (UHV). STM allows not only the real-space visualization of molecules in atomic resolution, but also highly controlled manipulation by the tip apex. Importantly, tip-induced chemical reactions (i.e. dehydrogenation) are key factors for constructing anchoring sites by application of local voltage pulses in our studies. In the following sections, we present three different chemical anchoring strategies on several molecules with different sizes, symmetries and classes on the Au(111) surface: (Sect. 2) by a methoxy-group, (Sect. 3) by a methylamine-group, and (Sect. 4) by a dehydrogenated dangling bond.

Table 1 Overview of different reported anchoring strategies of molecular rotors/gears on metal surfaces performed by STM experiments [4–6]

Axle/anchoring site	Description	Functionality of anchoring and corresponding motion of molecule	References
Atomic defect	By STM tip pushing the molecule on an atomic defect	Stepwise and stable rotation by lateral manipulation	Manzano et al. [1]
Herringbone elbow	By STM tip pushing the molecule on a herringbone elbow caused by surface reconstruction i.e. Au(111)	Stepwise rotation and small lateral displacement by lateral manipulation	Manzano et al. [1]
Metal adatom	By STM tip pushing the molecule on a metal adatom formed by tip dipping or deposition (i.e. Pd adatom on Cu surface)	Stepwise and stable rotation by lateral manipulation; rotations between train of gears reported	Soe et al. [2]
Molecular stator/propeller	Deposition of firstly molecular stator on surface, then the deposition of molecular rotor with radical cvclopentadiene that leads to anchoring on the stator with a ruthenium atomic ball axle	Stable stepwise or continuous rotation depends on temperature; chirality dependence for directionality induced by voltage pulse	Perera et al. [3], Zhang et al. [4]
Oxygen-/Sulfur-metal bond	Precursor molecules deposited on metal surface bound by oxygen or sulfur groups	Stable rotation excited by tip-induced voltage pulse; chirality dependence for directionality induced by voltage pulse	Tierney et al. [5], Eisenhut et al. [6]
Dangling bond	By tip-induced or surface annealing to trigger dehydrogenation/dehalogenation of the molecules	Stepwise and stable rotation of dehydrogenated/dehalogenated molecule by lateral manipulation; structure of molecule dependence: Up to three transmission of stable rotations	Au Yeung et al. (in preparation)

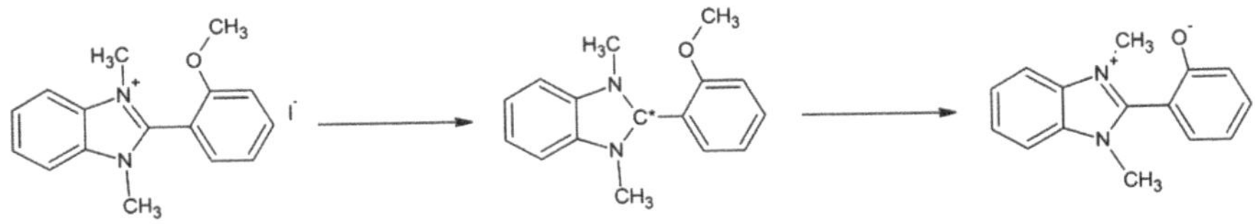

Fig. 2 Schematic diagram for expected products after on-surface synthesis by annealing. Iodine is first cleaved and then C–O bond cleavage occurs, forming a radical DMBI-P molecule

2 Anchoring Molecular Rotor by a Methoxy Group

Originally synthesized as an air-stable n-doping precursor molecule for the application in organic electronics [7, 8], we investigated the properties of single o-MeO-DMBI-I (2-(2-Methoxyphenyl)-1,3-dimethyl-1H-benzoimidazol-3-ium iodide) molecules deposited on the Au(111) surface. Our recent study shows that during thermal evaporation, the precursor is first reduced by iodine dissociation resulting in the radical form o-MeO-DMBI, and at temperatures around 490 K a C–O bond cleavage occurs, giving rise to DMBI-P and MeI as a byproduct. The resulted single DMBI-P molecules can be rotated stepwise and stable between six stable positions unidirectionally by application of local voltage pulses, depending on the molecule's chirality when adsorbed on Au(111) [9]. Hence, it is worth to investigate the surprisingly strong anchoring properties of the DMBI-P molecules and verify with motion analysis (Sect. 2.2). In this section, the quality of the anchoring strategy with the DMBI-P molecules (o-MeO-DMBI-I as precursors) will be discussed.

2.1 *On-Surface Synthesis*

We deposited the o-MeO-DMBI-I precursors at submonolayer coverage on a clean Au(111) surface kept at room temperature. After the C–O bond cleavage, the radical DMBI-P molecule chemisorbed on the Au(111) is anchored via back-donation (Fig. 2).

2.2 *Anchoring and Rotation*

The presence of a methoxy-group and of the radical form suggests a strong molecule-surface interaction that can be tested by pushing the single molecules with lateral manipulation. Upon manipulations, the molecules cannot be translated and no lateral displacements are observed. Manipulation trials with reduced tip height and consequently stronger tip-molecule interaction never lead to any lateral displacements. On the other hand, the orientation is clearly changed following the manipulation

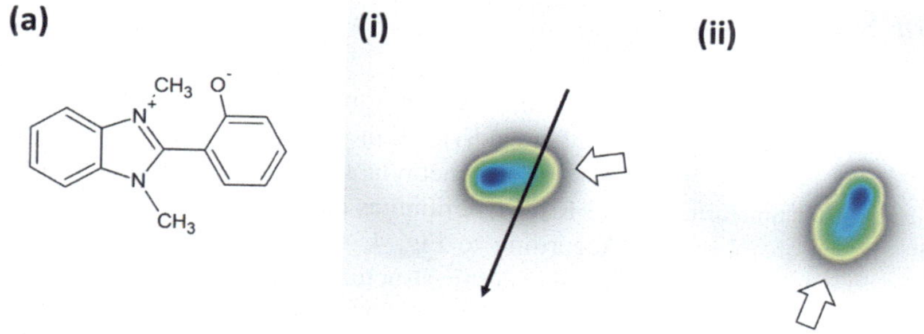

Fig. 3 The expected anchoring molecule DMBI-P adsorbed on the Au(111) surface. **a** Chemical structure of a DMBI-P molecule. (i), (ii) Sequential STM images of rotating a DMBI-P molecule on Au(111) before and after lateral manipulation. The black arrow indicates the manipulation trajectory. The white arrows indicate the orientation of the molecule on the surface (Parameters: I = 4.0 nA; V = 10 mV) (All image sizes: 5 nm × 5 nm; V = 0.2 V; I = 50 pA)

trajectory. This finding indicates that the rotation occurs around a fixed anchoring point at the side of the methoxy-group. In addition, the manipulation experiments demonstrate that the molecules are not physisorbed but chemically anchored to the Au(111) surface, as expected from the similar case of thioether-molecules which are known to strongly bind to surface atoms on coinage metals [10]. We therefore conclude that the single DMBI-P molecules are anchored to a gold atom through one of the two lone pairs of electrons of the oxygen (Fig. 3).

3 Tetracene-Based Rotor with a Methylamine Pin

The studies of poly-acenes on surface have been driving large attentions not only because of their instability starting from seven or more phenyl rings in solution, but more importantly to gain insights into their interesting electronic properties as extended π-electron structures in the one-dimensional limit. Recent advance regarding the on-surface synthesis of poly-acenes successfully reached to dodecacene consisting of twelve fused phenyl rings [11]. Other than the electronic properties, it is intriguing to understand the other physical and chemical properties, for instances, the adsorption on surface and the mechanical properties in terms of molecular mechanical machines with its extendable length in one dimension. More interestingly, large acenes show a very mobile characteristic, sometimes even a challenge for STM imaging from our observations (i.e. dodecacene). One could expect an anchoring or adsorption strategy for studying larger acenes in the future. The on-surface formation of tetracene and the corresponding electronic properties were reported in our earlier reports [12, 13]. In this section, we present a tetracene-based (Tn) rotor with a methylamine pin.

3.1 On-Surface Synthesis

After evaporating the tetracene precursors with two methylamine groups (Fig. 4), STM images reveal the topography of the individual molecules as a dumbbell shape. The protruding parts can be assigned to the methylamine groups of the molecule. Interestingly, after annealing at 443 K for five minutes on the Au(111) surface, four different species are observed. According to Fig. 4, we assign Cases **1–4** to the species for convenience. To confirm the atomic structures and the conformations of the resulted molecules, CO-functionalized tips were used to enhance the resolution of the STM images taken in constant height mode. For Cases **1** and **2**, the STM images show two additional side groups on the same or opposite sides and for Case **3**, only one side group is visible. Our findings suggest that the methylamine group can be either cleaved from one carbon atom leading to the formation of a benzene ring with a methylamine group at the side, or can be detached completely from the molecule leading to the formation of an acene. Note that the side groups can also be cleaved by voltage pulses above 3.0 V [12, 13].

3.2 Rotational Motion

As one can expect that a single anchoring site can work as an axle allowing for a stable rotation, we accordingly used the lateral manipulation mode of the STM to verify this. Indeed, it is not possible to move or rotate the molecules in Cases **1** and **2,** while the ones in Case **3** can be rotated. Figure 5 shows a sequential full rotation in six steps by lateral manipulation of the molecules from Case **3**. No lateral displacements are observed during manipulations. The results indicate that a strong molecule-surface interaction was generated due to the nitrogen of the side group as an anchor. On the contrary, the Tn molecule without any side groups (Case **4**) can be translated easily with a relatively low tunneling resistance of 5 MΩ. Our results reveal that it is possible to anchor a mobile tetracene molecule by a methylamine pin.

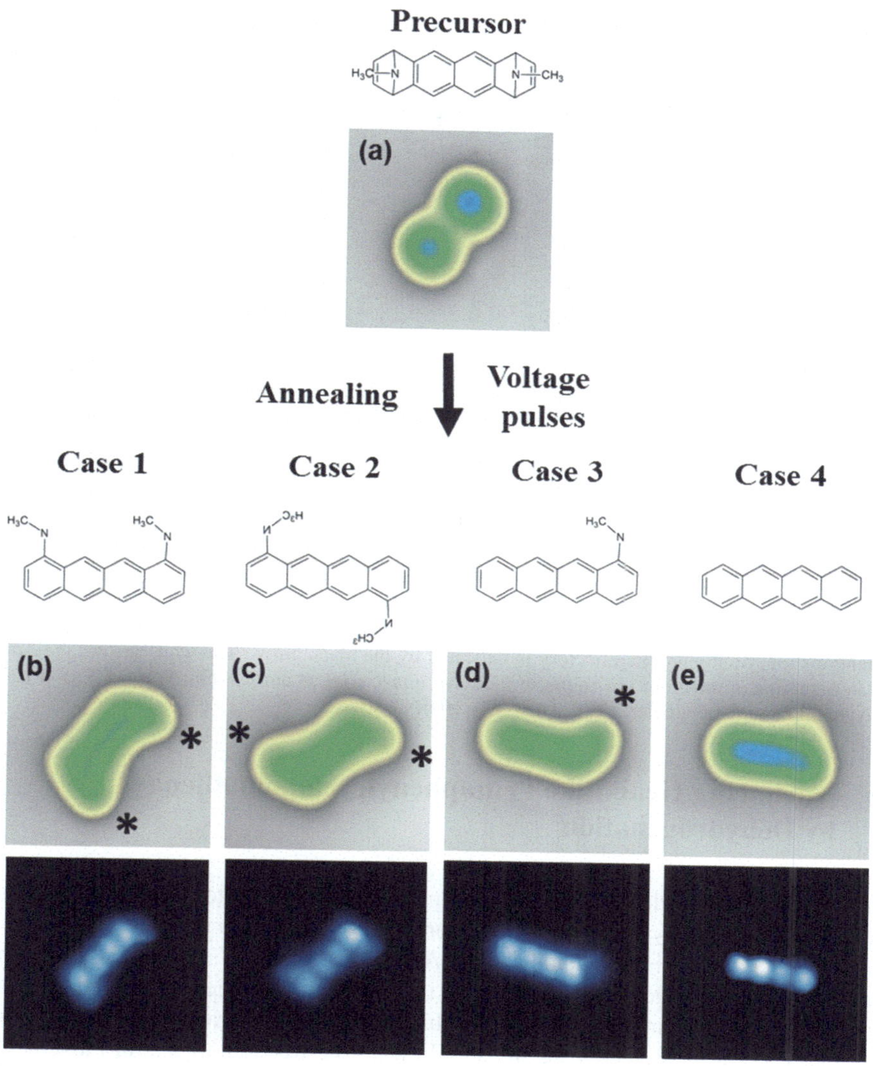

Fig. 4 The tetracene-based Tn(NCH$_3$)$_2$ molecular rotor. Top row: the proposed chemical structures of the precursor molecule and the resulting molecules after surface annealing at 170 °C or voltage pulses. **a** STM image of the precursor molecules on Au(111) (Image size: 2.5 nm × 2.5 nm; V = −560 mV; I = 73 pA). **b–e** Middle row: constant-current STM images of the different species after annealing (Image sizes: 2.5 nm × 2.5 nm; V = 500 mV; I = 100 pA); bottom row: corresponding constant-height STM images of the molecules achieved with a CO-functionalized tip (Image sizes: 2.5 nm × 2.5 nm; V = 10 mV)

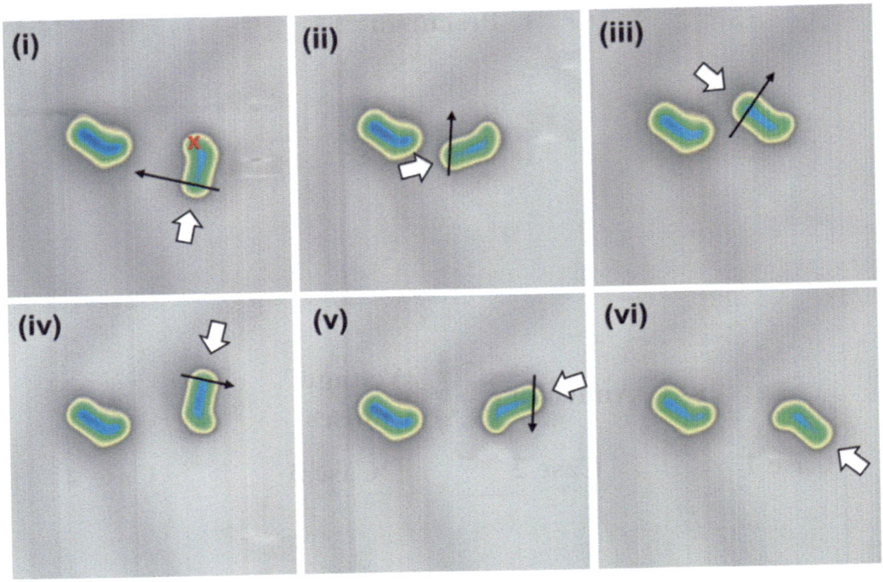

Fig. 5 Rotation of a reacted $Tn(NCH_3)_2$ molecule presented in Fig. 4 assigned as Case **3** on Au(111). (i)–(vi) Stepwise rotation of a tetracene-molecule with one methylamine-group by lateral manipulation. Manipulation trajectory is indicated by the black arrow. The white arrows indicate the orientation of the molecule on the surface (All image sizes: 8 nm × 8 nm; V = 0.5 V; I = 100 pA; manipulation parameter: I = 2.0 nA; V = 10 mV)

4 Anchoring Gear-Like Pentaphenylcyclopentadiene by Dehydrogenation

In this section, we will explore the opportunity of dehydrogenated dangling bonds as an anchoring strategy within the framework of star-shaped pentaphenylcyclopentadiene (PPCP; C_5Ph_5H) molecules on the Au(111) surface. As schematically shown in Fig. 6, the hydrogen atom (C–H bond) at the cyclopentadiene core is designed to work as a pin or an anchoring site of this molecule, taking a mechanical gear as an analogy. An ideal anchoring site should limit the freedom of lateral movement of the molecule. Hence, as a proof-of-concept, by inducing movements from the STM lateral manipulation, PPCP molecules have an advantage for us to analyze the pristine and radical states of the molecules on a metal surface. To implement the anchoring by dehydrogenated dangling bonds, the molecules are deposited by thermal evaporation on Au(111) and local STM tip-induced voltage pulses are used to trigger the dehydrogenation reaction for a concrete anchoring between the PPCP molecule and the Au(111) surface by a dangling bond.

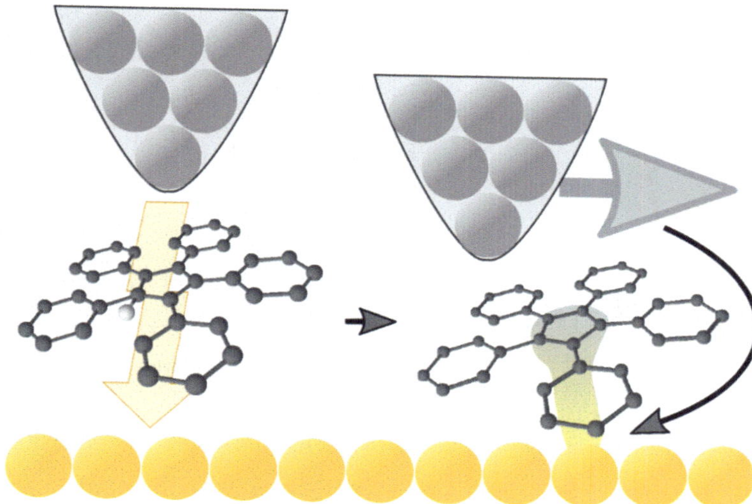

Fig. 6 Schematic diagram to illustrate the anchoring strategy for the gear-like PPCP molecule. The ball-stick model shows the chemical structure. After locally applying a STM tip-induced voltage pulse on a PPCP molecule at pristine state (left), a dehydrogenated dangling bond (a radical state) is formed with the Au(111) surface (right). To explore the properties of the anchor/pin, a tip-induced lateral manipulation is performed, usually at one of the phenyl rings of the molecule, to realize a rotational motion. STM images are taken before and after each manipulation for further motion analysis

4.1 *On-Surface Dehydrogenation*

After evaporating PPCP molecules at 448 K for 30 s, the Au(111) surface was covered with a sub-monolayer film. Interestingly, clusters of PPCP molecules were found along the step edges or some small islands on the surface consisting of three or four molecules. This observation suggests that PPCP molecules are physisorbed on the Au(111) surface. In the STM images (Fig. 7), a clear feature of this molecule is the bulge (higher apparent height) near its center. If the five lobes that delineate the contour of the molecule are the five phenyl rings, this bulge should correspond to the hydrogen atom. Indeed, the application of a voltage pulse of about 3 V right above the bulge alters the conformation of the molecule. Repeated pulsing can flatten this bulge to a lower apparent height even if the first pulse does not change the conformation. Figure 7 shows the STM image of two PPCP molecules where the upper one was flattened (For simplicity, we call the flat/pulsed molecule as "radical" and the other as "pristine".). Peak-to-peak apparent height difference is about 0.7 Å. The flattening of the molecule, i.e. the removal of the central bulge, is a strong evidence for a successful dehydrogenation reaction. However, in some cases we observe different conformations from the dehydrogenated molecule because of the many degrees of freedom of the phenyl rings.

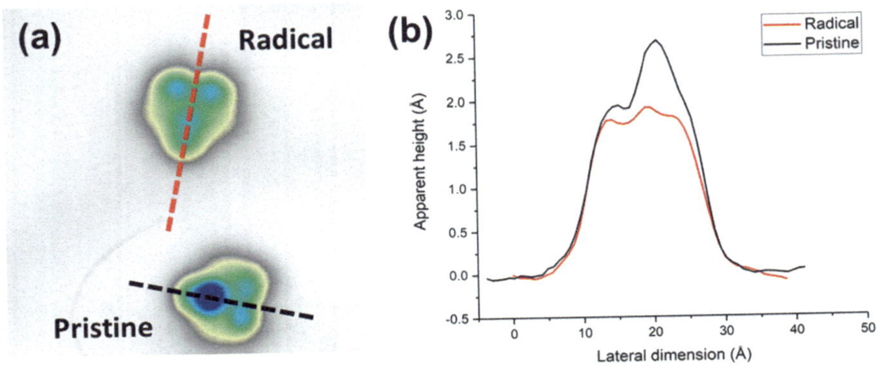

Fig. 7 **a** STM image and **b** corresponding height profiles of the pristine and radical PPCP molecules. Peak-to-peak apparent height difference is about 0.7 Å (Image size: 6.3 nm × 6.3 nm; V = 500 mV; I = 20 pA)

4.2 Movement Analysis

If a pristine PPCP molecule is dehydrogenated, it is reasonable to deduce that the remaining dangling bond can act as a pinning center so that the radical molecule rotates concentrically. Following this logic, a series of lateral manipulation experiments was performed. As shown in Fig. 8, it is possible to rotate the molecule step-by-step by lateral manipulation (Tunneling resistance = 25 MΩ). Note that some tiny lateral displacements occur after each manipulation, indicating that the anchoring is not strong enough for pure concentric rotations. To further understand the characteristics of the dehydrogenated dangling bond, we attempt to fix the molecule on one or two Au adatoms, however the bonding seems weaker than that on the flat gold surface.

4.3 Modified Design from the Framework of Pentaphenylcyclopentadiene

A modification of the PPCP molecule with five further *tert*-butyl groups is shown in Fig. 9. Each additional *tert*-butyl group is connected to each phenyl ring (namely PPCP–2 in short). The molecule was designed to keep the single benzene rings in a fixed conformation, providing a more symmetric gear. When adsorbed on the Au(111) surface, the molecules can be found at the kink of the herringbone, usually individual ones instead of clusters. This suggests that the interactions between molecules are relatively weak. From the STM images, one can easily identify the five lobes around the center part which represent the *tert*-butyl-phenyl rings. The adsorbed PPCP-2 molecules can be rotated by STM lateral manipulation (Fig. 9i–iii) without the need of

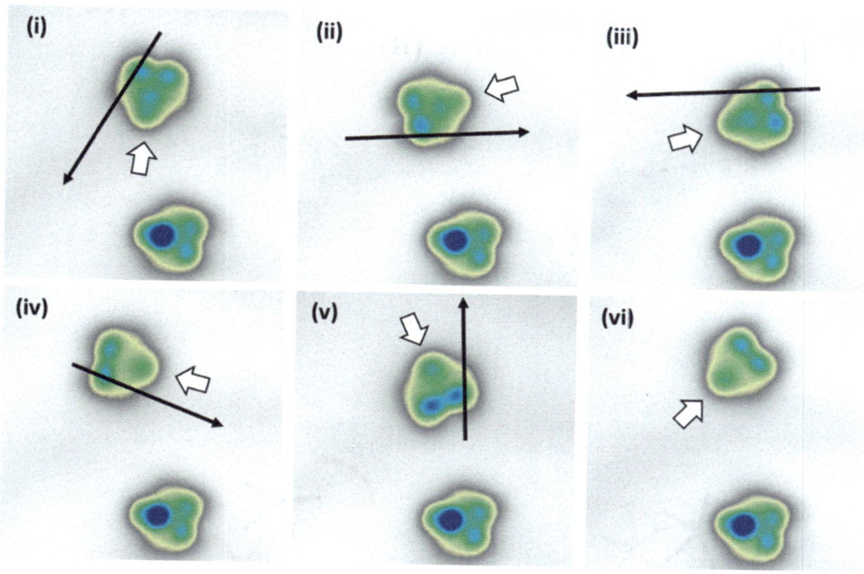

Fig. 8 Tip-induced movements of a radical PPCP molecule. (i)–(vi) The upper one is dehydrogenated by a 3.1 V voltage pulse while the lower one remains at the pristine state as a reference. The black arrows indicate the lateral tip manipulation events at V = 10 mV and I = 400 pA. The white arrows indicate the orientation of the molecule on the surface (All image size: 6.3 nm × 6.3 nm; V = 500 mV; I = 20 pA)

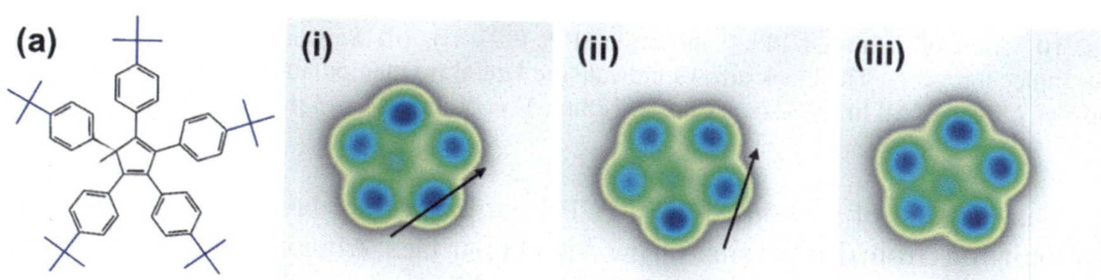

Fig. 9 Modified PPCP molecule (PPCP-2) with five *tert*-butyl groups attached to each phenyl ring. **a** Chemical structure of a PPCP-2 molecule. (i)–(iii) Sequence of rotations. The black arrows indicate the lateral tip manipulation events at V = 10 mV and I = 500 pA (All image sizes: 5 nm × 5 nm; V = 500 mV; I = 5 pA)

voltage pulses, suggesting that the H atom can be already cleaved during evaporation, allowing also in this case a week anchoring of the molecule via the radical 5-member ring. Note that it is still unclear to what extent a *tert*-butyl group can alter the tip manipulation (tip-molecule interactions) compared to a planar chemical structure (i.e. phenyl ring). In general, most of the cases suggest that a *tert*-butyl group is beneficial for lateral manipulation, which works similar to a “handle” [14, 15].

Figure 10 shows a train of PPCP-2 molecules that we formed to investigate collec-

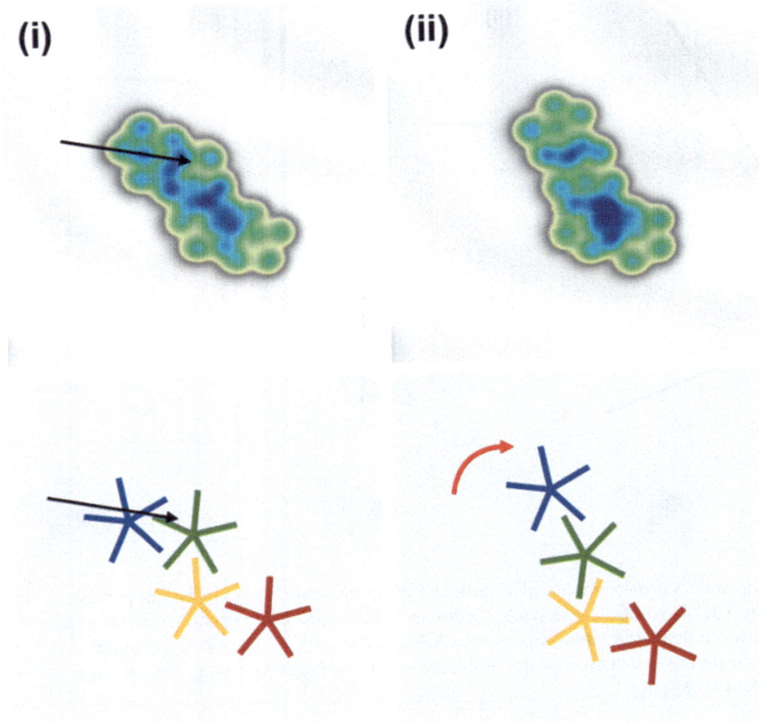

Fig. 10 Train of modified PPCP molecule (PPCP-2). (i), (ii) Sequence of rotation between four molecular gears. The black arrows indicate the lateral tip manipulation events at V = 10 mV and I = 2200 pA (All image size: 12 nm × 12 nm; V = 500 mV; I = 3 pA)

tive transmission of motions. Four PPCP-2 molecules are brought together by tip manipulations from distinct sites on the Au(111) surface. After manually interlocking all the four molecules, it is not possible to translate any motions by manipulating on the side of the molecule (unlike the case from Fig. 9). Interestingly, driving the tip across the center lobe of the molecule can easily lead to lateral displacement. Such manipulation strategy is used for rotating three or more interlocked molecules. As shown in Fig. 10, although it is not a stable and concentric rotation between molecules, a step of collective motion is clearly observed for the lower three molecules (green, yellow and red from the schematic drawing) transmitted from the uppermost molecule (blue). This indicates that the anchoring strategy using dehydrogenated dangling bonds in a radical state can provide both mobile and vertically stable motions, depending on the manipulation strategy.

5 Summary and Perspectives

In this Chapter, we focus on chemical anchoring strategies for different types of molecules by on-surface synthesis as an essential equipment of a molecular mechanical machine. Three different classes of molecules, from one-dimensional tetracene and o-MeO-DMBI molecules to a two-dimensional gear-like PPCP molecule, are discussed. Among the proposed molecules and anchoring strategies, indeed, one cannot directly compare the quality of the anchors because of the differences between molecular structures (including the symmetries) that lead to different tip-molecule and molecule-surface interactions (even inter-molecular interactions for transmission of motion). Nonetheless, more importantly as a fundamental understanding of molecular anchors and rotors, it is crucial to explore the opportunity from in-solution chemical synthesis to on-surface synthesis and STM manipulation as a complex workflow. The promising strategies of anchoring are beneficial for future constructions of not only molecular mechanical machines, but also organic electronics and two-dimensional materials on surface in terms of the insights of the characteristics and functionalities of metal-organic complexes.

Acknowledgements This work has received funding from the European Union's Horizon 2020 research and innovation program under the project MEMO, grant agreement No 766864. The authors thank Oumaima Aiboudi, Franziska Lissel, Claire Kammerer, Gwénaël Rapenne, and Diego Peña for providing part of the MeO-DMBI-I precursor molecules used in these studies.

References

1. Manzano, C., et al.: Step-by-step rotation of a molecule-gear mounted on an atomic-scale axis. Nat. Mater. **8**(7), 576–579 (2009)
2. Soe, W.-H., Srivastava, S., Joachim, C.: Train of single molecule-gears. J. Phys. Chem. Lett. **10**(21), 6462–6467 (2019)
3. Perera, U.G.E., et al.: Controlled clockwise and anticlockwise rotational switching of a molecular motor. Nat. Nanotechnol. **8**(1), 46–51 (2013)
4. Zhang, Y., et al.: A chiral molecular propeller designed for unidirectional rotations on a surface. Nat. Commun. **10**(1), 3742 (2019)
5. Tierney, H.L., et al.: Experimental demonstration of a single-molecule electric motor. Nat. Nanotechnol. **6**(10), 625–629 (2011)
6. Eisenhut, F., et al.: Inducing the controlled rotation of single o-MeO-DMBI molecules anchored on Au(111). Surf. Sci. **678**, 177–182 (2018)
7. Naab, B.D., et al.: Mechanistic study on the solution-phase n-doping of 1,3-dimethyl-2-aryl-2,3-dihydro-1H-benzoimidazole derivatives. J. Am. Chem. Soc. **135**(40), 15018–15025 (2013)
8. Bin, Z., Duan, L., Qiu, Y.: Air stable organic salt as an n-type dopant for efficient and stable organic light-emitting diodes. ACS Appl. Mater. Interfaces **7**(12), 6444–6450 (2015)
9. Eisenhut, F., et al.: Exclusive one-way rotation of a single molecule-rotor on a gold surface. Submitted (2020)
10. Murphy, C.J., et al.: Impact of branching on the supramolecular assembly of thioethers on Au(111). J. Chem. Phys. **142**(10), 101915 (2015)
11. Eisenhut, F., et al.: Dodecacene generated on surface: reopening of the energy gap. ACS Nano **14**(1), 1011–1017 (2020)

12. Krüger, J., et al.: Tetracene formation by on-surface reduction. ACS Nano **10**(4), 4538–4542 (2016)
13. Krüger, J., et al.: Molecular self-assembly driven by on-surface reduction: anthracene and tetracene on Au(111). J Phys Chem C **121**(37), 20353–20358 (2017)
14. Otero, R., et al.: Lock-and-key effect in the surface diffusion of large organic molecules probed by STM. Nat. Mater. **3**(11), 779–782 (2004)
15. Schunack, M., et al.: Long jumps in the surface diffusion of large molecules. Phys. Rev. Lett. **88**(15), 156102 (2002)

Transmission of Rotational Motion Between Molecule-Gears

W.-H. Soe, S. Srivastava, and C. Joachim

Abstract A molecule-gear rotating without a lateral jittering effect is constructed using a single copper adatom as a physical axle on a lead superconducting surface. The molecule-gear has a diameter of 1.2 nm with 6 tert-butyl-teeth. It is mounted on this Cu axle using the atom/molecule manipulation capability of a low temperature scanning tunneling microscope (LT-STM). Transmission of rotational motions between 2 molecule-gears, whose axles have to be exactly 1.9 nm separated, is functioning when this train of molecule-gears is completed with a molecule-handle. To manipulate the molecule-handle laterally, the first molecule-gear of the train directly entangled with the molecule-handle is step by step rotated around its Cu adatom axle. It drives the second molecule-gear mechanically engaged with the first gear to rotate like along a train of macroscopic solid-state gears. Such rotation transmission is one of the most basic function for the future construction of a complex molecular machinery.

Keywords Molecule-gear transmission · Molecule-machinery · LT-UHV STM · Molecule manipulation · Single atom axle

1 Introduction

Gears are essential elementary devices for constructing mechanical machinery like clocks, for transmitting and converting motor motive power (instead of rigid pull-outs) and to construct calculators (working in hostile environment like in nuclear power plant or in space). Gears are also essential to measure the motive power of

W.-H. Soe (✉) · S. Srivastava · C. Joachim
Centre d'Elaboration de Matériaux et d'Études Structurales (CEMES), Centre National de la Recherche Scientifique (CNRS), Université de Toulouse, 29 Rue J. Marvig, BP 4347, 31055 Toulouse Cedex, France
e-mail: we-hyo.soe@cemes.fr

W.-H. Soe · C. Joachim
International Center for Materials Nanoarchitectonics (WPI-MANA), National Institute for Material Sciences (NIMS), 1-1 Namiki, Tsukuba, Ibaraki 305-0044, Japan

C. Joachim (ed.), *Building and Probing Small for Mechanics*, Advances in Atom and Single Molecule Machines, https://doi.org/10.1007/978-3-030-56777-4_9

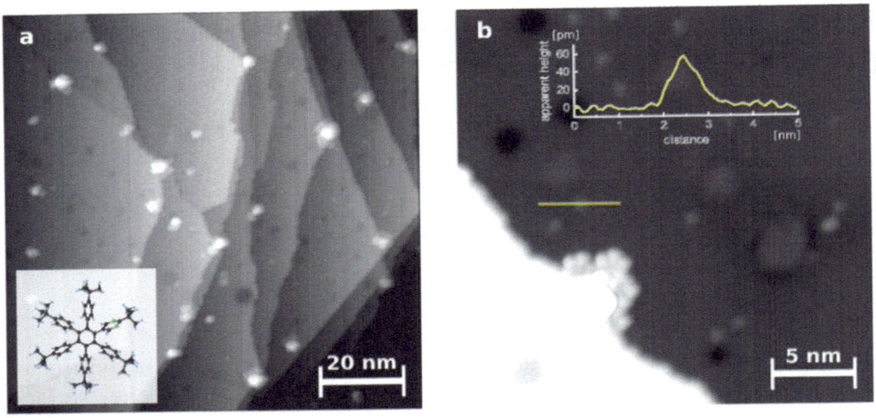

Fig. 1 Typical STM images after deposition of HB-NBP molecules and Cu atoms on a Pb(111) surface. **a** The molecules were formed large 2D clusters and also **b** stabilized attaching to atomic-sized impurities. A line scan profile measuring apparent height of Cu adatom was also shown in **b**. Chemical structure of HB-NBP molecule was illustrated in inset of **a**

a motor. For minimizing the energy required for those machinery to function and to reduce their weight for portability, it is continuously important to miniaturize as much as possible gears depending on the available technology. After hand fabrication, machine tools, optical and nanolithography's (see chapter "From the Anthycytera Astronomical Clock to Single Molecule Scale Machinery" of this volume), the next challenge at the beginning of our XXI century is to construct mechanical machinery with single molecule-gears that is gears whose size is between 1 and 10 nm in diameter and about 1 nm in height.

Currently the smallest functioning known molecule-gear is a hexa-tert-butyl-pyrimidopentaphenylbenzene ($C_{64}N_2H_{76}$, HB-NBP, Fig. 1a inset) molecule mounted on a native atomic scale impurity [1]. Other smaller molecules like PF_3 have also been explored for gearing effect but are not rigid enough for such functioning (see chapter "A Simple Example of a Molecule-Gear Train: PF3 Molecules on a Cu(111) Surface" of this volume). The native impurities are found at the herringbone-kink of the reconstructed Au(111) surface [1, 2]. But those herringbone-kinks are arranged around a 8 nm period on large Au(111) atomically flat terraces. This separation is too large as compared with the HB-NBP molecule-gear lateral size (about 1.2 nm) and there is no way to engage two of them together in a mechanical interaction at such 8 nm distance. The control and the stabilization of atomic scale rotation axles are the key for constructing mechanical machinery based on molecule-gears elementary devices. As described in this volume, a direct chemical bonding to the supporting surface [3] or a molecular tripod like support [4] can be used to define the molecule-gear mechanical axle of rotation (see respectively chapters "Prototypes of Molecular Gears with an Organo-metallic Piano-Stool Architecture", "Chemical Anchoring of Molecular Rotors" and "Anchoring Molecular Rotors by On-Surface Synthesis" of this volume).

We present here the construction of a functioning train of molecule-gears using a metal adatom axle per gear. We have selected Cu adatoms in combination with a Pb(111) surface among several possible candidates because a Cu adatom apparent height is equal to the impurity height on the herringbone-kink of Au(111) surface. The lead substrate was selected to minimize the electronic friction during the motion of the molecule on a superconducting metal surface due to cancelling the surface image charge (electron-hole quantum coupling) term in the electronic friction of the molecule [5]. HB-NBP molecules were sublimated at room temperature in UHV right after the UHV Pb(111) surface cleaning process. Then, the Cu adatoms were deposited in a very sub-monolayer amount on the surface mounted on the LT-STM sample stage of our LT-UHV 4-STM maintained at LHe temperature during this deposition.

2 Single Molecule Gear with Handle

Most of HB-NBP molecules sublimated thermally on the Pb(111) surface are forming 2D islands (Fig. 1a), but a few of them are also stabilized as individuals attached to atomic scale impurities existing at step edges on this surface as shown in Fig. 1b. Notice also that the Cu adatoms and the hexagonal Ar bubbles shadows can be identified.

The apparent height of Cu adatoms LT-UHV STM imaged is 60 pm on Pb(111) using standard tunneling conditions i.e. about several tens GΩ of tunneling resistance with a low bias voltage set-up (Fig. 1b). Unfortunately the pyrimidine chemical tag so useful to observe the molecule-gear rotation cannot be discriminated in the STM molecule-gear images when adsorbed on the Pb(111) surface unlike the molecule LT-UHV STM imaged on Au(111) and Cu(111) surfaces [1, 6]. Fortunately and on the Pb(111) surface, the tooth height difference can be imaged: one tooth over 6 tert-butyl teeth has always a lower apparent height than the 5 others in the ground state conformation. This apparent height difference can be recognized in the all the isolated molecules LT-UHV STM images through all this chapter.

STM image calculations using elastic scattering quantum chemistry (ESQC) [7] also confirm that the two methyl groups of one tert-butyl tooth are oriented slightly downward the Pb(111) surface producing a relatively dark STM contrast while the 5 other teeth face only one methyl group downward regardless of the adatom rotation axle. In this chapter, this lower height tooth was used to follow up the step by step molecule-gear rotation.

To construct a molecule-gear having its Cu adatom axle, LT-UHV STM molecule manipulation protocol was the following:

1. The selection of one molecule-gear natively stabilized at a mono-atomic step edge on the Pb(111) surface and of a target Cu adatom on the lower terrace using normal STM imaging conditions (a GΩ range for STM tunneling resistance),
2. The positioning the STM tip apex in the center of the selected molecule-gear,

3. The STM tip apex end atom is approached to the molecule-gear by reducing the STM tunneling resistance to 5 MΩ. The tip apex end atom must be closed enough to be able to manipulate the molecule-gear,
4. The tip apex trajectory must be real time recorded while direct positioning atop the targeted Cu adatom.

When the molecule-gear is successfully mounted atop the Cu adatom, the tip apex end atom height is jumping up exactly by 60 pm. Then, a protrusion is clearly STM imaged at the molecule-gear centre as shown in Fig. 2.

After the mounting of the molecule-gear on its Cu adatom axle and its centring, the same protocols used to step by step rotate an HB-NBP molecule-gear on Au(111) without any lateral movement [1] were tested on the Pb(111) surface. But both constant current and constant height LT-UHV STM mechanical manipulation were not successful. On Pb(111), we have observed that in most of the case, the molecule-gear is dismounted from its Cu axle (Fig. 3a) and in some cases, the molecule-gear and its Cu adatom axle are transferred laterally on the Pb(111) surface (Fig. 3b). This indicates that a molecule-gear centred atop its Cu axle on Pb(111) is in a metastable surface conformation. The STM tip apex to molecule-gear interactions are always destabilizing this atop configuration during an STM mechanical manipulation.

Rotations using a bias voltage increase were also performed. Regardless of the tip apex position on the molecule-gear, permutation between a lower apparent height tooth and one of higher teeth occurred above a positive bias voltage of 2.5 V. In that case, two methyl groups with a downward tert-butyl tooth conformation switch into one methyl group's downward tooth conformation. At the same time, one over five of one methyl group's downward teeth become two methyl group's downward tooth.

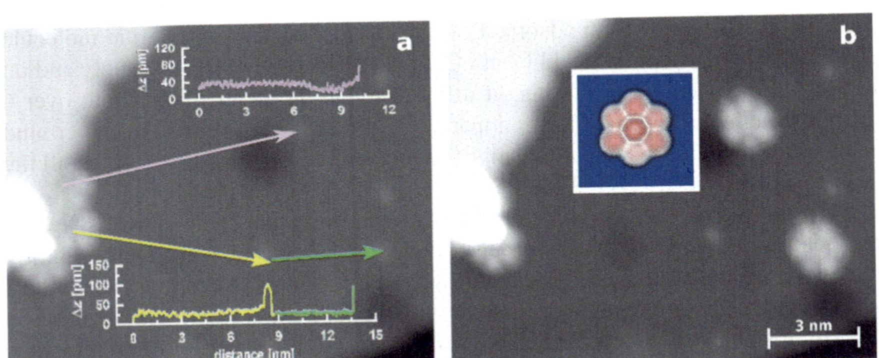

Fig. 2 STM images showing the HB-NBP molecules **a** before and **b** after manipulations. The arrows drawn on **a** display the tip trajectories during manipulations and corresponding tip height profiles are also presented using same colored lines. When the molecule is about to be mounted atop the Cu adatom, the tip height increases (Yellow and green jumps, **a** Δz inserted curve). After successful construction of the molecule-gear, the central protrusion coming from its Cu axle is appeared as an ESQC-calculate STM image shown in the inset of **b**. Redrawn from Ref. [12] with permission: copyright 2019, American Chemical Society

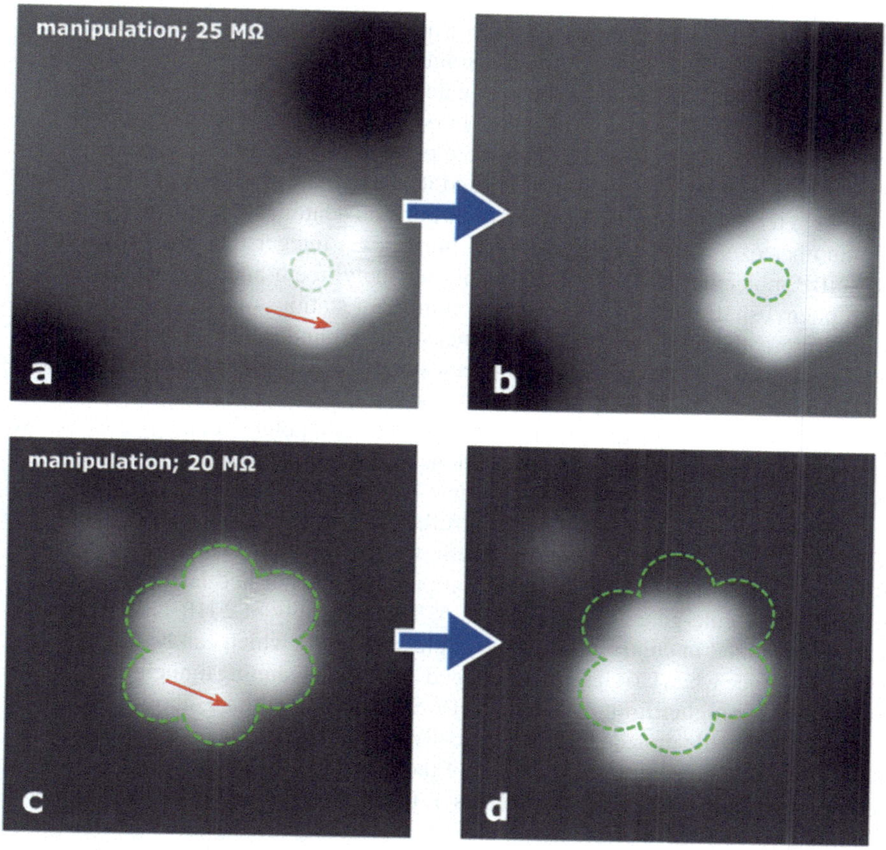

Fig. 3 Trials of the molecule-gear rotation using constant current STM mechanical manipulation for standalone molecule-gear on its Cu axle. Images **a** and **c** were taken before manipulations and **b** and **d** are after manipulations performed for tip trajectories drawn by red arrows on **a** and **c**. After manipulation in this way, most molecules are dismounted from Cu adatom as shown in **b**. Here green dotted circles are indicating Cu adatom position before trial. Cu adatom is rarely translated together with molecule as resulted in **d**. Reused from our Ref. [12] with permission: copyright 2019, American Chemical Society

Here the chirality of the HB-NBP molecule-gear on the surface and the excited state potential energy surfaces acceded during an inelastic tunnelling excitation at large bias voltage are not efficient to make one-way directional collective conformation change for rotation due to the micro-reversibility principle. As a result, this structural permutation leads a random rotation in both direction and angle of the molecule-gear. Here, we have also to notice that an inelastic tunnelling excitation is a quantum process which by no means results in complete occupation of those excited states. As a consequence, such large voltage inelastic tunnelling excitation involving excited states is less efficient as compared to a vertical UV like optical excitation used for example to activate molecular motor in solution.

Semi-empirical molecular mechanics calculations [8] for a stand-alone molecule-gear mounted on its Cu adatom on a Pb(111) surface lead after energy optimization to a total 15 stable stations on its ground state potential energy surface when exploring a full rotation of the molecule-gear around its Cu axle as presented in Fig. 4. Since a Pb(111) fcc surface has C3v symmetry around an adatom located in a hollow-site, the number of stations for a rigid body must be an integer multiple of 3. For the HB-NBP molecule-gear case, it is mainly tert-butyl teeth and their angles around each phenyl shaft which determines the number of stations on the Pb(111) surface. Furthermore, the rotation angle dependence of the ground state energy indicates that the conformational barrier heights between stations 1 & 2 (repeated at 6 & 7 and 11 & 12 due to a C3 symmetry) and 3 & 4 (likewise, 8 & 9 and 13 & 14) are below 50 meV. After a detail exploration of this ground state potential energy surface, we have identified only 9 possible stable stations instead of 15 possible for this molecule on a Pb(111) fcc surface because of the tert-butyl teeth relaxation.

To stabilize and be able to rotate controllably a Cu atop mounted molecule-gear, a second molecule-gear (hereinafter referred to the 'handle') which had lost one of its tert-butyl teeth during the sublimation process was meshed laterally to this Cu mounted molecule-gear as shown in Fig. 5. This assembly of 2 HB-NBP molecules is now stabilized in contrast with a molecule-gear alone which can be dragged on a Pb(111) terrace when LT-UHV STM imaged even if the scanning tunneling resistance is set up at the higher limit of our LT-UHV STM for imaging i.e. about 1 pA.

Once the handle molecule is mechanically entangled with the molecule-gear, The LT-UHV STM molecular manipulation of the handle molecule (and no more the Cu mounted molecule-gear directly) enables a step by step rotation of the atop centered Cu molecule-gear. In total, 9 stable stations were observed for this molecule gear-handle machinery while rotating only the handle (Fig. 5). The detail LT-UHV STM molecular manipulation steps are the following:

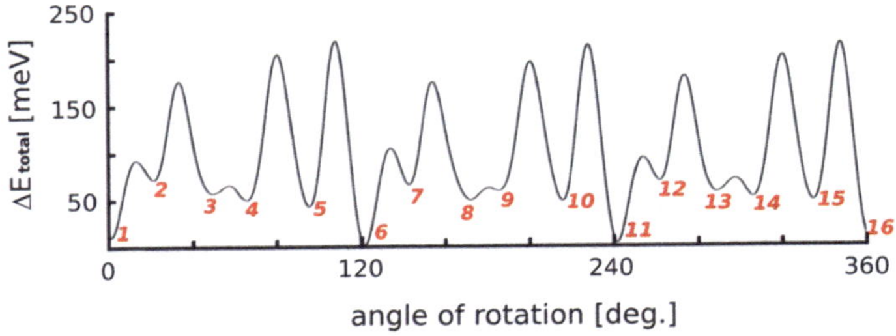

Fig. 4 Possible 15 stable stations during a full rotation of the HB-NBP molecule mounted on a Cu adatom on Pb(111) surface evaluated from a total energy calculation using a semi-empirical molecular mechanics method. Here 1st, 6th, 11th and 16th stations are identical (likewise, station groups {2nd, 7th, 12th}, {3rd, 8th, 13th}, {4th, 9th, 14th}, and {5th, 10th, 15th}) even if these energies are not exactly the same due to conformation change during rotation. Reused and transformed from Ref. [12] with permission, copyright 2019, American Chemical Society

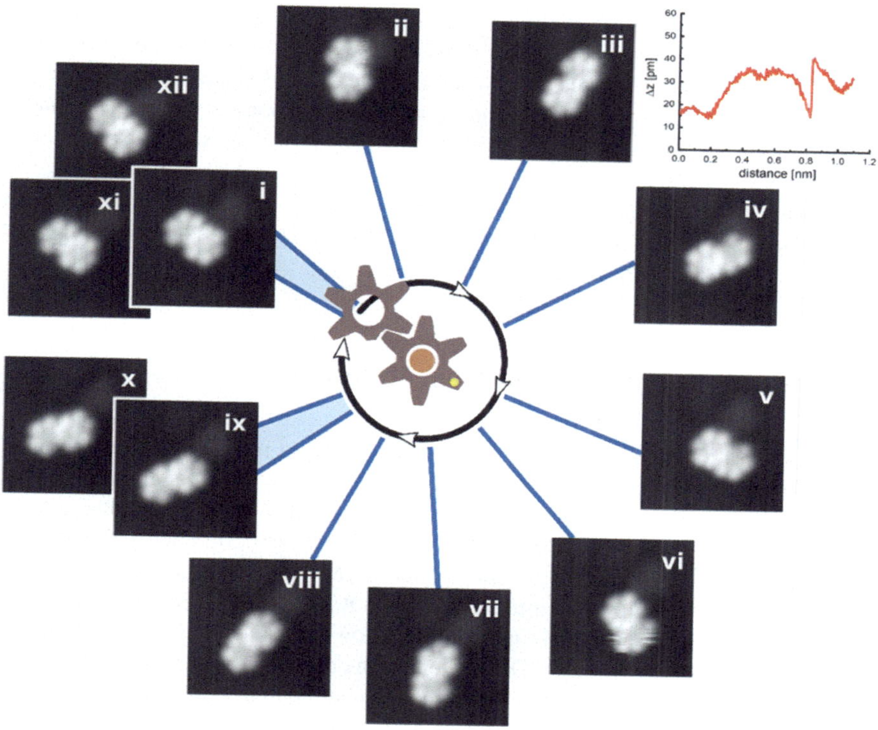

Fig. 5 A series of experimentally observed stable stations during a step by step full clockwise-rotation of a single molecule-gear using ancillary molecule-handle. Blue colored lines are indicating molecule orientations at each station extrapolating from the line segment between lower apparent height tooth and the molecule-gear center. A typical tip height profile during handle manipulation is also presented on the top right corner. Redrawn using the LT-STM images from Ref. [12] with permission: copyright 2019, American Chemical Society

1. positioning the STM tip apex in the center of handle molecule,
2. approaching the tip by reducing tunneling resistance to 20 MΩ close enough to be able to manipulate this handle molecule,
3. tracing the tip apex trajectory about 1 nm along a tangent to a circle passing through the center of the handle molecule.

When succeeding in a one-step rotational motion, a discontinuous point appears in the variation of the relative tip height during the handle molecule manipulation as presented in Fig. 5. For a complete 360° rotation of the Cu mounted molecule-gear, the above procedure must be repeated until the handle molecule reaches again its initial position. The handle molecule has many possible degrees of freedom for its entangling angle with its molecule-gear partner (each ±12° in a rigid body model). Furthermore and according to its own possible stable conformations on the surface, the number of stations of the gear-handle molecular machinery must be reduced in reference to the unstable 15 stations of the molecule-gear alone. It is also down to 9 stations because of 3σv specular symmetry in C3v system. Experimentally, some fluctuations were also observed around some stable stations (defined by their angles as illustrated in Fig. 5) because of small conformation changes of the gear-handle machinery during rotations and the stations i (xi) & xii or ix & x can be counted as a single station each (see Fig. 5).

The main advantage to use the handle is to weaken the direct interaction between the STM tip apex and the Cu mounted molecule-gear and to avoid this molecule-gear to stray out of its Cu adatom rotary axis.

3 Transmission of Molecule Gear Rotation to Another Gear

To perform a molecule-gear transmission experiment, a train of two Cu mounted molecule-gears was constructed also by LT-UHV STM molecular manipulations. Both molecule-gears were mounted and centered each on a single Cu adatoms positioned 1.9 nm away from each other. This distance was experimentally determined by trial and error. Notice that in general, a direct STM manipulation of a small radius metal adatom is difficult on a large radius metal atom surface because small single adatoms (Cu in the present case) sink into the hollow-sites surrounded by large atoms (Pb in the present case). In fact, DFT calculations are also showing that Cu and/or Co adatoms adsorbed on the Pb(111) surface may buried under the topmost Pb layer.

To adjust Cu adatom axle position on the Pb(111) surface, a lateral diffusion barrier lowering effect of Cu adatoms was found when manipulating an HB-NBP molecule on a Cu(111) surface [9]. We have applied the same manipulation procedure here. Even with this procedure, only a few atomic step manipulations are possible on the Pb(111) surface (Fig. 3). But it was sufficient for exploring the best interatomic distance between two Cu adatom axles to construct a train of 1.2 nm in diameter molecule-gears since the nearest-neighbor hollow-site distance of the Pb(111) surface is 0.20 nm.

When the Cu–Cu distance is too short, the second molecule-gear cannot be centered around its Cu adatom after the construction of the first molecule-gear because the inter-teeth van der Waals repulsion forces between the two molecules rapidly increases upon a decrease of molecule-molecule distance. When the Cu–Cu distance permits a Cu atop configurations for both molecules but with a distance slightly shorter than ideal distance, the two molecule-gears are assembled in a stable HB-NBP entangled molecular dimer by a forced conformation change applied through the handle molecule manipulation. As a result, the two molecule-gears are stuck together and only the molecule-handle rotates around the first Cu mounted molecule-gear. When now the Cu–Cu distance is larger than 1.9 nm, a first molecule-gear rotation is not transmitted to the second mounted molecule-gear, which just quivering its teeth facing to the first gear and keeps its initial orientation conformation on the Pb(111) surface (angle/position). This behavior was also well reproduced by semi-empirical molecular mechanics calculations (Fig. 6) (see also

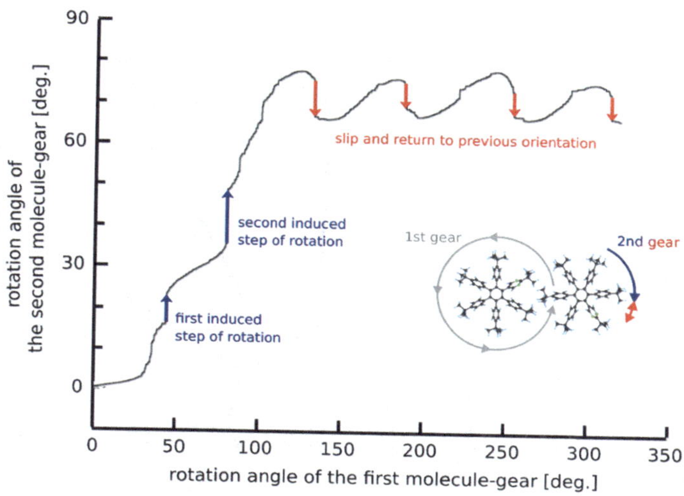

Fig. 6 Enforced rotation angle to the first molecule-gear versus induced angle to the second gear by mechanical transmission simulated using the semi-empirical molecular mechanics method. After two steps of successful transmission of rotation, the second molecule-gear starts backward slip roughly every 60° of the first gear rotation. Redrawn and transformed from Ref. [12] with permission, copyright 2019, American Chemical Society

chapter "Mechanical Transmission of Rotation for Molecule Gears and Solid-State Gears" of this volume for a more complete discussion).

Once the optimized distance between the two Cu centers of the molecule-gears is found, the first molecule-gear of the train is rotated by manipulating the handle using the same protocol than in Sect. 2. The induced step by step rotational motion of the first molecule-gear is now transmitted successfully to the second molecule-gear as shown in Fig. 7. At each step, the stable conformations of this three molecules molecular machinery are the result of a competition between the transmission of the rotation from molecule-gear to molecule-gear and the self-stabilization of each molecule-gear. However, in this rather complex molecular machinery as compared with the single molecule-gear with its handle, the observed stable conformations are mainly coming from the competition between the two molecule-gears attached to their axles because the handle molecule has more degrees of freedom to be stabilize by the Pb(111) surface. Furthermore, a completely stabilized molecule-gear can support and maintain the other molecules in a metastable surface conformation. As a consequence, the rotational motion is transmitted from molecule to molecule by steps of about 30°. The exact rotation angle at each step cannot be 30° because this HB-NBP molecule trimer must be accommodated to its ground state at each rotation step on the quite rigid highly corrugated atomic lattice surface. The rotation angle of the second molecule-gear alternates between 27 and 33°. In contrast, the first molecule-

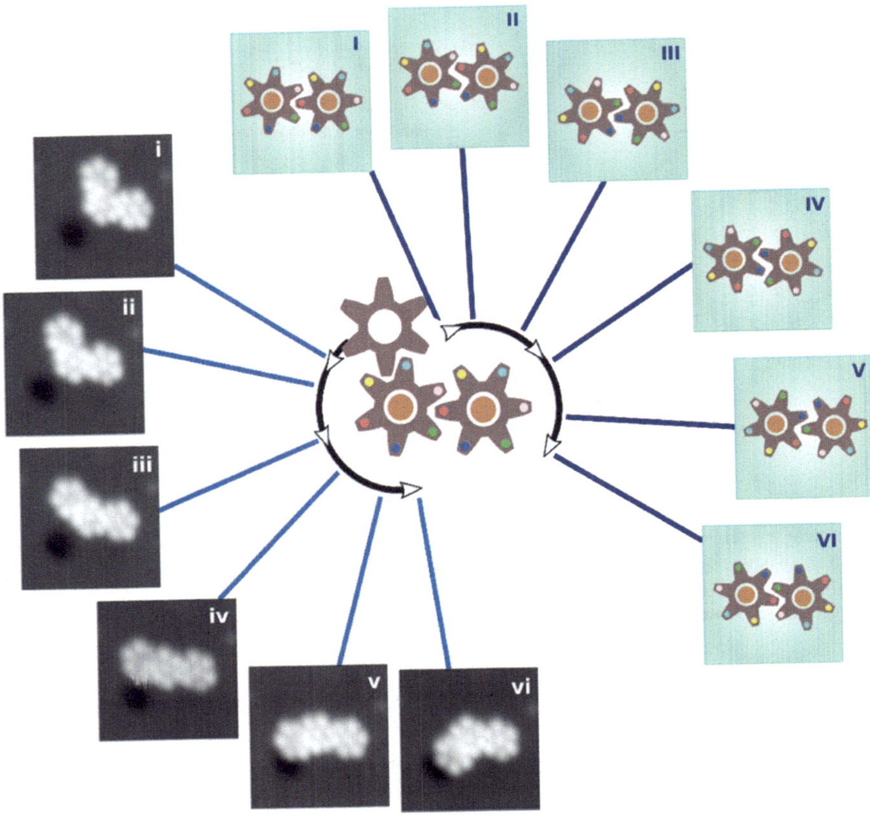

Fig. 7 A series of experimentally succeeded transmission of step by step rotation along a molecule-gear train using molecule-handle. Lower and upper cases roman numerals are used for experimental STM images and corresponding schematic diagrams, respectively. Light blue colored lines are indicating the first molecule-gear (gear molecule in the middle) orientations at each step, and dark blued lines for the second gear (molecule on the right). Redrawn using the Ref. [12] LT-UHV STM images with permission: copyright 2019, American Chemical Society

gear rotation is between 20 and 35° due to stabilization effect coming from the handle molecule, which is better mechanically coupled with the first molecule-gear as compared with the coupling between the two molecule-gears.

4 Conclusion

The transmission of a rotation along a molecule-gear train results from a more complex mechanics than along a rigid macroscopic train of solid-state gears [10] or of mesoscopic machinery constructed with solid state nanogears [11]. For the transmission of a rotational motion to occur from the first molecule-gear to the next, a precise optimization of the distance between their rotation atomic axles is first required. Second, the molecules used here in the construct of the train of molecule-gear machinery are not only flexible but are also changing their conformations at each stationary station during their functioning. This renders the optimization of their functioning on the ground state potential energy surface of this molecular machinery more challenging than their rigid body millimeters macroscopic or 100 nm mesoscopic in diameter counterparts.

Acknowledgements We thank Dr. A. Gourdon for having kindly provided to us a few milligrams of the remaining $C_{64}N_2H_{76}$ molecule-gears, and the European Union Horizon 2020 FET open project "Mechanics with Molecule(s)" (MEMO, grant 766864) for financial support.

References

1. Manzano, C., Soe, W.-H., Wong, H.S.J., Ample, F., Gourdon, A., Chandrasekhar, N., Joachim, C.: Step by step rotation of a molecule-gear mounted on an atomic-scale axis. Nat. Mater. **8**, 576–579 (2009)
2. Gao, L., Liu, Q., Zhang, Y.Y., Jiang, N., Zhang, H.G., Cheng, Z.H., Qiu, W.F., Du, S.X., Liu, Y.Q., Hofer, W.A., Gao, H.-J.: Constructing an array of anchored single-molecule rotors on gold surfaces. Phys. Rev. Lett. **101**, 197209 (2008)
3. Tierney, H.L., Murphy, C.J., Jewell, A.D., Baber, A.E., Iski, E.V., Khodaverdian, H.Y., McGuire, A.F., Klebanov, N., Sykes, E.C.H.: Experimental demonstration of a single-molecule electric motor. Nat. Nanotechnol. **6**, 625–629 (2011)
4. Perera, U.G.E., Ample, F., Kersell, H., Zhang, Y., Vives, G., Echeverria, J., Grisolia, M., Rapenne, G., Joachim, C., Hla, S.-W.: Controlled clockwise and anticlockwise rotational switching of a molecule motor. Nat. Nanotechnol. **8**, 46–51 (2013)
5. Persson, B.N.J., Tosatti, E.: The puzzling collapse of electronic sliding friction on a superconductor surfaces. Surf. Sci. **411**, L855–L857 (1998)
6. Chiaravalloti, F., Gross, L., Rieder, K.H., Stojkovic, S., Gourdon, A., Joachim, C., Moresco, F.: A rack and pinion device at the molecular scale. Nat. Mater. **6**, 30–33 (2007)
7. Sautet, P., Joachim, C.: Calculation of benzene on rhodium STM images. Chem. Phys. Lett. **185**, 23–30 (1991)
8. Ample, F., Joachim, C.: A semi-empirical study of polyacene molecules adsorbed on a Cu(110) surface. Surf. Sci. **600**, 3243–3251 (2006)
9. Gross, L., Rieder, K.H., Moresco, F., Stojkovic, S., Gourdon, A., Joachim, C.: Trapping and moving metal atoms with a six-leg molecule. Nat. Mater. **4**, 892–895 (2005)
10. Howe, R.T., Muller, R.S., Gabriel, K.J., Trimmer, W.S.N.: Silicon micromechanics: sensors and actuators on a chip. IEEE Spectr. **27**(4), 29–35 (1990)
11. Yang, J., Deng, J., Troadec, C., Ondarcuhu, T., Joachim, C.: Solid state SiO_2 nanogears AFM tip manipulation on HOPG. Nanotechnology **25**, 465305 (2014)
12. Soe, W.-H., Srivastava, S., Joachim, C.: Train of single molecule-gears. J. Phys. Chem. Lett. **10**, 6462–6467 (2019)

A Simple Example of a Molecule-Gear Train: PF_3 Molecules on a Cu(111) Surface

S. Srivastava, W.-H. Soe, and C. Joachim

Abstract A train of molecule gears consisting of PF_3 molecules was studied using semi-empirical ASED+ method to explore the mechanism of rotational transmission along this train. It was observed that a unidirectional rotational transmission occurs between only the first two PF_3 molecules for a PF_3 molecule train up to six molecule-gears, the four PF_3 molecules at the end of the train being used to rigidify the rotation axle of the first two PF_3. This demonstrates that in a train of molecule-gears, the rotation of each molecule is resulting from a collective action of many degrees of freedom per molecule. This collective motion is rather fragile against many others possible minimum energy trajectories which can develop on the multidimensional ground state potential energy surface of a molecule-gear train to respond to the increase of the potential energy required to rotate the first molecule-gear of the train.

Keyword Single molecule-gear transmission

1 Introduction

The transmission of mechanical movements from one part to another distant part of a mechanical machinery is essential for its functioning. Such a transmission can be found in mechanics at any scale from large trucks and rockets to nanoscale macro-molecular machinery [1, 2]. Often employed in macroscopic machines, the transmission of a rotation at the nanoscale with a train of molecule-gears is largely preferable to the use of a molecular chain pull-tab [3] because long single molecule rods are rather flexible [3]. Well before the first reported single molecule-gear experiments

S. Srivastava (✉) · W.-H. Soe · C. Joachim
Centre d'Elaboration de Matériaux et d'Études Structurales (CEMES), Centre National de la Recherche Scientifique (CNRS), Université de Toulouse, 29 Rue J. Marvig, BP 4347, 31055 Toulouse Cedex, France
e-mail: Saurabh.mohit@gmail.com

W.-H. Soe · C. Joachim
International Center for Materials Nanoarchitectonics (WPI-MANA), National Institute for Material Sciences (NIMS), 1-1 Namiki, Tsukuba, Ibaraki 305-0044, Japan

C. Joachim (ed.), *Building and Probing Small for Mechanics*, Advances in Atom and Single Molecule Machines, https://doi.org/10.1007/978-3-030-56777-4_10

[4–6], single molecule to molecule gearing effects were called on board to explain the peculiar surface structure of some molecular monolayers self-assembled on a metal surface, more specifically the mechanical frustrations showing up at domain boundaries in these monolayers [7]. A well-known case is the molecular series NH_3, NF_3 or PF_3 molecules chemisorbed on Ag(111), Cu(111), Ni(111) or Ru(0001) surfaces [7–10]. Here molecular rotation hindering comes into play when increasing the molecular coverage density on those metallic surfaces. It creates mechanical disorders at the boundaries between well-ordered extended molecular domains. Among those molecules, PF_3 is an important example for single molecule to single molecule gearing effects because its phosphorus central atom is chemisorbed atop a given surface metal atom with a bond length larger than for the nitrogen in the NH_3 ammonia molecule case. This long bond length is minimizing the 3 fluorine top umbrella atoms electronic interactions with the metal surface (Fig. 1). Furthermore, a fluorine atom has a larger van der Waals radius than hydrogen. Therefore, and at a well-adapted low coverage, PF_3 molecule-gears are more adapted to a cogwheel like PF_3 to PF_3 molecules mechanical interactions than, for example, NH_3 [11]. This explains why PF_3 was selected as a starting example to simulate the mechanics of a linear PF_3 train of molecule-gears [12–14].

In this chapter, the mechanics of a PF_3 train of molecule-gears is revisited without using cyclic boundary conditions i.e. to study the mechanical transmission of rotation of a truly isolated train of molecule-gear. In Sect. 2, a single PF_3 molecule chemisorbed on the Cu(111) surface is presented. We analyze the native rotation energy barrier around the Cu–P bond axle. In Sect. 3, different strategies to rotate one-way a single PF_3 molecule around this axis are discussed including the possibility to push with an STM tip on a single P–F bond. In Sect. 4, a train of two PF_3 molecule-gears is studied and the difficulties to master a transmission of rotation

Fig. 1 The atomic model of a single PF_3 molecule chemisorbed atop a Cu surface atom on a Cu(111) surface. The Cu(111) surface is represented by a 3 layer Cu slab with 54 Cu atoms per layer. No periodic boundary conditions were used in the calculations results presented in this chapter explaining the large lateral size of this slab to avoid any slab border effects

from one PF_3 to the next are analyzed. In Sect. 5, ancillary PF_3 molecule are used to stabilize this transmission leading to the Sect. 6 conclusions.

2 A Single PF_3 Molecule on the Cu(111) Surface

On Ag(111), Cu(111), Ni(111) or Ru(0001) surfaces, an isolated PF_3 molecule chemisorbs atop a metal surface atom with its phosphorus directly bonded to this atom. The 3 fluorine atoms are forming a reverse trigonal pyramid umbrella (Fig. 1). In this chapter, we have chosen the PF_3/Cu(111) case because the Cu–P distance is larger than 0.2 nm which is maintaining the 3 fluorine atoms at a large distance from the metallic surface.

The sp hybridized lone pair of the highest occupied molecular orbital (HOMO) of the PF_3 is mainly localized on the phosphorus atom. It offers a σ-coupling to the 4s orbital of a Cu surface atom. This bonding is weaker than, for example, the Ni–PF_3 one leading to a less stable PF_3 chemisorption on Cu(111) as compared to Ni(111) [15]. As already noticed, this leads to a quite long 0.21 to 0.225 nm Cu–P bond length depending on the optimization calculation technique used. This is quite useful for surface molecular gearing effect because with a 0.157 nm P–F bond length and a 70° opening of the fluorine umbrella, the Cu–F surface distance is as large as 0.3 nm [10]. Taking also into account that the PF_3 HOMO and LUMO (the lowest unoccupied molecular orbital) weights on the 3 fluorine atoms are small as compared to their weight on phosphorus [16], the fluorine to Cu(111) surface interactions are in the meV range as discussed below.

Having a sp hybridized HOMO bonding to the 4s surface orbital of a top Cu atom is very convenient for a gearing effect based on a collective rotation of the 3 fluorine atoms around the Cu–P axis. Considering after R. Zwanzig a PF_3 molecule on a surface as a gear in an engineering sense [11], this umbrella rotation is one of the many degrees of freedom of a PF_3 molecule on the Cu(111) surface. But PF_3 is not exactly a rigid object as compared with a macroscopic solid-state gear. With its 4 atoms, a free single isolated PF_3 has 12 mechanical degrees of freedom. On a Cu(111) surface and below a 220 K surface temperature (to avoid its lateral diffusion on this surface [15]), PF_3 has 9 intrinsic mechanical degrees of freedom, to list a few of them: the Cu–P bond vibrations (distance d1), the deformation of the fluorine atoms umbrella, a possible inversion of this umbrella for large d1 before desorption from the surface and of course the collective umbrella rotation angle $\varphi 1$ around the Cu–P axle. Therefore and while chemisorbed on Cu(111), the PF_3 Born-Oppenheimer (BO) electronic E_g ground and E^* excited states potential energy surfaces (PES) are all a regular function of 9 variables. Among those variables, some are essential for the understanding of the PF_3 rotation mechanics like $\varphi 1$ and d1. Others like the eccentricity of the umbrella rotation and how each P–F bond stretching is mechanically responding to an external excitation (Thermal, optical and mechanical) will be discussed in the next section. Notice that in this chapter, we are considering the BO semi-classical mechanics of the PF_3 molecules while chemisorbed on Cu(111). This is supposing that the ro-vibronic mechanical parts of a PF_3 molecule quantum states on a surface are very rapidly decohering while chemisorbed and mechanically activated at temperature below 220 K. Freezing completely the PF_3/Cu(111) system at very extreme low temperature well below 4 K will certainly bring up some consequences of mechanical quantification [9].

In Fig. 2, the calculated rotation angle dependency of ground state potential energy E_g (φ1) are presented first by building up a completely rigid rotating PF_3 molecule trajectory on the ground PES and second by minimizing at each φ1 value along this trajectory, the ground state potential energy of the PF_3 chemisorbed molecule after a detail exploration of the PES on all the nine (φ1, d1, v1, v2, v3, ..., v7) dimensional space. Since the PES are defined using 9 coordinates, there is some arbitrariness on how to select those coordinates for optimization starting from the native Cartesian coordinates of the PF_3 constituent atoms relative to the Cu(111) atoms where an absolute Cartesian referential can be defined.

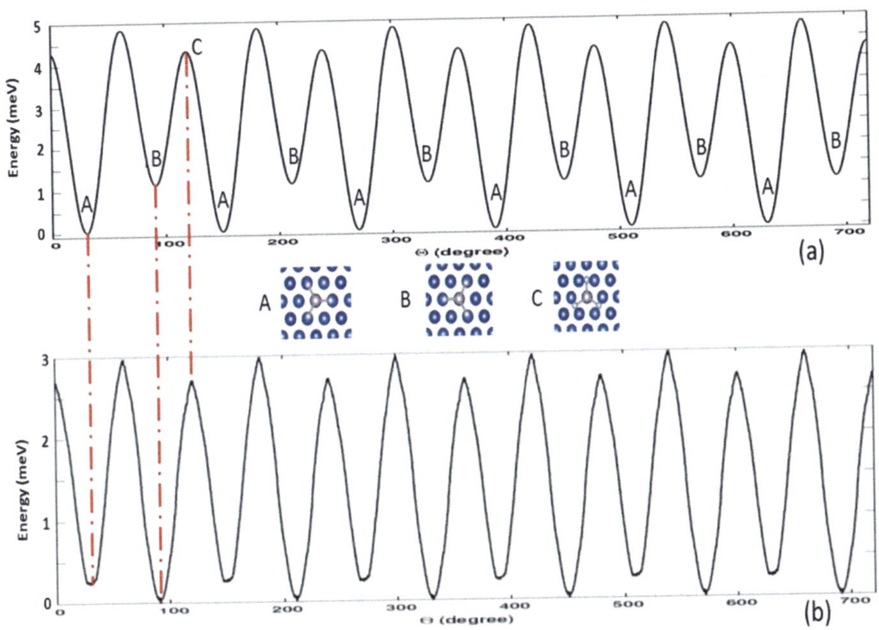

Fig. 2 The ASED+ calculated ground state potential energy E_g curves of the Fig. 1 slab with a chemisorbed PF_3 as a function of the Fluorine pyramid collective rotation angle φ1. **a** PF_3 is rigid while rotating around the Cu–P axis and **b** the total energy had been optimized in all the 9 mechanical degrees of freedom while rotating by step of 0.01°. The A–B minima sequence is respected in both cases. In **b** the B minima are recovering their absolute character. For (**a**), the starting d1 = 0.225 nm was ASED+ optimized for a starting C orientation. For **a** and **b**, the zero in energy is defined by the series of absolute minima found locally on the total ground state PES

For exploring the ground state PES and to determine the φ1 collective rotation trajectory of interest, a freezing of all the coordinates except φ1 is one solution guided by the gearing mechanical effect to be explored. Minimizing the E_g(φ1, d1, v1, v2, v3, ..., v7) function while distinguishing φ1 and d1 among all the nine directions on this PES is a more precise strategy. This is supposing, for example, that the main rotational mechanics of interest is first the rotation φ1 of the umbrella and then a change in the Cu–P bond length d1 when the fluorine atoms are interacting with the surface Cu atoms during a PF_3 rotation. Notice that PF_3 can also adopt an eccentric rotation to minimize those interactions. This is not problematic for a single PF_3 rotation but can lead to some difficulties when this first PF_3 is supposed to induce the rotation of a second PF_3.

The Fig. 2 potential energy curves were calculated using the semi-empirical atom superposition and electron delocalization (ASED+) molecular orbital method [17, 18] with the long-range van der Waals forces calculated using the MM4 parametrization [19]. This semi-empirical method was preferred to DFT-SCF calculations because of the large Cu(111) surface considered in this chapter (see Fig. 1). The total ground state energy was calculated at each φ1 value summing up twice all the mono-electronic energy of the Fig. 1 slab including the PF_3 molecule MO up to half of its mono-electronic states. The φ1 elementary step for those calculations was 0.01°.

The rigid PF_3's E_g(φ1) curve is giving about 5 meV energy barrier for a rotation and the fully optimized E_g(φ1) a 3 meV rotation energy barrier. A little ratchet effect is showing up in Fig. 2b at each maximum indicating that there are some mechanical constrains on the E_g(φ1, d1, v1, v2, v3, ..., v7) PES in particular on the Cu–P bond length. Notice that the use of the ASED+ semi-empirical method allows to explore two complete turns of the PF_3 rotation demonstrating also the stability of this rotation. In conclusion, there are 6 stations along the total ground state PES for a 360° rotating PF_3 molecule-gear on the Cu(111) surface. The fluorine atoms are far away enough from the Cu(111) surface that their surface interactions is accounting for 2 meV in the stabilization of this rotation as compared to the 3 meV rotation barrier.

3 Rotating a Single PF_3 Molecule on Cu(111)

The Fig. 2b potential energy curve is reflecting only a very local part of the total E_g(φ1, d1, v1, v2, v3, ..., v7) PES multidimensional surface topology. The small ratchet effect found at each maximum on this curve is an indication that it is not offering a complete description of this PES. For example, this φ1 rotation trajectory does not necessarily define the trajectory to be followed by a PF_3 molecule-gear for rotating at least a second PF_3 molecule along a train of PF_3 molecule-gears. As a consequence, and for such functioning to occur, the φ1 collective coordinate trajectory on the PES must be the result of an external excitation providing the required energy to pass over the rotation barrier. It must not be the result of a mathematical definition as it is usually done for φ1. In the case of a two PF_3 molecule-gears train that will be considered in the next section, the E_g(φ1, φ2, d1, d2, v1, v2, v3, …, v14) ground state PES is a regular function of $2 \times 9 = 18$ coordinates. Here, the synchronized expected (φ1, φ2) trajectory on this 18-dimensions surface for describing a train of 2 molecule-gears in rotation must result from a precise external excitation and not predefined mathematically.

Let us first return to the simple single PF_3 molecule-gear case. There are basically 4 different possible ways to try to induce a collective φ1 rotation trajectory on the E_g(φ1, d1, v1, v2, v3, …, v7) PES: a thermal excitation coming from the Cu(111) surface, an optical excitation involving some E^*(φ1, d1, v1, v2, v3, …, v7) electronic states, an inelastic tunneling effect using a voltage bias tip of a scanning tunneling microscope (STM) resulting generally in an incoherent mixing of the ground state E_g(φ1, d1, v1, v2, v3, …, v7) and of some low lying in energy E^*(φ1, d1, v1, v2, v3, …, v7) electronic states and finally a mechanical push (pulling) on a single P–F bond of the PF_3 molecule by repulsive (attractive) interactions with the STM tip apex end atom. Let us discuss in detail each of them in the following.

3.1 A Thermal Excitation?

Starting at a low Cu(111) surface sample temperature T_1 where a given isolated PF_3 molecule is not φ1 rotating i.e. for $T_1 \ll 30$ K, one can increase progressively the copper sample temperature in such a way to increase the internal PF_3 temperature. This temperature T_2 is defined by the ro-vibronic energy $E_g = kT_2$ (k: Boltzmann constant) relative to the quantum zero-point energy in one of the local minima identified by the calculations presented in Fig. 2. According to the classical equi-repartition theorem, each PF_3 mechanical degrees of freedom in interactions with a thermostat at temperature T_1 is capturing in average $1/2\ kT_1$ of energy leading for PF_3 to a total of $9/2\ kT_1$ of internal mechanical energy equi-distributed among the 9 directions on the ground state BO PES. Notice that the excited states are not impacted by this thermal excitation because the ground state to excited state energy difference is largely in the UV range for a PF_3 molecule [16].

The surface energy capture by a PF_3 molecule will be first delivered to the φ1 soft rotation mechanical mode before randomly exciting also the harder ones like the Cu–P chemical bond. Due to the micro-reversibility principle, there is no mechanical degrees of freedom favored during this PF_3 energy capture. There is an intramolecular energy equilibration with the surface thermostat temperature.

This thermalization process has one important consequence. When kT_1 is reaching the Fig. 2 rotation barrier height threshold, the PF_3 top fluorine umbrella is first fluctuating in its low temperature initial minimum energy (6 per 360°) and then rotates to reach a nearby energy minimum on the E_g(φ1, d1, v1, v2, v3, …, v7) PES. Since the Cu(111) surface thermostat temperature is constant, this leads to a random exploration of the φ1 rotation trajectory on this PES as characterized by the Fig. 2b curve. This was already observed with other single molecules rotating on a metal surface after a sample temperature increase [1, 20–23]. Such a random thermal exploration of the $E_g(\varphi 1, d1, v1, v2, v3, \ldots, v7)$ PES will not be able to drive the rotation of a second PF_3. This random process can also be easily simulated by a standard Langevin like equation [24].

An interesting point to notice is the role in such a random process of the little ratchet-like part of the potential energy curve calculated for each Fig. 2b maximum. When such a potential energy curve is used in a 1D Langevin equation, a deviation from a purely random rotation is calculated. But as already noticed, the micro-reversibility principle is quite robust in the sense that the Fig. 2b curve is just a 1D section of the total $E_g(\varphi 1, d1, v1, v2, v3, \ldots, v7)$ PES along the $\varphi 1$ rotation trajectory. Using the complete 9 dimensions potential energy function in a 9 coupled Langevin equations system will resume the random rotation process. This comes from the fact that the calculated ratchet effect using only the $E_g(\varphi 1)$ 1D potential energy is artificially created by the restriction of the $E_g(\varphi 1, d1, v1, v2, v3, \ldots, v7)$ PES thanks to the minimization procedure used in ASED+ [24]. As an indication, there is no ratchet effect using the Fig. 2a potential energy curve in a 1D Langevin equation.

3.2 A Vertical Optical Excitation?

Fixing now the PF_3/Cu(111) surface sample temperature T_l well below 30 K, the PF_3 molecules are each stable in one of their potential energy minima (6 minima per 360° per PF_3 molecule). A UV light source is illuminating this surface to trigger an optical transition from the E_g ground to the first E^* excited state of a given PF_3 molecule. Many PF_3 molecules on the Cu(111) surface will be illuminated by this light source because the UV light wavelength is too large to focus this illumination on a single PF_3 molecule even using a near field optical microscope instrument. Considering now one of those PF_3 under illumination, this vertical optical transition will bring this PF_3 molecule from one minimum of $E_g(\varphi 1, d1, v1, v2, v3, \ldots, v7)$ to the $E^*(\varphi 1, d1, v1, v2, v3, \ldots, v7)$ excited state surface. Located about 6 eV from the ground state [25, 26], the exact surface topology of this $E^*(\varphi 1, d1, v1, v2, v3, \ldots, v7)$ electronic state is not known in detail. What is known is that the LUMO entering in the composition of this state is built up from a degenerate pair of π^* orbitals with a very large molecular orbital weight on the phosphorus atom [16].

Therefore and along the possible $\varphi 1$ rotation trajectory, the rotation barrier on the $E^*(\varphi 1, d1, v1, v2, v3, \ldots, v7)$ surface will be well below 1 meV because also this LUMO is 6 eV up the HOMO entering in the ground state. This means that by reaching this E^* surface, there will be no rotation of PF_3 activated by reaching this excited state PES. By rapidly relaxing in its ground state, PF_3 can either remain as it is in original minimum or explore any of the lateral ones and not necessary the two nearest neighbor ones because of the flatness in energy of this $E^*(\varphi 1, d1, v1, v2, v3, \ldots, v7)$ PES. It can be even worse. For example and for the NH_3/Cu(100) prepared surface, the UV absorption reaching E^* is triggering a fast NH_3 trajectory on this 9D surface leading to a large vibration of the Cu–N bond after the spontaneous relaxation to E_g, then to an inversion of the H_3 umbrella on the Cu(100) surface and finally to a desorption of the NH_3 molecule because of this inversion [27].

3.3 An STM Inelastic Tunneling Current Excitation?

On a prepared PF_3/Cu(111) surface maintained at LHe temperature and with a very low coverage for the PF_3 molecules (to be electronically and mechanically independent from each other), the tip apex end atom of a LT-UHV-STM must at least be located on a single PF_3 molecule to trigger its rotation as it was first observed for the simple O_2/Pt(111) system [28].

During an STM scanning, at low bias voltages (a few 100 mV) and at low tunneling current intensities (below 100 pA), a PF_3 molecule will normally remain in one of its local $E_g(\varphi 1)$ minima reached after the PF_3/Cu(111) surface preparation. Such STM imaging will lead to the determination of the position of the fluorine umbrella. Increasing the bias voltage up to about a volt will normally destabilize PF_3 from its atop adsorption site leading to a lateral hopping mechanism. Increasing further this bias voltage will lead to a desorption of the PF_3 molecule. This was already observed experimentally for the NH_3/Cu(111) system where a lateral hopping happens at a 450 mV of bias voltage and a 2.1 nA of tunneling current [29].

For PF_3 and between 0.1 and 1.0 V of bias voltage, it seems difficult to find some voltage range to activate in one direction the PF_3 $\varphi 1$ rotation mode. The reason is that a tunneling current intensity through a molecule is mainly due to a random sequence of billions of virtual occupations of the E^* (of the E_g) states via a single electron (or a single hole) transfer process per second respectively. Among all of those transfer events, 0.01% will be inelastic delivering for example a small amount of ro-vibronic energy to E^*. In view of the flatness of the $E^*(\varphi 1, d1, v1, v2, v3, \ldots, v7)$ PES, there will be no preferential trajectory on this surface triggered by this increase of energy.

For the NH_3/Cu(100) case, it was already observed that this inelastic tunneling effect leads first to a stretching of the N–H bonds and for larger bias voltages to an inversion of the H_3 umbrella exactly as discussed above for a complete UV transition [30]. For PF_3/Cu(111), we anticipate a random lateral hopping on the PF_3 molecule on the Cu(111) surface instead of a rotation. There will be no reversal of the F_3 umbrella because the inversion energy barrier is too large for this molecule. To get a chance to observe a $\varphi 1$ random rotation, the end atom must at least be positioned off the P center of the PF_3 molecule. But it will be non directional.

3.4 A Local STM Mechanical Push on a Single P–F Bond?

The only way we have found to initiate a collective $\varphi 1$ fluorine umbrella rotation around the Cu–P axis by a local excitation is to push mechanically on a single P–F bond using the tip apex end atom of the STM. This is exactly the STM manipulation protocol which is used for molecule-gears larger than the PF_3 molecule (see the chapters "Anchoring Molecular Rotors by On-Surface Synthesis" and "Transmission of Rotational Motion Between Molecule-Gears" this volume, [5, 20, 31]). The advantage of a simple molecule like PF_3 is that the number of degrees of freedom is small. In that case, how a minimum energy trajectory on the E_g PES can result in a Fig. 2b like potential energy curve is easy to appreciate.

To simulate the STM tip apex push, one P–F bond was simply twisted by step of 0.1°, changing only one fluorine atom (x, y) Cartesian coordinates and freezing its z Cartesian coordinate. By this procedure, the collective φ1 angle is not mathematically predefined. It must result from the collective mechanical reaction of the three fluorine atoms umbrella and of all the intramolecular degrees of freedom of the PF_3 molecule. Only the φ11 torsion angle in the plan parallel to the Cu(111) surface of the selected P–F bond is defined by the change of the (x, y) coordinates starting from an initial minimum energy of the E_g(φ11, φ12, φ13, d1, w1, w2, w3, w4, w5) PES. Notice that the 9 (φ11, φ12, φ13, d1, w1, w2, w3, w4, w5) variables used now to define this function are different from the previous E_g(φ1, d1, v1, v2, v3, …, v7) PES because φ11 is first defined relative to the absolute Cu(111) surface referential. Then, φ12 and φ13 are calculated after each optimization step on this complete 9D $E_g(\varphi 11_{fix}$, φ12, φ13, d1, w1, w2, w3, w4, w5) PES. Not including the STM tip apex in the calculation largely limits the ASED+ minimization energy calculation time per φ11 step.

Figure 3 is demonstrating with a great detail how φ12 (and also φ13) are in fact almost exactly following step by step the φ11 P–F torsion. It is a purely adiabatic

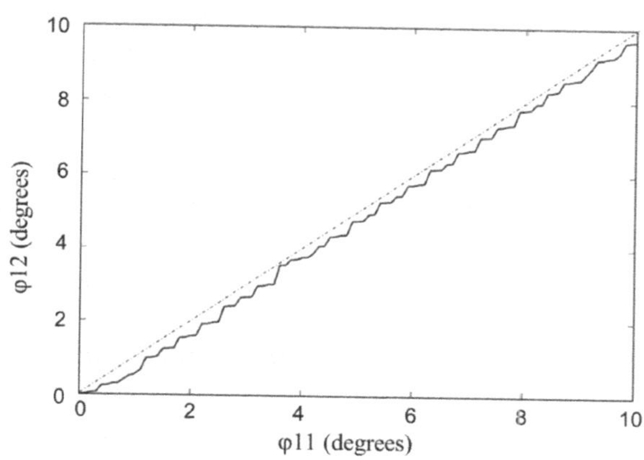

Fig. 3 The resulting φ12 torsion angle parallel to the Cu(111) plan of the P–F bond next to the φ11 twisted one in the plan P–F bond. This was calculated by first performing the φ11 twist by steps of 0.1° and then finding the minimum energy on the E_g(φ11, φ12, φ13, d1, w1, w2, w3, w4, w5) surface while maintaining this constrain. The linear dotted line is the awaited ideal response of PF_3 umbrella to the φ11 torsion. The φ13 is following similar variations as a function of φ11

molecular mechanics calculations simulating how a tip apex STM molecular manipulation will step by step work. No molecular dynamics here since a one-step molecular manipulation push on a single molecule is a very slow process, about 10 ms per push for STM molecular manipulation. The molecular dynamics part of the STM manipulation process is accommodated by the ASED+ energy minimization algorithm after each P–F bond torsion parallel to the Cu(111) plan.

Once engaged in a trajectory on the $E_g(\varphi 11, \varphi 12, \varphi 13, d1, w1, w2, w3, w4, w5)$ PF_3 ground state PES where $\varphi 12$ and $\varphi 13$ are following $\varphi 11$ almost linearly as presented in Fig. 3, it is possible to continue to push step by step on the same P–F bond following the rotation of the umbrella or to change of P–F bond for twisting. This will lead to the same result i.e. a trajectory on the E_g surface where the F_3 umbrella is step by step following a collective trajectory almost equivalent to the mathematical trajectory leading Fig. 2b $\varphi 1$ potential energy curve and this with no lateral hoping of the molecule on the Cu(111) surface. It comes from the fact that a 1° twist performed on a single P–F bond is increasing the PF_3 internal energy only by 0.7 meV, not enough to destabilized the PF_3 molecule from its adsorption site. The corresponding potential energy curve is presented in Fig. 4b extracted from the ($\varphi 11$, $\varphi 12$) energy contour plot presented in Fig. 4a.

This STM manipulation protocol leading to the definition of a $\varphi 1$ collective rotation trajectory of the complete fluorine atoms umbrella is different from the three previously possible way of triggering such a rotation. With a well applied local STM tip apex push on a single P–F bond, one is looking for the best way to benefit from the complete E_g PES topology for an umbrella rotation. With a thermal excitation, an optical transition or its intermediate inelastic tunneling current effect, the provided energy is distributed among all the 9 degrees of freedom. The thermal equi-distribution cannot define a coherent trajectory on the E_g PES. The inelastic tunneling and the optical ones are delivering energy to the E_g ground state via the excited state (respectively partially or totally). There is no reason why a trajectory initiated partially or totally on the E^* PES be adapted to the target umbrella collective trajectory on the E_g PES especially for a flat in energy E^* where the ro-vibronic energy can trigger mechanical fluctuations starting from many directions on this E^* surface.

4 Rotating a Train of 2 PF_3 Molecule-Gears with One

On the ground state PES, the driving conditions to rotate step by step and in one-way a single PF_3 molecule-gear can now be applied to a train of two identical PF_3 molecule-gears whose optimized surface molecular dimer structure is presented in Fig. 5. The first step was to design this train for the $(PF_3)_1$ driver to be chemisorbed a good distance away from the $(PF_3)_2$ follower molecule. On the Cu(111) surface, there is not a lot of choice since the only chemisorption site for a PF_3 molecule is an on-top site of Cu(111) surface. As presented in Fig. 5 and after a few trials, we have selected for this dimer chemisorption, two Cu surface atoms separated only by one Cu

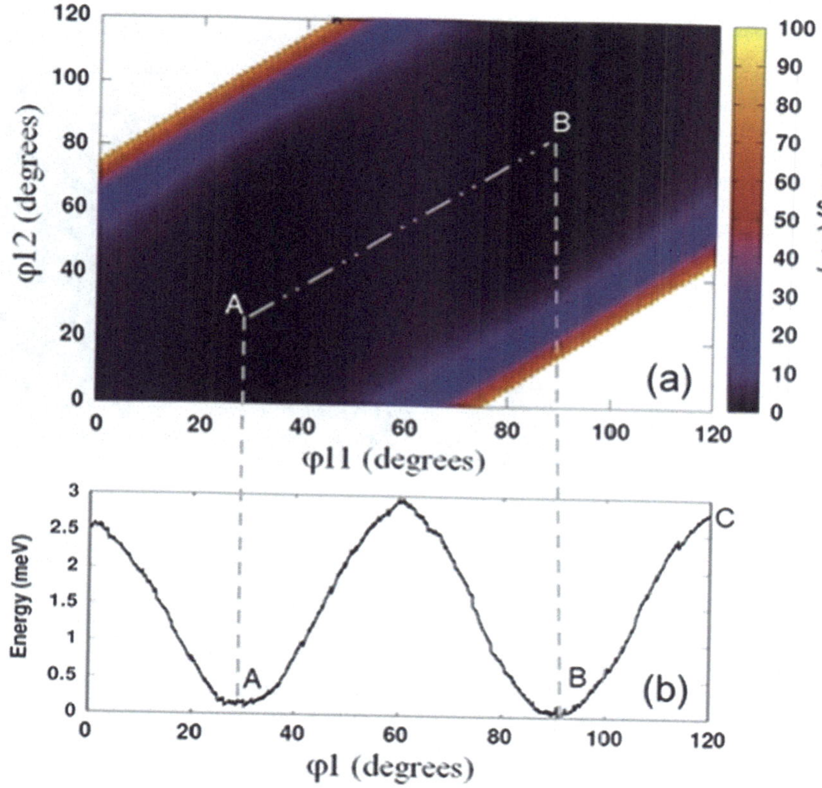

Fig. 4 **a** The minimized $E_g(\varphi 11, \varphi 12)$ potential energy contour plot while twisting step by step $\varphi 11$. **b** The extracted $E_g(\varphi 11 = \varphi 12)$ potential energy curve plotted as a function of the collective angle $\varphi 1$ where $\varphi 1 = \varphi 11 = \varphi 12 = \varphi 13$. This (**b**) curve must be compared with the Fig. 2b curve where the first two stations A and B are identical in molecular conformation. The saddle energy point at 60° is the position on the E_g PES where the PF$_3$ collective umbrella rotation occurs following the dashed double dotted central line on **a**

atom along $[01\bar{1}]$ crystallographic surface orientation leading to the Fig. 5 presented configuration with an 0.455 nm P–P surface distance. Such a surface configuration can normally be experimentally constructed PF$_3$ molecule per PF$_3$ molecule using a now standard STM tip lateral molecular manipulation protocol.

As already discussed above, the ground state potential energy surface of the Fig. 5 PF$_3$/Cu(111) dimer is now a regular function of 18 variables: $E_g(\varphi 11, \varphi 1, \varphi 2, d1, d2, w1, w2, w3, \ldots, w13)$ where $\varphi 11$ is the twisting angle of the $(PF_3)_1$ P–F bond driver as indicated also in Fig. 5. Notice that we have changed the P–F bond on the driver $(PF_3)_1$ when the one used for a given rotation is starting to interact with the $(PF_3)_2$ follower molecule. Each 0.1° elementary twist is followed by an ASED+ energy minimization to determine the new adiabatic stable configuration of the PF$_3$ dimer/Cu(111) on this 18D E_g PES.

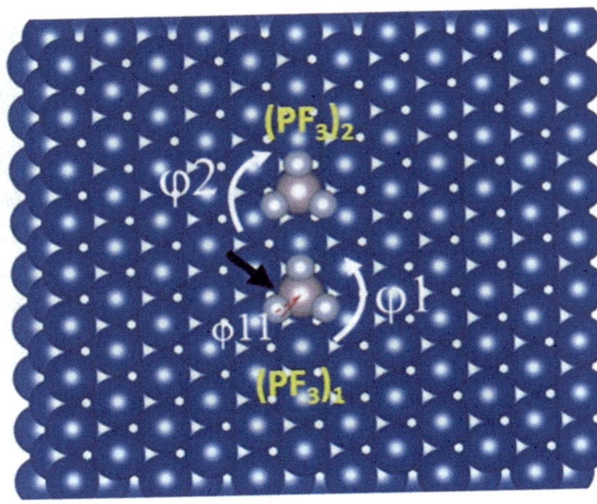

Fig. 5 The PF_3/Cu(111) designed surface structure with its two PF_3 molecule-gears positioned to what is expected to be their best surface conformation for a gearing effect. $(PF_3)_1$ is the driver and $(PF_3)_2$ the follower molecule-gears. The black arrow is indicating the location of the P–F bond step by step twisted by 0.1° at the beginning of the driving simulation. The Cu(111) surface is represented by a 3 layer Cu slab with 54 Cu atoms per layer and no boundary conditions

Figure 6 is presenting the results of this optimization for a complete 360° step by step rotation of the $(PF_3)_1$ driver where the twisted P–F bond had been changed 2 times at each 120° rotation loaded by the $(PF_3)_2$ follower molecule, the $(PF_3)_1$ driver is still able to nicely rotate 360°. The step by step P–F bond twists are able to maintain a φ1 coherent umbrella rotation angle with a linear relation between φ1 and φ11 as presented in Fig. 6a (The starting details of such rotation was given in Fig. 3). Interestingly, the motive power of the $(PF_3)_1$ driver is not large enough to rotate the $(PF_3)_2$ follower. More exactly, the driver increase of potential energy is consumed in the deformation of the Fig. 5 PF3 dimer as discussed below.

Following the first 10° φ11 variation, a first 60° rotation of the $(PF_3)_2$ molecule is obtained (Fig. 6b). Then, $(PF_3)_2$ stops and enters in a slippering oscillation mode with about a 10° oscillation amplitude. This is better characterized Fig. 7 by a plot

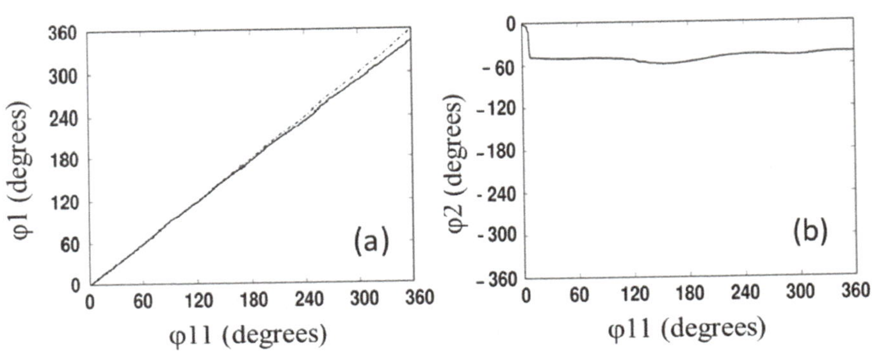

Fig. 6 Variation of the umbrella collective rotation angle **a** φ1 for $(PF_3)_1$ and **b** φ2 for $(PF_3)_2$ as a function of the φ11 twisting angle step by step driving $(PF_3)_1$

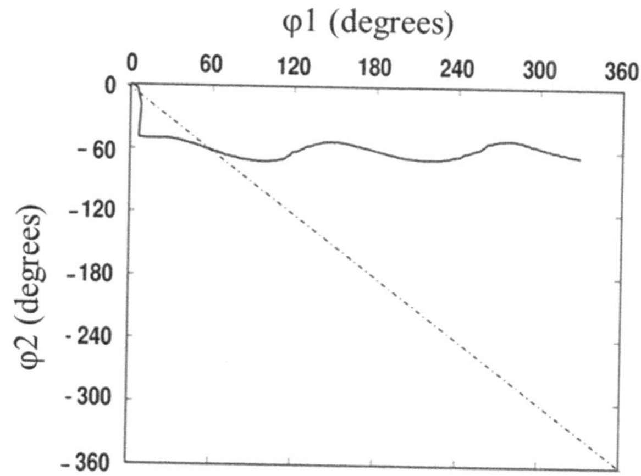

Fig. 7 Variation of the φ2 umbrella collective rotation angle for $(PF_3)_2$ as a function of the φ1 $(PF_3)_1$ driver umbrella rotation angle. The expected φ2 rotation supposed to be driven by $(PF_3)_1$ is indicated by the dotted line

of the $\varphi 2 = \varphi 2(\varphi 1)$ function since the $\varphi 1(\varphi 11)$ and $\varphi 2(\varphi 11)$ functions are the 2 mechanical response curves of the same variable φ11.

To analyze in detail this slippering mechanical motion of the supposed follower $(PF_3)_2$ molecule-gear, we have plotted in Fig. 8 the energetics and the umbrella center of mass distance between the two PF_3 molecule-gears as a function of the φ11 driver $(PF_3)_1$ twisting angle. On the 18D E_g PES, the first driving event induced by the step by step P–F twisting on $(PF_3)_1$ result in a 7 meV increase of the PF_3 dimer potential energy (Fig. 8a). This provides a mechanical constrain induced by $(PF_3)_1$ on $(PF_3)_2$. Then, $(PF_3)_2$ releases this constrain by an abrupt almost 60° rotation (see Fig. 7) resulting for the PF_3 dimer to reach a new absolute energy minimum on the 18D E_g PES. As a consequence, the center of mass distance is reduced by 0.015 nm indicative of a van der Waals attraction between the two PF_3 (Fig. 8b).

After this first $(PF_3)_2$ rotation event, the next 110° $(PF_3)_1$ rotation is forcing the PF_3 dimer to reach a second 120° energy minimum on the 18D E_g PES (see Fig. 8a). But this was not followed by the second expected $(PF_3)_2$ 60° rotation. The reason is that the two PF_3 are now too close for $(PF_3)_2$ to rotate following the $(PF_3)_1$ rotation (see Fig. 8b). This stop of $(PF_3)_2$ is corresponding to a minimum energy position and there is no decomposition of the molecule nor a lateral jump on the surface because the P–F twist is very small in amplitude. This can also be followed on the plot of slowly increasing center of mass when rotating further $(PF_3)_1$ from this low energy position. After this 120° energy minimum, the next $(PF_3)_1$ rotation step is even more problematic for $(PF_3)_2$ since now the mechanical constrain in the PF_3 dimer had increased with an intramolecular dimer energy larger than 30 meV (Fig. 8a). This leads to a large umbrella center of mass distance increase creating the slippering motion of $(PF_3)_2$ while $(PF_3)_1$ can still continue to rotate almost regularly as indicated by the third 245° energy minimum found by the ASED + minimizer on the 18D E_g PES.

The sequence of events which can be followed on the Fig. 8 curves with the two PF_3 umbrella center of mass deviation and the consequent slippering effect of the

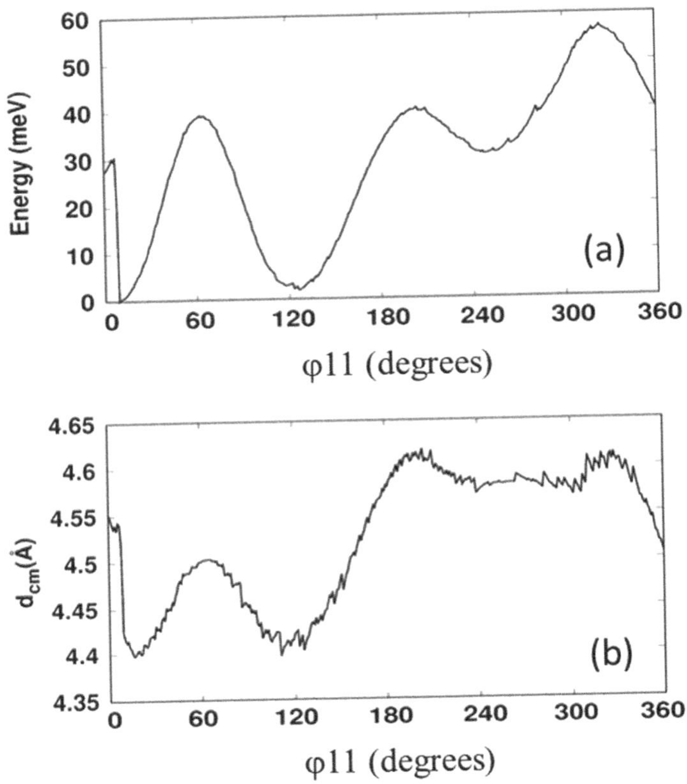

Fig. 8 Variation **a** of the potential energy and **b** of the umbrella center of mass distance between the two PF_3 as a function of a complete rotation of the $(PF_3)_1$ molecule-gear driven by the φ11 P–F twisting angle. During this 360° sequence, $(PF_3)_1$ was rotating as described in Fig. 6a. The potential energy fall at 10° is caused by the 60° rotation of $(PF_3)_2$ induced by $(PF_3)_1$. This leads to a net attraction between the two PF_3 molecule-gears which still remains both bonded to their original Cu surface atom

$(PF_3)_2$ molecule-gear preventing its rotation is well known in macroscopic gears mechanics where there is sometime looseness between the rotation axle and the solid-state gear itself. For the PF_3 molecule-gear train, the Cu–P bond rotation axle per molecule-gear is natively not rigid enough to support a rotation as compared to the different intramolecular possible ways of dispersing the potential energy increase induced by the P–F twisting. For example, the two PF_3 can bend to respond from attractive or repulsive interactions between the two fluorine-umbrella gearing parts. This looseness was also observed in the case of larger molecule-gears than the PF_3 molecules as described in detail in the chapter "Transmission of Rotational Motion Between Molecule-Gears" of this volume.

Contrary to solid state gears down to a diameter of about 5 nm (see chapter "Fabricating Solid State Gears at the Nanoscale: A Top–Down Approach" of this volume), the intramolecular cohesion of a PF_3 molecule-gear is not strong enough for the PF_3 molecule to response mechanically collectively to a strong driving force. The 18D ground state energy E_g of the PF_3 dimer is full of sideways minima as compared to the expected rotation trajectory. Therefore, many directions on this multidimensional potential energy surface can enter into play to release the constrain introduced in the PF_3 dimer by a very well localized push of the tip apex of an STM. On this 18D surface, the expected coherent rotation trajectory represented by the dotted line in Fig. 7 is effectively very difficult to construct. This is not the case of a solid state nanogear where upon a potential energy increase of the driver gear, all the atoms of the follower gear will intrinsically follow the same rotating trajectory.

The importance of the solid state gears mechanical cohesion was already explored experimentally using surface nanoscale NaCl cantilevers (for example 22.9 nm $\times$ 4.9 nm $\times$ 0.56 nm surface cantilevers) grown on a Cu(111) surface. A threshold number of atoms (about 3000 atoms) constitutive of such a cantilever is necessary to recover the classical mechanics Hooke spring law [32]. Below this limit, each atom of a NaCl nanoscale cantilever can play its own atomic scale mechanics (A lateral diffusion on the Cu(111) surface, a diffusion at the cantilever border, an interchanged with internal cantilever atoms). Passing over this limit, a cohesive motion result [32]. Well below this limit, the PF_3 single molecule-gear is only constituted of 4 atoms. As described in the next section (and also reported in chapter "Transmission of Rotational Motion Between Molecule-Gears" of this volume), it is possible to refrain the looseness of small number of atoms molecule-gears by using ancillary molecules. Let us insist on the fact that a 1 nm in thickness and 5 nm in diameter solid state nanogear will normally show up the mechanical cohesive rotation of a macroscopic solid-state gear when its cohesion is coming from electrostatic, ionic of covalent bonds between its constitutive atoms.

5 Ancillary PF_3 to Stabilize a PF_3 Molecule-Gear Train

To freeze the two PF_3 umbrella centers of mass deviation induced by the driver $(PF_3)_1$ molecule, the train of PF_3 molecule-gears have to be stabilized laterally. Demonstrated in this section, one solution is to add up ancillary PF_3 molecules at the end of the train to refrain the deformation of the rotation axles noticed with two PF_3 molecules.

As presented in Fig. 9, a third chemisorbed PF_3 molecule was atop chemisorbed on line with the two other PF_3 molecules to form a train of three PF_3 molecule-gears. As compared to the Fig. 5 PF_3/Cu(111) dimer structure, the molecule-gear train orientation of the Fig. 9 PF_3/Cu(111) trimer on the Cu(111) surface is set along $[11\bar{2}]$ orientation to ensure a better mechanical interaction between the three PF_3 molecules.

The $(PF_3)_1$ P–F driver was twisted as already described in Fig. 5. The result is first a $(PF_3)_2$ $\varphi 2 = -70°$ rotation and at the same time a $\varphi 3 = +70°$ rotation while $(PF_3)_1$

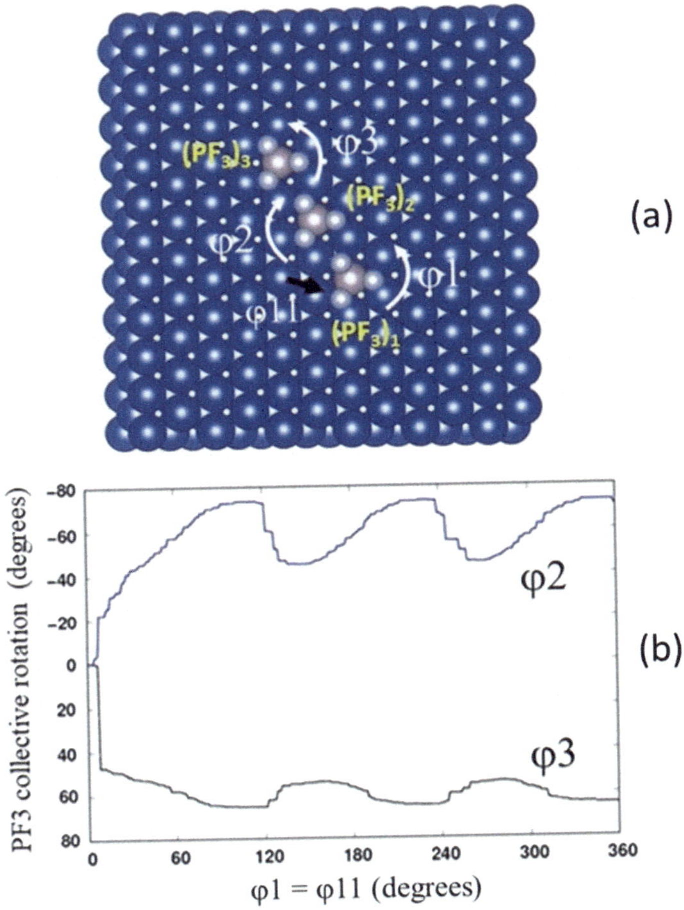

Fig. 9 **a** The PF_3/Cu(111) designed surface structure with its three PF_3 molecule-gears positioned to what is expected to be their best surface conformation for a gearing effect. $(PF_3)_1$ is the driver, $(PF_3)_2$ the follower and $(PF_3)_3$ the ancillary follower blocking lateral displacement of molecule-gears. The black arrow is indicating the location of the P–F bond step by step twisted by 0.1° at the beginning of the driving simulation. The Cu(111) surface is represented by a 3 layer Cu slab with 66 Cu atoms per layer and no boundary conditions. **b** The resulting $\varphi 2(\varphi 11)$ and $\varphi 3(\varphi 11)$ rotation curve while $\varphi 1(\varphi 11)$ (not shown) is perfectly 360° rotating

is rotated 120°. Then, both $(PF_3)_2$ and $(PF_3)_3$ stop rotating and start to slip with a periodic rotation angle of 30° for $(PF_3)_2$ and less than 10° for $(PF_3)_3$. The $(PF_3)_3$ blocking power of the $(PF_3)_2$ deformation is not strong enough. On the now 27D E_g PES, the $(PF_3)_3$ mechanical constrain is not strong enough to bring the $(PF_3)_1$–$(PF_3)_2$ dimer on the trajectory corresponding to $\varphi 2(\varphi 11) = \varphi 1(\varphi 11)$.

For the above ancillary PF_3 blocking strategy to be effective, it was necessary to complete the Fig. 9 PF_3 molecule-gears train by a fourth PF_3 molecule leading to

PF_3/Cu(111) tetramer structure presented above in the Fig. 10. Now, while $(PF_3)_1$ is still driving, $(PF_3)_2$ is now step by step rotating, $(PF_3)_3$ is absorbing a part of the constrain imposed by $(PF_3)_1$ for rotation and $(PF_3)_4$ is almost immobile. Only a first $\varphi 11$ initial instability is preserved at $\varphi 11 = 10°$. Then, $\varphi 2$ is linearly following $\varphi 11$ with long its E_g PES trajectory, a $-25°$ of retro-rotation at $\varphi 11 = 180°$, and followed again by a linear variation able to recover the initial trajectory until $\varphi 11 = 310°$. On the Fig. 10 presented 360° sequence for $(PF_3)_1$, it comes from a retro-rotation of $(PF_3)_3$ also at $\varphi 11 = 180°$ and $\varphi 11 = 310°$.

Increasing once more the length of the PF_3 molecule-gear train reaching a PF_3/Cu(111) pentamer structure open the way to stabilize the rotation of the $(PF_3)_2$ and $(PF_3)_3$ molecule-gears rotation along this train. Following the ASED+ optimization procedure used for the Figs. 5, 9 and 10 trains of molecule-gear, the molecular structure of this PF_3 pentamer was optimized at each $\varphi 11 = 0.1°$ corresponding to an exploration of an 90D E_g potential energy surface. As presented in Fig. 11a, the $\varphi 2(\varphi 11)$ and $\varphi 3(\varphi 11)$ functions are following a linear trend reaching each an 150° rotation for a 360° variation of $\varphi 1(\varphi 11)$. It was 170° for the $(PF_3)_2$ molecule in the case of the tetramer Fig. 10. In Fig. 11a, the $(PF_3)_4$ and $(PF_3)_5$ molecules are almost no more rotating playing their ancillary role of restraining the deformation of the $(PF_3)_2$ and $(PF_3)_3$ molecule during their step by step rotation.

Unfortunately, for the next step i.e. a PF_3/Cu(111) hexamer surface, this stabilization is not working anymore. As presented in Fig. 11b, the $(PF_3)_2$ molecule is rotating, the $(PF_3)_3$ molecule almost rotating and the three ancillary PF_3 molecules at the end of the train are not rotating aside from an initial instability at $\varphi 11 = 70°$. At $\varphi 11 = 360°$, the $(PF_3)_2$ rotation is now only 100°. With this hexamer, the ground state potential energy surface is now of dimension 108 and it becomes now difficult to explore this 108 surface to step by step minimize the PF_3/Cu(111) hexamer structure at each $\varphi 11 = 0.1°$ of the P–F bond twisting on the first $(PF_3)_1$ of the train.

6 Conclusion

In this chapter and using small PF_3 molecules supposed to be chemisorbed on a Cu(111) surface, we have analyzed in detail their train of molecule-gears functioning using the ASED+ semi-empirical molecular mechanics calculations. DFT calculations were not practicable in view of the very large number of atoms involved. We have studied only an isolated train of molecule-gears with no through surface electronics and no through space mechanical interactions between the trains contrary to what is considered when using periodic boundary conditions.

Chemisorbed on a Cu(111) surface, a single PF_3 molecule-gear has 9 classical mechanical degrees of freedom. We have first determined the best excitation for this PF_3 to rotate its fluorine top umbrella 3 atoms in one direction to prepare a gearing effect. By a single small 0.1° by 0.1° P–F bond twist of this umbrella, we have simulated the adiabatic molecular manipulation which can be performed on such a single molecule by an STM tip apex. A well optimized STM manipulation will be

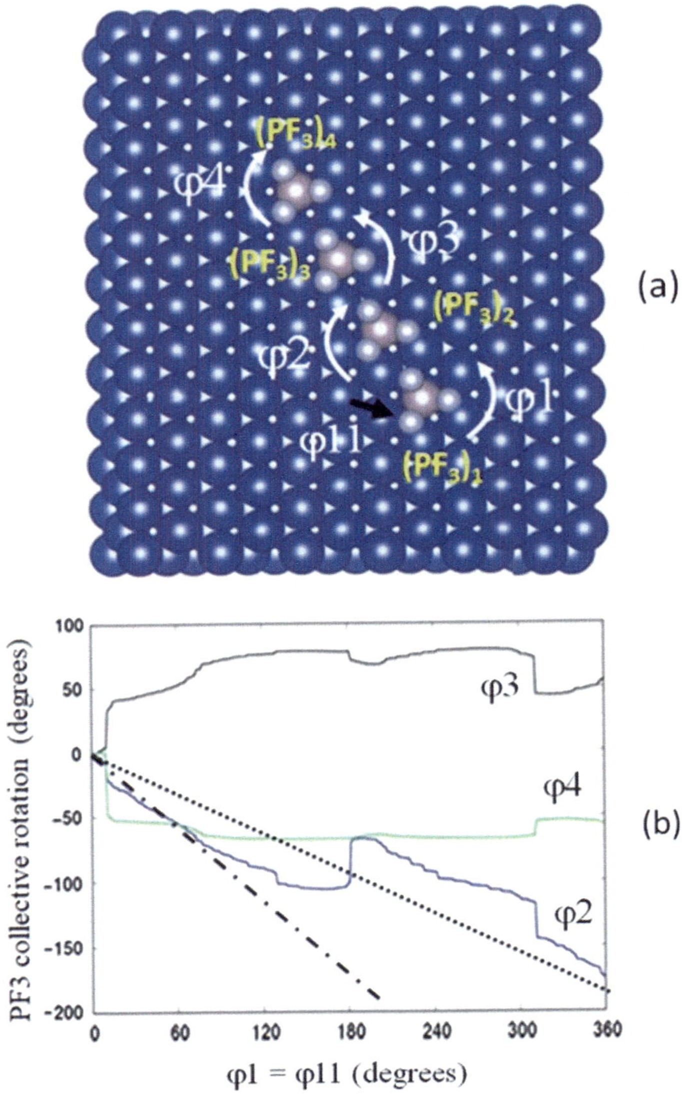

Fig. 10 **a** The PF_3/Cu(111) designed surface structure with its four PF_3 molecule-gears positioned to what is expected to be their best surface conformation for a gearing effect. $(PF_3)_1$ is the driver, $(PF_3)_2$ the follower, $(PF_3)_3$ and $(PF_3)_4$ not only followers but ancillary molecule-gears. The black arrow is indicating the location of the P–F bond step by step twisted by 0.1° at the beginning of the driving simulation. The Cu(111) surface is represented by a 3 layer Cu slab with 72 Cu atoms per layer and no boundary conditions. **b** The resulting $\varphi 2(\varphi 11)$, $\varphi 3(\varphi 11)$ and $\varphi 4(\varphi 11)$ rotation curve while $\varphi 1(\varphi 11)$ (not shown) is perfectly 360° rotating. The dot line is indicating the average linear rotation slope of the $\varphi 2(\varphi 11)$ function and the dashed-dotted line what would be the ideal $\varphi 2(\varphi 1)$ curve for a perfect 360° rotation of $(PF_3)_1$ and counter rotation of $(PF_3)_2$

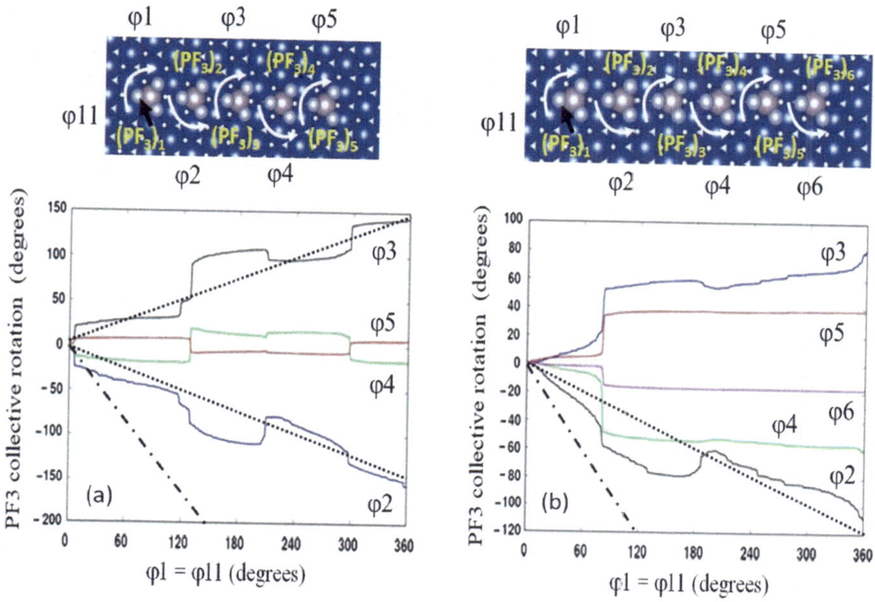

Fig. 11 The PF_3/Cu(111) designed surface structure with its **a** five and **b** six PF_3 molecule-gears with $(PF_3)_1$ the driver. In (**a**) $(PF_3)_2$ and $(PF_3)_3$ are following the $(PF_3)_1$ rotation. The $(PF_3)_4$ and $(PF_3)_5$ ancillaries are stabilizing the transmission of rotation between $(PF_3)_1$ and $(PF_3)_3$. In (**b**) $(PF_3)_2$ is following the $(PF_3)_1$ rotation, $(PF_3)_3$ is starting to be blocked together with the 3 other ones while $\varphi 1(\varphi 11)$ (not shown) is perfectly 360° rotating. The dot line is indicating the average linear rotation slope of the $\varphi 2(\varphi 11)$ function and the dashed-dotted line what would be the ideal $\varphi 2(\varphi 1)$ curve for a perfect 360° rotation of $(PF_1)_1$ and counter rotation of $(PF_3)_2$. The black arrows are indicating in **a** and **b** the location of the P–F bond step by step twisted by 0.1°

able to rotate step by step the fluorine umbrella of a PF_3 molecule defining a nice umbrella collective rotational trajectory on its 9-dimensional ground state potential energy surface. Then, this first PF_3 was loaded by a second PF_3 to construct a train of two PF_3 molecule-gear. It results a surface molecular structure whose conformation and mechanics is determined on a ground state Born-Oppenheimer potential energy surface of dimension 18. We have demonstrated that among all the possible minimum energy mechanical motion of the PF_3 dimer described by a trajectory on this 18D surface, a transmission of rotation from the first to the second PF_3 is not the main minimum energy path. Step by step driven by a 0.1° twist of one of the P–F bonds of the PF_3 fluorine umbrella, the minimum energy trajectory results in an almost periodic deviation of the center of mass distance between the 2 umbrellas. No transfer of rotation and a slippery motion of the second PF_3 results. To stop this center of mass deviation, we have first added a third PF_3 in mechanical interactions with the second PF_3 of the train. This was not enough to maintain the two rotational axles of the PF_3 molecules vertical in the train. A fourth one was necessary to cancel this center of mass potential energy release resulting from the P–F bond twist input. With a train of 4 PF_3 molecules with the end two PF_3 in charge of maintaining the rotation axles almost vertical to the Cu(111) surface, a transmission of rotation occurs. The first PF_3 rotating 360° leads to a rotation of the second by almost 200°. The two ancillary PF_3 at the end of the train are not rotating.

For relatively flexible molecule (or for small molecules like PF_3), those results are pointing out the delicate balance between the tendency of each molecule along a molecule-gears train to release itself the initial potential energy input and the expected transmission of rotation along this train. This collective behavior (the transmission of rotation) is one of the many possible minimum energy paths on generally a large dimensional potential ground state energy surface. Predefining mathematically the collective rotational degrees of freedom inside a molecule-gears train without considering the experimental way of rotating the first molecule-gear of the train can really lead to a wrong prediction and therefore wrong molecular designs. In more cohesive nanogears like solid state material ones with a gear diameter down to 5 nm, the material cohesion ensures that all the atoms of a given gear are following the motion of the other atoms of the same gear. In that case, the ground state potential energy surface of a solid-state nanogear train can be reduced to one dimensional per nanogear. This is not the case for molecule-gear where a detail knowledge of the complete ground state potential energy surface is required to identify how the molecule-gear collective rotation angle build-up.

Acknowledgements We thank the European Union Horizon 2020 FET open project "Mechanics with Molecule(s)" (MEMO, grant 766864) for financial support.

References

1. Freeth, T., Bitsakis, Y., Moussas, X., Seiradakis, J.H., Tselikas, A., Hangou, H., Zafeiropoulou, M., Hadland, R., Bate, D., Ramsey, A., Allen, M., Crawley, A., Hockley, P., Malzbender, T., Gelb, D., Ambrisco, Edmunds, M.G.: Decoding the ancient Greek astronomical calculator known as the Antikythera mechanism. Nature **444**, 587–591 (2006)
2. Lanyi, J.K., Schobert, B.: Local-global conformational coupling in a heptahelical membrane protein: transport mechanism from crystal structures of the nine states in the bacteriorhodopsin photocycle. Biochemistry **43**, 3–8 (2004)
3. Merkle, R.C.: Two types of mechanical reversible logic. Nanotechnology **4**, 114–131 (1993)
4. Gimzewski, J.K., Joachim, C., Schlittler, R.R., Langlais, V., Tang, H., Johanson, J.: Rotation of a single molecule within a supramolecular bearing. Science **281**, 531–533 (1998)
5. Manzano, C., Soe, W.-H., Wong, H.S.J., Ample, F., Gourdon, A., Chandrasekhar, N., Joachim, C.: Step by step rotation of a molecule-gear mounted on an atomic-scale axis. Nat. Mater. **8**, 576–579 (2009)
6. Frantz, D.K., Linden, A., Baldridge, K.K., Siegel, J.S.: Molecular spur gears comprising triptycene rotators and bibenzimidazole-based stators. J. Am. Chem. Soc. **134**, 1528–1535 (2012)
7. Alvey, M.D., Yates Jr., J.T., Uram, K.J.: The direct observation of hindered rotation of a chemisorbed molecule: PF3 on Ni(111). J. Chem. Phys. **87**, 7221 (1987)
8. Kerkart, M., Woodrufft, D.P., Avilat, J., Asensioi, M.C., Fernindez-Garcia, M., Conesas, J.C.: Structure determination for PF3 absorption on Ni(111). J. Phys.: Condens Matter **4**, 6509–6522 (1992)

9. Kaji, H., Kakitani, K., Yoshimori, A.: Quantum rotational states of PF3 molecule pair on Ru(001) surface. Surf. Sci. **493**, 305–313 (2001)
10. Jones, R.G., Abrams, N.E., Jackson, G.J., Booth, N.A., Butterfield, M.T., Cowi, B.C.C., Woodruff, D.P., Crapper, M.D.: The structure of PF3 adsorbed on Cu(111). Surf. Sci. **414**, 396–408 (1998)
11. Zwanzig, R.: Rotational relaxation in a gear network. J. Chem. Phys. **87**, 4870 (1987)
12. Zhao, R., Zhao, Y.-L., Qi, F., Hermann, K.E., Zhang, R.-Q., Van Hove, M.A.: Interlocking mechanism between molecular gears attached to surfaces. ACS Nano **12**, 3020–3029 (2018)
13. Zhao, R., Qi, F., Zhao, Y.-L., Hermann, K.E., Zhang, R.-Q., Van Hove, M.A.: Interlocking molecular gear chains built on surface. J. Phys. Chem. Lett. **9**, 2611–2619 (2018)
14. Zhao, R., Qi, F., Zhang, R.-Q., Van Hove, M.A.: How does the flexibility pf molecules affect the performance of molecular rotors? J. Phys. Chem. C **122**, 25067–25074 (2018)
15. Nitschké, F., Ertl, G., Küppers, J.: Coordination chemistry of metal surfaces: chemisorption of PF3. J. Chem. Phys. **74**, 5911–5920 (1981)
16. Edith Chan, A.W., Hoffmann, R.: PF3, PF2 and PF on Ni(111): theoretical aspects of their chemisorption. J. Chem. Phys. **92**, 699–708 (1990)
17. Ample, F., Joachim, C.: a semi-empirical study of polyacene molecules adsorbed on a Cu(110) surface. Surf. Sci. **600**, 3243–3251 (2006)
18. Ample, F., Joachim, C.: The Chemisorption of poly-aromatic molecules on Si(100)H dangling bond. Surf. Sci. **602**, 1563–1571 (2008)
19. Bowen, J.P., Allinger, N.L.: Molecular mechanics: the art and science of parametrization. In: Lipkowitz, K.B., Boyd, D.B. (eds.) Reviews in Computational Chemistry, pp. 81–97. New York, NY, VHC (1991)
20. Soe, W.H., Srivastava, S., Joachim, C.: Train of single molecule-gears. J. Phys. Chem. Lett. **10**, 6462–6467 (2019)
21. Tierney, H.L., Murphy, C.J., Jewell, A.D., Baber, A.E., Iski, E.V., Khodaverdian, H.Y., McGuire, A.F., Klebanov, N., Sykes, E.C.H.: Experimental demonstration of a single-molecule electric motor. Nat. Nanotechnol. **6**, 625–629 (2011)
22. Gao, L., Liu, Q., Zhang, Y.Y., Jiang, N., Zhang, H.G., Cheng, Z.H., Qiu, W.F., Du, S.X., Liu, Y.Q., Hofer, W.A., Gao, H.J.: Constructing an array of anchored single-molecule rotors on gold surfaces. Phys. Rev. Lett. **101**, 197209 (2008)
23. Eisenhut, F., Meyer, J., Krüger, J., Ohmann, R., Cuniberti, G., Moresco, F.: Inducing the control rotation of single o-MeO-DMBI molecules anchored on Au(111). Surf. Sci. **678**, 177–182 (2018)
24. Echevaria, J., Moturet, S., Joachim, C.: One way rotation and efficiency of a single molecule motor driven by a shot noise. Nanoscale **6**, 2793–2799 (2014)
25. McAdams, M.J., Russel, B.R.: The vacuum ultraviolet spectra of fluorine chlorine substituted trivalent phosphorus. Chem. Phys. Lett. **18**, 402–404 (1973)
26. Zlatar, M., Allan, M., Fedor, J.: Excited states of $Pt(PF_3)_4$ and their role in focused electron beam nanofabrication. J. Phys. Chem. C **120**, 10667–10674 (2016)
27. Hertel, T., Wolf, M., Ertl, G.: UV photostimulated desorption of ammonia from Cu(111). J. Chem. Phys. **102**, 3414 (1995)
28. Stipe, B.C., Rezai, M.A., Ho, W.: Inducing and viewing the rotational motion of a single molecule. Science **279**, 1907–1909 (1998)
29. Bartels, L., Wolf, M., Klamroth, T., Saalfrank, P., Kühnle, A., Meyer, G., Rieder, K.H.: Atomic-scale Chemistry: desorption of ammonia from Cu(111) induced by tunneling electrons. Chem. Phys. Lett. **313**, 544–552 (1999)
30. Pascual, J.I., Lorente, N., Song, Z., Conrad, H., Rust, H.P.: Selectivity in vibrationally mediated single-molecule chemistry. Nature **423**, 525–528 (2003)
31. Chiaravalloti, F., Gross, L., Rieder, K.H., Stojkovic, S., Gourdon, A., Joachim, C., Moresco, F.: A rack and pinion device at the molecular scale. Nat. Mater. **6**, 30–33 (2007)
32. Bombis, C., Ample, F., Meilke, J., Mannsburger, M., Villagomez, C., Roth, Ch., Joachim, C., Grill, L.: Mechanical behaviour of nanocrystalline NaCl islands on Cu(111). Phys. Rev. Lett. **104**, 185502 (2010)

Mechanical Transmission of Rotation for Molecule Gears and Solid-State Gears

Huang-Hsiang Lin, Jonathan Heinze, Alexander Croy, Rafael Gutierrez, and Gianaurelio Cuniberti

Abstract The miniaturization of gears towards the nanoscale is a formidable task posing a variety of challenges to current fabrication technologies. In context, the understanding, via computer simulations, of the mechanisms mediating the transfer of rotational motion between nanoscale gears can be of great help to guide the experimental designs. Based on atomistic molecular dynamics simulations in combination with a nearly rigid-body approximation, we study the transmission of rotational motion between molecule gears and solid-state gears, respectively. For the molecule gears under continuous driving, we identify different regimes of rotational motion depending on the magnitude of the external torque. In contrast, the solid-state gears behave like ideal gears with nearly perfect transmission. Furthermore, we simulate the manipulation of the gears by a scanning-probe tip and we find that the mechanical transmission strongly depends on the center of mass distance between gears. A new regime of transmission is found for the solid-state gears.

Keywords Classical molecular dynamics · Quaternions · Molecule gears · Solid state gears

1 Introduction

The miniaturisation of gears down to the atomic scale, in order to transmit mechanical motion, represents a major challenge, with trains of molecule gears being the ultimate target [1]. To guide ongoing experiments, it is of crucial interest to shed light on the microscopic features that govern the mechanics of molecule gears. In addition to fabrication technologies based on a bottom-up approach [2], the production of solid-state gears using top-down methods (e.g. focused ion beam [3] or electron beam [4, 5]) may yield a viable path towards miniaturization.

H.-H. Lin · J. Heinze · A. Croy · R. Gutierrez · G. Cuniberti (✉)
Institute for Materials Science and Max Bergmann Center of Biomaterials, TU Dresden, 01069 Dresden, Germany
e-mail: g.cuniberti@tu-dresden.de

C. Joachim (ed.), *Building and Probing Small for Mechanics*, Advances in Atom and Single Molecule Machines, https://doi.org/10.1007/978-3-030-56777-4_11

To manipulate molecule gears in cutting-edge experiments, typically the tip of a scanning tunneling microscope (STM) is used [6–11]. In those experiments, the molecules are deposited on a suitable substrate and moved onto nearby adatoms, whenever possible. For example, the scheme in Fig. 1 illustrates the experimental setup reported in Ref. [11]. There, up to four hexa-*t*-butyl-hexaphenylbenzene (HB-HPB) molecules were mounted on copper atoms (in yellow) on top of a lead-(111) surface (in green). In this situation, the molecules interact only weakly with each-other and with the substrate via van-der-Waals interactions. As it was demonstrated, by pushing one of the gears, its rotation can be transmitted to the others.

It is interesting to compare the situation with molecule gears to the behavior of solid-state gears. For the latter, one expects perfect transmission of rotation for suitable distances between the gears. For such gears, with mesoscopic dimensions (few nm), the number of atoms is large enough to manifest classical behavior [12]. The main difference to the molecular case is the softness of the molecules, which influences the conditions for observing collective rotations [13]. For the molecule gears, several atomistic calculations based on density-functional theory (DFT) and classical molecular dynamics (MD) have been carried out to investigate the transmission properties between gears. For instance, DFT has been used to study a cyclopentadienyl ring with cyano groups mounted on a manganese atom above graphene [14], as well as PF_3 molecules on a Copper(111) surface [15] (see also the chapter by Srivastava et al. in this volume). In particular, the influence of the flexibility of gears and the slippage between gears has been investigated. MD simulations have been performed for carbon nanotube, fullerene-based and molecule gears [13, 16, 17]. But at the moment, a direct comparison of different gears in terms of the mechanical transmission between them is still missing. Therefore, a systematic analysis for different type of gears, separation distance and external driving is of particular interest.

In this chapter, we focus on the mechanical transmission of motion in both molecule-based gears and nanoscale solid-state gears, and investigate the conditions under which collective rotation is possible. In particular, we compare the results for molecule gears based on HB-HPB with those for solid-state gears (diamond) using the same model for the substrate and the same temperature. We use all-atom molecular dynamics (MD) simulations to investigate the problem, since it allows to reach relevant timescales of about 100ps to 1ns, even for solid-state gears. The simulations also yield trajectories longer than the surface relaxation time, which is on the order of few picoseconds [18]. The trajectories are analyzed using a nearly-rigid body approximation (NRBA), which enables a separation of the rigid-body motion and the internal deformation of the gears [13].

The chapter is organized as follows: in Sect. 2, we introduce the nearly rigid-body approximation (Sect. 2.1) and the details of the MD simulations (Sect. 2.3) for molecule and solid-state gears. In Sect. 3, we show and discuss the results for a train of molecule and solid-state gears driven by an external torque (Sect. 3.1) and under tip manipulation (Sect. 3.2). Finally, in Sect. 4, we summarize our results and provide an outlook.

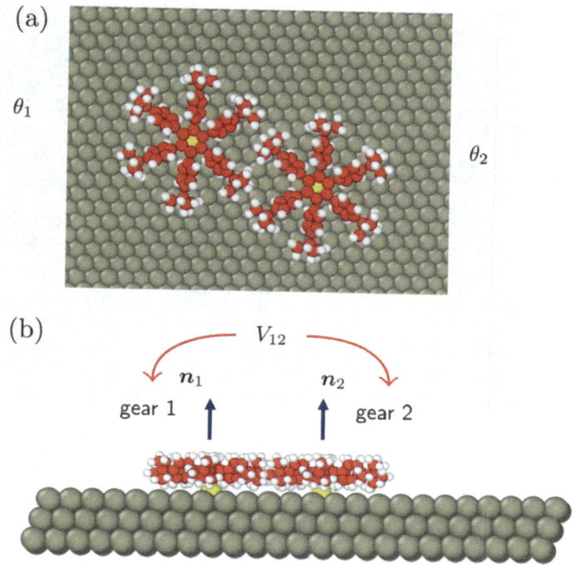

Fig. 1 Schematic plots of **a** top-view (with rotation angles θ_1 and θ_2) and **b** side-view (with rotation axes $\boldsymbol{n}_1$ and $\boldsymbol{n}_2$) for the setup of a train of molecule gears HB-HPB mounted above Cu (yellow) atoms on top of Pb (brown) (111) surface. The interaction V_{12} mediates the transmission of rotation between the gears

2 Modelling

In this section, we will introduce the NRBA to define the orientation vectors of individual gears and describe the setup of the MD simulations for a train of molecule gears and solid-state gears.

2.1 *Nearly Rigid-Body Approximation*

In order to define the orientation-vector of the molecule and solid-state gears, we use the NRBA as introduced in Ref. [13]. First, we consider a train of gears as shown in Fig. 1. For each gear, we define an appropriate reference structure represented by a set of Cartesian coordinates $\boldsymbol{r}^0_{\alpha k}$, where α denotes the gear index and k is running over all atoms in the αth gear. For instance, we can choose the initial frame $\boldsymbol{r}_{\alpha k}(t_0)$ of the MD simulation, which corresponds to the optimized geometry of the molecule. Secondly, the gear geometry at a later time t is given by $\boldsymbol{r}_{\alpha k}(t)$ (the structure on the right in Fig. 2). Next, we assume that the deformation of the gear during the simulation is sufficiently small, so that we can always find a unique set of rotational axes $\boldsymbol{n}_\alpha$ and angles θ_α. Those define the best-fitting rigid-body rotation transformation $\boldsymbol{R}^\alpha(\theta_\alpha, \boldsymbol{n}_\alpha)$ of the reference structure (the thinner structure on the right panel in Fig. 2) to the current structure. At the same time, the deviation from the best-fitting transformation is defined as deformation.

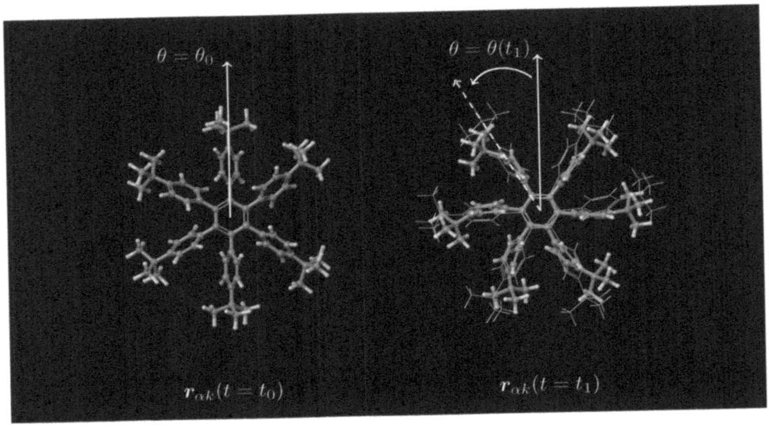

Fig. 2 Demonstration of the nearly rigid-body approximation. On the left, the initial molecular geometry at $t = t_0$ is set as a reference frame with orientation $\theta = \theta_0$; on the right, the deformed structure at $t = t_1$ is mapped to the best-fitting rigid-body transformation (with thinner sticks) from the reference frame with the uniquely-defined orientation $\theta = \theta(t_1)$

To be specific, the *deformation* for a given structure $\boldsymbol{r}_{\alpha k}$ in the NRBA is defined as:

$$\boldsymbol{\varepsilon}_{\alpha k} = \boldsymbol{R}^{\alpha}(\theta_\alpha, \boldsymbol{n}_\alpha)\boldsymbol{r}^0_{\alpha k} - \boldsymbol{r}_{\alpha k} \, . \tag{1}$$

The weighted sum of squared deformation for all atoms of the αth gear is given by:

$$\varepsilon^{\alpha}_{tot} = \sum_k w_{\alpha k} |\boldsymbol{\varepsilon}_{\alpha k}|^2 \, , \tag{2}$$

where the positive weight is taken to be $w_{\alpha k} = m_{\alpha k} / \sum_k m_{\alpha k}$, i.e. the ratio between the individual mass $m_{\alpha k}$ of the kth atom and the total mass $\sum_k m_{\alpha k}$ of gear α. This implies a larger contribution to the total deformation for heavier atoms.

Technically, the best-fitting transform $\boldsymbol{R}^{\alpha}(\theta_\alpha, \boldsymbol{n}_\alpha)$ can be found by using quaternions [19]. The latter are defined by four numbers, $\underline{q}^{\alpha} = (q_0^{\alpha}, q_1^{\alpha}, q_2^{\alpha}, q_3^{\alpha})$ (for simplicity, we suppress the α index in what follows). The rotation matrix is then related to the quaternions via:

$$\boldsymbol{R}^{\alpha}(q) = \begin{pmatrix} q_0^2 + q_1^2 - q_2^2 - q_3^2 & 2(-q_0q_3 + q_1q_2) & 2(q_0q_2 + q_1q_3) \\ 2(q_0q_3 + q_1q_2) & q_0^2 + q_2^2 - q_1^2 - q_3^2 & 2(-q_0q_1 + q_2q_3) \\ 2(-q_0q_2 + q_1q_3) & 2(q_0q_1 + q_2q_3) & q_0^2 + q_3^2 - q_1^2 - q_2^2 \end{pmatrix} . \tag{3}$$

Accordingly, the quaternion components are related to the rotation axes $\boldsymbol{n}_\alpha = (n_x^{\alpha}, n_y^{\alpha}, n_z^{\alpha})$ and the rotation angle θ_α by:

$$\begin{aligned} q_0 &= \cos(\theta_\alpha/2) \,, \\ q_1 &= \sin(\theta_\alpha/2) n_x^\alpha \,, \\ q_2 &= \sin(\theta_\alpha/2) n_y^\alpha \,, \\ q_3 &= \sin(\theta_\alpha/2) n_z^\alpha \,. \end{aligned} \tag{4}$$

In order to obtain the best rigid-body transform $\boldsymbol{R}^\alpha(q)$, we insert Eq. (3) into Eq. (1), and subsequently minimize Eq. (2) with respect to $\underline{q}^\alpha$, and subject to the normalization condition $\underline{q}^\alpha \cdot \underline{q}^\alpha = 1$. Equivalently, the quaternion $\underline{q}^\alpha$ can be obtained by minimizing the following function via the method of Lagrange multipliers:

$$f(\underline{q}^\alpha, \lambda^\alpha) = \varepsilon_{tot}^\alpha(\underline{q}^\alpha) - \lambda^\alpha(\underline{q}^\alpha \cdot \underline{q}^\alpha - 1) \,. \tag{5}$$

This results in the eigenvalue problem:

$$\boldsymbol{M}_\alpha \underline{q}^\alpha = \lambda^\alpha \underline{q}^\alpha \qquad \text{with} \qquad \underline{q}^\alpha \cdot \underline{q}^\alpha = 1 \,, \tag{6}$$

where the matrices $\boldsymbol{M}_\alpha$ can be shown to depend directly on $\boldsymbol{r}_{\alpha k} = (x_{\alpha k}, y_{\alpha k}, z_{\alpha k})$ and $\boldsymbol{r}^0_{\alpha k} = (x^0_{\alpha k}, y^0_{\alpha k}, z^0_{\alpha k})$ [19]. More explicitly,

$$\boldsymbol{M}_\alpha = \sum_k w_{\alpha k} \boldsymbol{M}_{\alpha k} \tag{7}$$

with the independent components of the symmetric matrix $\boldsymbol{M}_{\alpha k}$ given by:

$$\begin{aligned}
(\boldsymbol{M}_{\alpha k})_{11} &= x^2_{\alpha k} + y^2_{\alpha k} + z^2_{\alpha k} + (x^0_{\alpha k})^2 + (y^0_{\alpha k})^2 + (z^0_{\alpha k})^2 - 2(x_{\alpha k} x^0_{\alpha k} + y_{\alpha k} y^0_{\alpha k} + z_{\alpha k} z^0_{\alpha k}) \,, \\
(\boldsymbol{M}_{\alpha k})_{12} &= 2(y_{\alpha k} z^0_{\alpha k} - z_{\alpha k} y^0_{\alpha k}) \,, \\
(\boldsymbol{M}_{\alpha k})_{13} &= 2(-x_{\alpha k} z^0_{\alpha k} + z_{\alpha k} x^0_{\alpha k}) \,, \\
(\boldsymbol{M}_{\alpha k})_{14} &= 2(x_{\alpha k} y^0_{\alpha k} - y_{\alpha k} x^0_{\alpha k}) \,, \\
(\boldsymbol{M}_{\alpha k})_{22} &= x^2_{\alpha k} + y^2_{\alpha k} + z^2_{\alpha k} + (x^0_{\alpha k})^2 + (y^0_{\alpha k})^2 + (z^0_{\alpha k})^2 - 2(x_{\alpha k} x^0_{\alpha k} - y_{\alpha k} y^0_{\alpha k} - z_{\alpha k} z^0_{\alpha k}) \,, \\
(\boldsymbol{M}_{\alpha k})_{23} &= -2(x_{\alpha k} y^0_{\alpha k} + y_{\alpha k} x^0_{\alpha k}) \,, \\
(\boldsymbol{M}_{\alpha k})_{24} &= -2(x_{\alpha k} z^0_{\alpha k} + z_{\alpha k} x^0_{\alpha k}) \,, \\
(\boldsymbol{M}_{\alpha k})_{33} &= x^2_{\alpha k} + y^2_{\alpha k} + z^2_{\alpha k} + (x^0_{\alpha k})^2 + (y^0_{\alpha k})^2 + (z^0_{\alpha k})^2 + 2(x_{\alpha k} x^0_{\alpha k} - y_{\alpha k} y^0_{\alpha k} + z_{\alpha k} z^0_{\alpha k}) \,, \\
(\boldsymbol{M}_{\alpha k})_{34} &= -2(y_{\alpha k} z^0_{\alpha k} + z_{\alpha k} y^0_{\alpha k}) \,, \\
(\boldsymbol{M}_{\alpha k})_{44} &= x^2_{\alpha k} + y^2_{\alpha k} + z^2_{\alpha k} + (x^0_{\alpha k})^2 + (y^0_{\alpha k})^2 + (z^0_{\alpha k})^2 + 2(x_{\alpha k} x^0_{\alpha k} + y_{\alpha k} y^0_{\alpha k} - z_{\alpha k} z^0_{\alpha k}) \,.
\end{aligned}$$

Finally, the quaternion, which is minimizing the deformation, is given by the eigenvector of Eq. (6) with the smallest eigenvalue. The degree of deformation is directly given by the corresponding eigenvalue.

In summary, the NRBA allows us to extract the rigid-body transformation and the deformation connecting two arbitrary configurations as illustrated in Fig. 2.

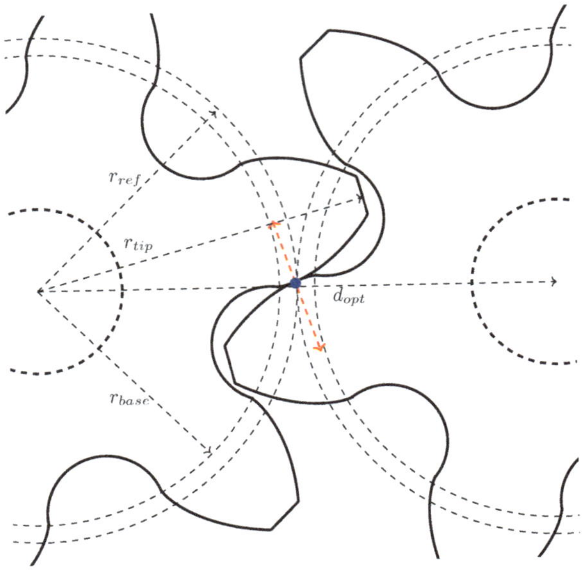

Fig. 3 Demonstration of two interlocked solid-state gears. The single-point contact (pressure point) is marked by a blue dot in the center. As the gears rotate, the pressure point moves along the line of action (red line) which will stay tangent to the base circles with radius r_{base} at all times for optimal transfer of angular momentum. Since we use multiple gears with equal dimensions, the optimal distance is given by the center of mass distance between two gears $d_{opt} = 2 \cdot r_{ref}$. The gear size r_{tip} is defined as the distance between the center of mass and the gear tip

2.2 *Solid-State Gear Meshing*

In order to create the solid state gears, we follow a general algorithm for creating involute spur gears [20]. We then use the Open Visualization Tool (OVITO) [21] to cut the gears from a bulk diamond crystal. To be specific, the typical structure to define gears is shown in Fig. 3. The figure shows the contact between two ideal involute gears; they touch each other at a single point (pressure point) marked by the blue dot in the center. As the gears rotate, the pressure point moves along the line of action (red line) which will stay tangent to the base circles with radius r_{base} at all times for optimal transfer of angular momentum. Since we use multiple gears with equal dimensions, the optimal distance is given by the center of mass distance between two gears $d_{opt} = 2 \cdot r_{ref}$. For later discussion, we define the gear size by r_{tip}, which is the distance between the center of mass and a gear tip.

2.3 Molecular Dynamics

Since we focus on the rotational transmission between gears, we model our problem based on the following general assumptions for both solid-state gears and molecule gears:

1. The gears are weakly coupled to the surface. Therefore, charge transfer effects between the two systems can be neglected and the specific atomic position on the surface is not relevant.
2. The gears are well anchored, which can be mimicked by fixing the centers of mass.
3. The gears are initially in thermal equilibrium with the surface.

To be specific, we use the Large-scale Atomic/Molecular Massively Parallel Simulator (LAMMPS) [22] for implementing the MD simulations. Based on the first assumption above, we use an artificial van-der-Waals surface, which interacts with the molecules via a 9-3 Lennard-Jones-Potential:

$$V_{LJ}(r) = \varepsilon \left[\frac{2}{15} \left(\frac{\sigma}{r} \right)^9 - \left(\frac{\sigma}{r} \right)^3 \right] . \tag{8}$$

Here, we use $\varepsilon = 0.1$ eV, $\sigma = 5$ Å and an initial distance of 5 Å between the surface and the gears. Then, according to the third assumption, we use a Langevin thermostat with the relaxation time $\tau = 1$ ps [18]. For the interatomic potentials describing the molecular gear and diamond-based solid-state gear, respectively, we use the adaptive intermolecular reactive empirical bond order (AIREBO) potential [23], which is suitable for simulations of hydrocarbons. In all simulations, we set the temperature to $T = 10$ K, mimicking the typical conditions of a low-temperature STM experiment. Before running the simulation, the total system undergoes a geometry optimization by the conjugate gradient method built into LAMMPS.

3 Results and Discussion

In this section, we treat two different methods to rotate gears by either applying an external torque to one of the gears in a train, or by moving a gear via a realistic tip manipulation. We compare the locking coefficients and transmission coefficients, which provide a measure for the transmission quality and which will be defined below, for both a train of molecule gears and solid-state gears (with radius $r = 3$ nm and 4932 atoms). In the following, we discuss the two methods in more detail.

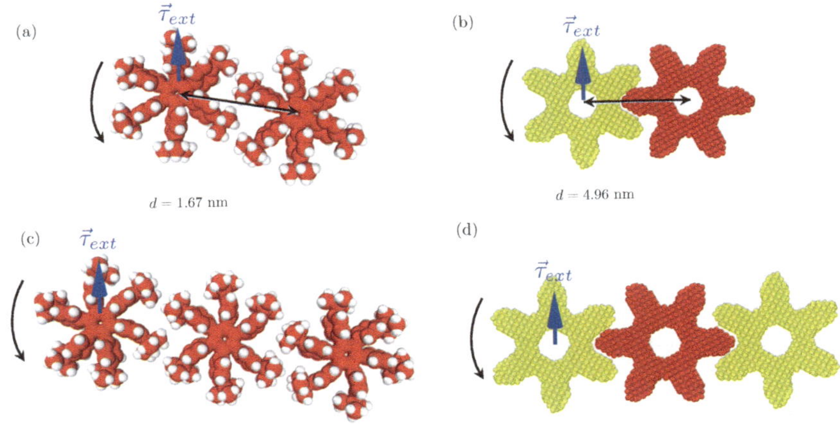

Fig. 4 Schematic plots for driving gears with external torque $\boldsymbol{\tau}_{ext}$ (orientations are indicated by blue arrow) applied to the first gear on the left. **a** Two and **c** three molecule gears with center of mass distance $d = 1.67$ nm; **b** two and **d** three diamond-based solid-state gears with distance $d = 5$ nm

3.1 Rotation Driven by an External Torque

First, we consider the scenario shown in Fig. 4, where we apply an external torque τ_{ext} (with the blue arrow pointing to $+z$ direction) to the first gear on the left, which would in turn drive the neighboring gear counterclockwise. Moreover, depending on molecule gear or solid-state gear, one has to decide the center of mass distance d between gears. For HB-HPB gears, the maximal distance with interlocking is shown to be between 1.67 and 1.74 nm. Here we use $d = 1.67$ nm. For solid-state gears, since we use the standard spur gear, the optimal distance for the 3 nm gear is 4.5 nm. However, in reality, the atoms cannot be arbitrarily close due to the strong repulsion, therefore we adjust the distance to 5 nm.

Once the distances are specified, we run the MD simulations to study the response of the gears to the external torque. The results are shown in Fig. 5. In order to characterize the transmission of motion across gears, we define the *locking coefficient* as follows:

$$L_j = \frac{\langle \omega_j \rangle}{\omega_R} \,. \tag{9}$$

Here, $\langle \omega_j \rangle$ denotes the average angular velocity of the jth gear and ω_R represents the terminal angular velocity of perfectly interlocked rigid-body gears in a train with N gears. The terminal angular velocity is given by:

$$\omega_R = \frac{\tau_{ext}}{N\gamma} = \frac{\tau_{ext}\tau}{NI} \,, \tag{10}$$

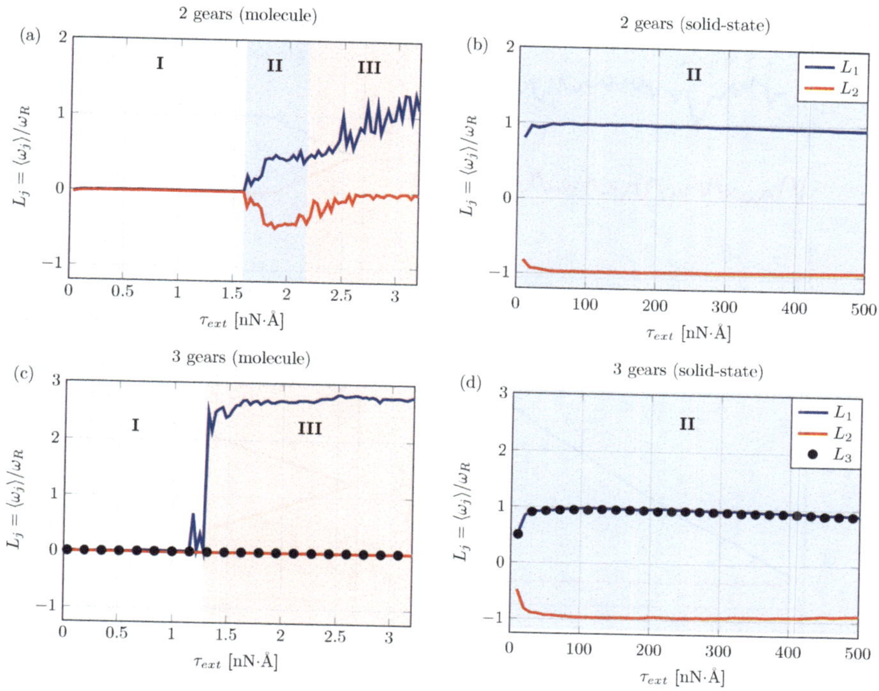

Fig. 5 The locking coefficient $L_j = \langle\omega_j\rangle/\omega_R$ for **a** two and **c** three molecule gears with center of mass distance $d = 1.67$ nm; and **b** two and **d** three solid-state gears, where $\langle\omega_j\rangle$ denotes the average angular velocity of jth gear and ω_R represents the terminal angular velocity of rigid-body gear. The region I (white), II (blue) and III (red) indicate the underdriving ($|L_i| \approx |L_j| \approx 0$), driving ($0 < |L_i| \approx |L_j| \leqslant 1$) and overdriving ($0 \approx |L_i| \approx |L_j| \ll L_1$) regimes, respectively

where γ is the damping coefficient given by $\gamma = I/\tau$ with $N = 2$ or 3, the moment of inertia is $I = 2.13 \times 10^{-41}$ kg m^2 [13] and the relaxation time is set to $\tau = 1$ ps [18].

The locking coefficient provides a measure for the ability to transfer rotations between gears. For perfectly interlocked gears, the coefficient is equal to ± 1. In Fig. 5a, we show results for a MD simulation of two molecule gears (as in in Fig. 4a) within 100 ps. We compute the dependence of the locking coefficients to the external torque, which is ramped up from 0 to 3.2 nNÅ. One can see that there are three different regions of motion for gears (highlighted in white, blue and red) [13].

Region I:

$$|L_1| \approx |L_2| \approx 0\,. \tag{11}$$

For $0 < \tau_{ext} < 1.6$ nNÅ, both locking coefficients are vanishing, meaning that the gears barely rotate. The corresponding trajectories θ_1 and θ_2 with $\tau_{ext} = 1.12$ nNÅ

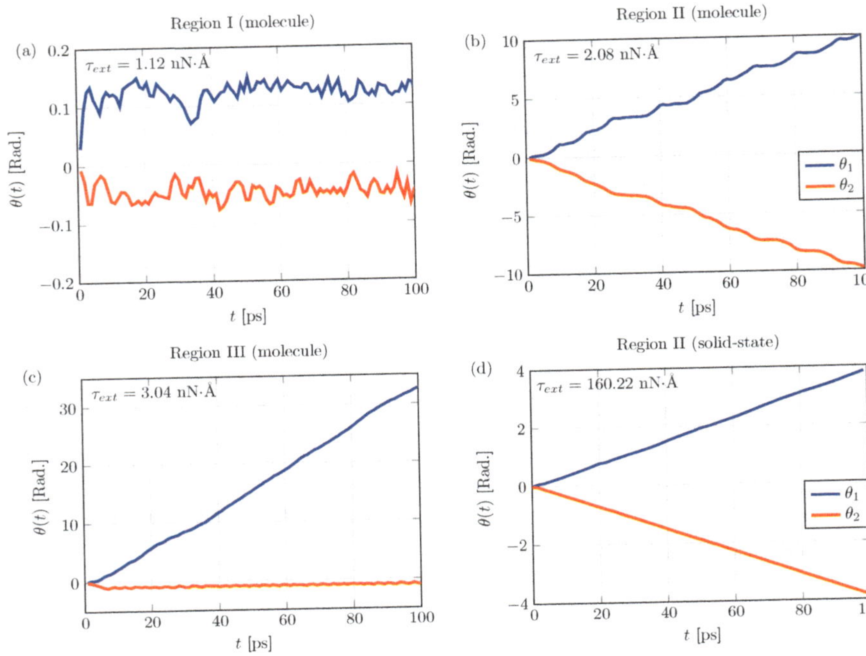

Fig. 6 The trajectories within 100 ps for **a** region I (underdriving) with external torque $\tau_{ext} = 1.12$ nN Å, **b** II (driving) with $\tau_{ext} = 2.08$ nN Å and **c** III (overdriving) with $\tau_{ext} = 3.04$ nN Å of two molecule gears. **d** The trajectories for two solid-state gears with $\tau_{ext} = 160.22$ nN Å (only region II exists)

are shown in Fig. 6a, which correspond to typical Brownian rotations at finite temperature. In this case, we say that the two gears are in the *underdriving phase*.
Region II:

$$0 < | L_1 | \approx | L_2 | \leqslant 1 \ . \tag{12}$$

For $1.6 < \tau_{ext} < 2.2$ nN Å, the locking coefficients are approximately opposite to each other, which means that the gears are interlocked. The corresponding trajectories θ_1 and θ_2 with $\tau_{ext} = 2.08$ nN Å are shown in Fig. 6b, which represent the pattern of step-by-step collective rotation. We denote this case as the *driving phase*. One can see that, for this type of molecule gear, the locking coefficient $| L_j |$ is around 0.5 in the driving phase, which indicates that the gears are rather soft and some energy is dissipated into the internal degrees of freedom in the form of deformations.
Region III:

$$0 \approx | L_2 | \ll L_1 \leqslant N \ . \tag{13}$$

For $\tau_{ext} > 2.2$ nN Å, the locking coefficient L_1 is much larger than all the others, so that only the first gear rotates. The corresponding trajectories θ_1 and θ_2 with $\tau_{ext} = 3.04$ nN Å are shown in Fig. 6c, and represent the pattern of a single gear rotation. In this case, we say that the two gears are in the *overdriving phase*. Note that in the overdriving phase L_1 may be larger than one but it has to be bounded by the terminal velocity of free single gear rotation, namely:

$$L_1 = \frac{\omega_1}{\omega_R} \leqslant \frac{\tau_{ext}/\gamma}{\tau_{ext}/(N\gamma)} = N \, . \tag{14}$$

We can do a similar analysis for three molecule gears as shown in Fig. 4c. In this case, one immediately sees that there are only two regions (I and III): underdriving phase for $0 < \tau_{ext} < 1.2$ nN Å and overdriving phase for $\tau_{ext} > 1.2$ nN Å. This implies that no matter how hard the first gear is driven, it is not possible to have a collective rotation due to the softness of the molecules.

For comparison, we move on to the solid-state gears as shown in Fig. 4b, d. Since the gear is based on diamond, which is the hardest material, one expects a rather stiff or rigid behavior. As one can see in Fig. 5b, d, only region-II behavior appears, so that gears are always in the driving phase. On the other hand, the locking coefficients L_j are close to one, indicating a rigid-body interlocked rotation. This is also consistent with the trajectories obtained for $\tau_{ext} = 160.22$ nN Å in Fig. 6d.

Since the solid-state gears are rather rigid, one can see that the collective rotation happens even in the ten gears case as shown in Fig. 7. Besides the collective rotation, there is a delay time between the gear response. For instance, the total propagation time from the first gear to the last one lies approximately between 15 ps $\leq \Delta t \leq 50$ ps.

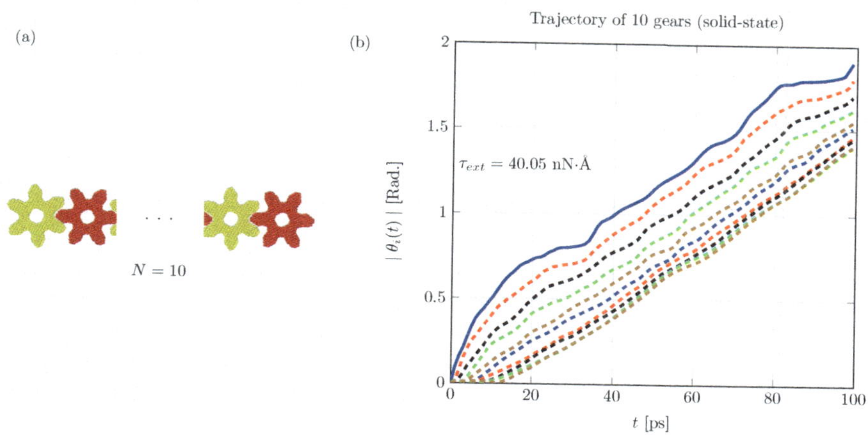

Fig. 7 **a** Schematic configuration of a gear train consisting of $N = 10$ gears driven by an external torque $\tau_{ext} = 40.05$ nN Å. **b** The trajectories (solid (dashed) lines represent counterclockwise (clockwise)) in *driving phase*

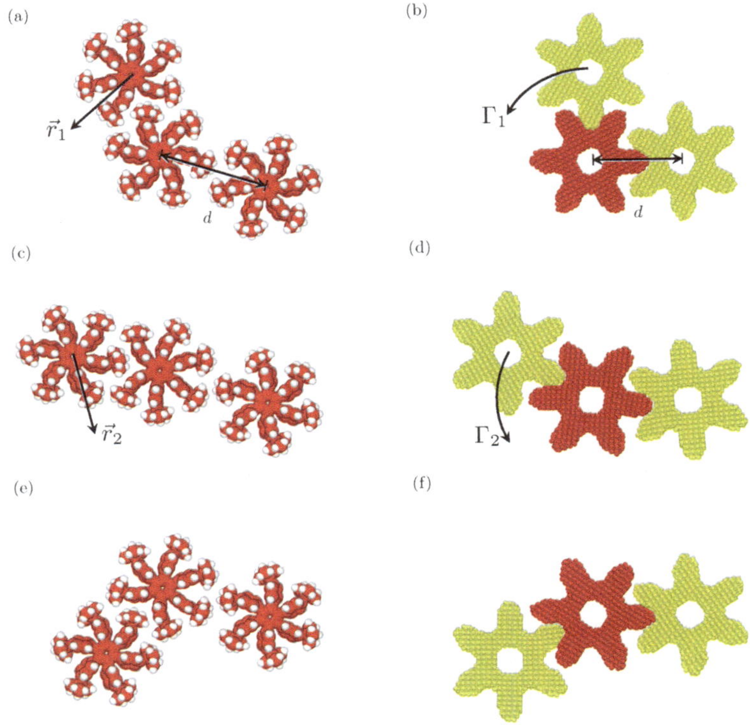

Fig. 8 Schematic plots for manipulating the first gear on the left for both molecule gears and solid-state gears with constant speed $v = 3$ m/s. **a**, **c** and **e** show three different conformations during the *linear* two-step manipulation (with manipulation vectors $\mathbf{r_1}$ and $\mathbf{r_2}$ and center of mass distance $d = 1.80$ nm); **b**, **d** and **f** indicate three conformations during the *circular* two-step manipulation (along two circular segments Γ_1 and Γ_2 and with distance $d = 4.725$ nm)

3.2 Rotation via Tip Manipulation

In a typical STM experiment, the torque cannot be applied to the gears directly. Instead, a handle gear is introduced as a mediator between STM tip and target gear [11]. To mimic this situation, we manipulate the handle gear along two specific trajectories as shown in Fig. 8, which will in turn drive the second gear counterclockwise.

For the molecule gears, we use a *linear* two-step manipulation along two vectors $\boldsymbol{r}_1$ and $\boldsymbol{r}_2$ with a waiting time of 30 ps between the two steps. For the solid-state gears we use a *circular* two step manipulation path along the trajectories Γ_1 and Γ_2 due to large deformations occurring when using *linear* paths.

Both manipulations are done with a fixed distance between the first and the second gear (before and after moving along the respective trajectory): for molecule gears we take $d = 1.5$ nm and for solid state gears $d = 4.725$ nm. The distance between

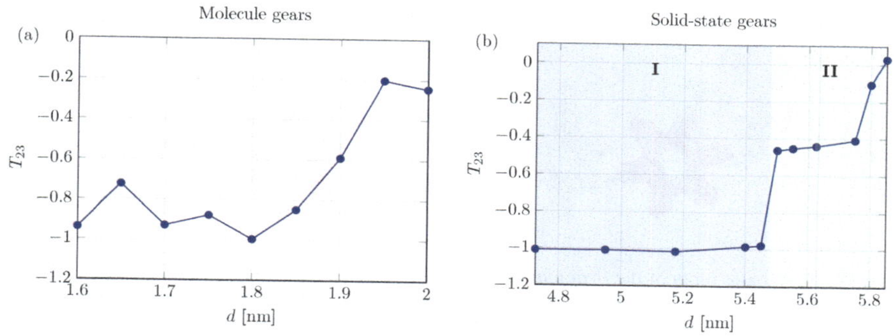

Fig. 9 The transmission coefficient $T_{23} = \theta_3/\theta_2$ between the second and third gear for different center of mass distances d with constant speed for **a** molecule gears during the *linear* two-steps manipulation and **b** solid-state gears during the *circular* two-steps manipulation, where θ_2 and θ_3 are the total angular displacements for the second (middle) and third (right) gear. For solid-state gears we can distinguish between region I (blue) and II (green) indicating the driving ($| T_{23} | \approx 1$) and dragging ($0 \leq | T_{23} | \leq 0.5$) regimes, respectively

the second and the third gear is varied. From the perspective of the second gear, the first gear moves 60° per step, amounting to a total of 120°.

The results of the MD simulations are shown in Fig. 9. For molecule gears, the movement took 1 ns (excluding relaxation time) and covered a total distance of 3 nm. For the solid-state gears it took 3.3 ns and covered an arc length of 5 nm. In order to compare the results, we define the transmission coefficient as follows:

$$T_{23} = \frac{\theta_3}{\theta_2} , \tag{15}$$

where θ_2 and θ_3 are the total angular displacements of the second and third gears, respectively. This quantity describes how well both gears are interlocked, even though the angular velocity cannot be obtained directly. For instance, when the handle gear moves two circular-steps (120° with respect to the second gear), the third gear will also rotate two steps in opposite direction. One can use the NBRA to estimate the corresponding angle and to compute the transmission coefficient, which gives a value in the range $-1 \leq T_{23} \leq 0$.

Figure 9a shows the average transmission coefficient of 20 simulations for different center of mass distances d during the *linear* two-step manipulation. For $d \leq 1.9$ nm, we have similar interlocked rotations with $0.6 \leq | T_{23} | \leqslant 1$. The optimal collective rotation can be found at $d = 1.8$ nm with $| T_{23} | \approx 1$. For larger distances, we see a quick decay for transmission to around $| T_{23} | \approx 0.2$.

In Fig. 9b, the average transmission coefficient of solid-state gears for different center of mass distance d is shown. Here we can distinguish two different regions (highlighted in blue and green):

Region I:

$$| T_{23} | \approx 1 \tag{16}$$

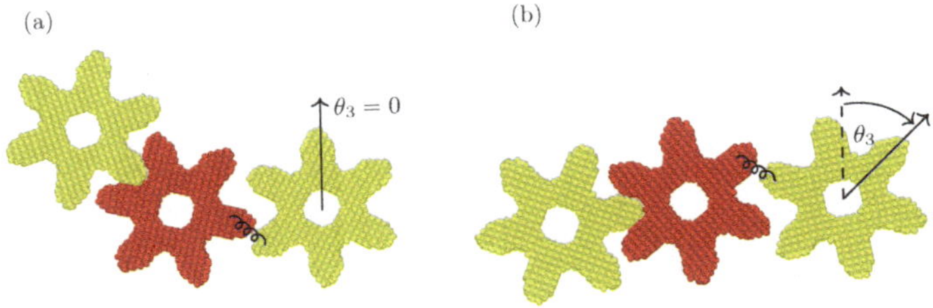

Fig. 10 Schematic plots of the *dragging phase*. **a** and **b** show two different conformations during the *circular* two-step manipulation for center of mass distances $d \geq 5.50$ (Region II). **a** While the first and the second gear rotates as shown in Fig. 8, the third gear does not interlock; it is still in its starting conformation ($\theta_3 = 0$). As the distances between the teeth become smaller, the middle gear starts to drag the right gear (represented by the string. **b** As the motion continues, the distance between the teeth becomes too big to sustain the drag and they lose contact. θ_3 is the total angle covered by the motion of the right gear

For $d \leq 5.45$ nm, the gears are in *driving phase*, with almost perfectly interlocked rotation.
Region II:

$$0 \leq | T_{23} | \leq 0.5 \tag{17}$$

For larger distances, we see a plateau between $5.50 \leq d \leq 5.75$ followed by another sudden decrease in $| T_{23} |$. We call this region *dragging phase*, the gears barely touch at their respective tips, and the rotation of the third gear is mainly driven by the attractive force between the atoms of the tips, as shown in Fig. 10a, b.

In Fig. 10a, while the first two gears undergo the first step in Fig. 8, the third does not interlock and stays in its starting conformation ($\theta_3 = 0$). When the distance between the teeth becomes sufficiently small, the middle gear starts to drag (highlighted by a spring) the right one (see Fig. 8). This motion will then continue until the distance between the teeth becomes too large to sustain the drag. The angle covered by the right gear due to the drag is θ_3. In the end, this results in a decrease of $| T_{23} |$. For distances $d \geq 5.85$ nm, the gears are too far apart for any collective rotation to occur.

While there are two regions for the solid-state gears, the molecule gears do not show such a distinct pattern for changes in the center of mass distance. In comparison, their transmission coefficient is subject to much higher fluctuations for every change in the center of mass distance, whereas for solid-state gears significant changes only occur in the transition between the regions.

4 Conclusions and Outlook

In this chapter, we have carried out, using atomistic Molecular Dynamics simulations, a comparative study of the transmission of rotational motion across molecule gears as well as solid-state gears. Our approach is based on a nearly rigid-body approximation, which helps to define the orientation vector of the gear for weakly deformed structures. We discussed two possible strategies to induce a rotational motion of the leading gear: either by (i) applying an external torque or (ii) by mimicking the manipulation with an STM tip. In the first case (i), the introduction of *locking coefficients* allowed to clearly identify different rotational regimes, denoted as underdriving, driving and overdriving phases. It turns out that for molecule gears, collective rotations are possible only up to two gears, a result related to the dissipation of energy into internal molecular degrees of freedom. In contrast, the solid state gears largely preserve the rigid-body like character, so that collective rotations become possible up to ten gears. Concerning case (ii), we found out that transmission of rotational motion across more than two molecule gears is feasible and it critically depends on the center-of-mass distance between the gears. For for solid-state gears, driving and dragging phases were identified, in dependence of the center of mass distance between the gears.

Future computational studies will need to include the influence of a real substrate in order to address additional energy dissipation channels, which may hamper the efficient transmission of motion across a gear train. This problem is closely connected with the more general problem of the theoretical description of friction processes at the nanoscale [24]. Elucidating the interaction mechanisms between nanoscale gears and various substrates builds an integral part of the understanding of the working principles of nanoscale machinery [25]. Looking beyond the classical regime, the possibility of studying quantum effects in mechanical gears provides a fascinating perspective [26, 27].

Acknowledgements We would like to thank C. Joachim, A. Kutscher, A. Mendez, A. Raptakis, T. Kühne, D. Bodesheim, S. Kampmann, R. Biele, D. Ryndyk, A. Dianat, and F. Moresco for very useful discussions and suggestions. This work has been supported by the International Max Planck Research School (IMPRS) for "Many-Particle Systems in Structured Environments" and also by the European Union Horizon 2020 FET Open project "Mechanics with Molecules" (MEMO, grant nr. 766864).

References

1. Soong, R.K., Bachand, G.D., Neves, H.P., Olkhovets, A.G., Craighead, H.G., Montemagno, C.D.: Science (80-.) **290**(5496), 1555 (2000). https://doi.org/10.1126/science.290.5496.1555
2. Gisbert, Y., Abid, S., Bertrand, G., Saffon-Merceron, N., Kammerer, C., Rapenne, G.: Chem. Commun. **55**(97), 14689 (2019). https://doi.org/10.1039/C9CC08384G
3. Ju Yun, Y., Seong Ah, C., Kim, S., Soo Yun, W., Chon Park, B., Han Ha, D.: Nanotechnology **18**(50), 505304 (2007). https://doi.org/10.1088/0957-4484/18/50/505304
4. Deng, J., Troadec, C., Ample, F., Joachim, C.: Nanotechnology **22**(27), 275307 (2011). https://doi.org/10.1088/0957-4484/22/27/275307
5. Yang, J., Deng, J., Troadec, C., Ondarçuhu, T., Joachim, C.: Nanotechnology **25**(46), 465305 (2014). https://doi.org/10.1088/0957-4484/25/46/465305
6. Gimzewski, J.K., Joachim, C., Schlittler, R.R., Langlais, V., Tang, H., Johannsen, I.: Science (80-.) **281**(5376), 531 (1998). https://doi.org/10.1126/science.281.5376.531

7. Tierney, H.L., Murphy, C.J., Jewell, A.D., Baber, A.E., Iski, E.V., Khodaverdian, H.Y., McGuire, A.F., Klebanov, N., Sykes, E.C.H.: Nat. Nanotechnol. **6**(10), 625 (2011). https://doi.org/10.1038/nnano.2011.142
8. Ohmann, R., Meyer, J., Nickel, A., Echeverria, J., Grisolia, M., Joachim, C., Moresco, F., Cuniberti, G.: ACS Nano. **9**(8), 8394 (2015). https://doi.org/10.1021/acsnano.5b03131
9. Perera, U.G.E., Ample, F., Kersell, H., Zhang, Y., Vives, G., Echeverria, J., Grisolia, M., Rapenne, G., Joachim, C., Hla, S.W.: Nat. Nanotechnol. **8**(1), 46 (2013). https://doi.org/10.1038/nnano.2012.218
10. Gao, L., Liu, Q., Zhang, Y.Y., Jiang, N., Zhang, H.G., Cheng, Z.H., Qiu, W.F., Du, S.X., Liu, Y.Q., Hofer, W.A., Gao, H.J.: Phys. Rev. Lett. **101**(19), 197209 (2008). https://doi.org/10.1103/PhysRevLett.101.197209
11. Soe, W.H., Srivastava, S., Joachim, C.: J. Phys. Chem. Lett. 6462–6467 (2019). https://doi.org/10.1021/acs.jpclett.9b02259
12. Bombis, C., Ample, F., Mielke, J., Mannsberger, M., Villagómez, C.J., Roth, C., Joachim, C., Grill, L.: Phys. Rev. Lett. **104**(18), 185502 (2010). https://doi.org/10.1103/PhysRevLett.104.185502
13. Lin, H.H., Croy, A., Gutierrez, R., Joachim, C., Cuniberti, G.: Phys. Rev. Appl. **13**(3), 034204 (2020). https://doi.org/10.1103/PhysRevApplied.13.034024
14. Zhao, R., Qi, F., Zhao, Y.L., Hermann, K.E., Zhang, R.Q., Van Hove, M.A.: J. Phys. Chem. Lett. **9**(10), 2611 (2018). https://doi.org/10.1021/acs.jpclett.8b00676
15. Zhao, R., Zhao, Y.L., Qi, F., Hermann, K.E., Zhang, R.Q., Van Hove, M.A.: ACS Nano. **12**(3), 3020 (2018). https://doi.org/10.1021/acsnano.8b00784
16. Han, J., Globus, A., Jaffe, R., Deardorff, G.: Nanotechnology **8**(3), 95 (1997). https://doi.org/10.1088/0957-4484/8/3/001
17. Robertson, D., Dunlap, B., Brenner, D., Mintmire, J., White, C.: MRS Proc. **349**(Md), 283 (1994). https://doi.org/10.1557/PROC-349-283
18. Persson, B., Nitzan, A.: Surf. Sci. **367**(3), 261 (1996). https://doi.org/10.1016/S0039-6028(96)00814-X
19. Kneller, G.R., Yvette, F.G.S.: Mol. Simul. **7**, 113 (1991)
20. Norton, R.L.: Machine Design: An Integrated Approach, 4th edn. Prentice Hall (2010)
21. Stukowski, A.: Model. Simul. Mater. Sci. Eng. **18**(1), 015012 (2010). https://doi.org/10.1088/0965-0393/18/1/015012
22. Plimpton, S.: J. Comput. Phys. **117**(1), 1 (1995). https://doi.org/10.1006/jcph.1995.1039
23. Stuart, S.J., Tutein, A.B., Harrison, J.A.: J. Chem. Phys. **112**(14), 6472 (2000). https://doi.org/10.1063/1.481208
24. Panizon, E., Santoro, G.E., Tosatti, E., Riva, G., Manini, N.: Phys. Rev. B **97**(10), 1 (2018). https://doi.org/10.1103/PhysRevB.97.104104
25. Roegel, D., Ann, I.E.E.E.: Hist. Comput. **37**(4), 90 (2015). https://doi.org/10.1109/MAHC.2015.79
26. MacKinnon, A.: Nanotechnology **13**(5), 678 (2002). https://doi.org/10.1088/0957-4484/13/5/328
27. Liu, Z., Leong, J., Nimmrichter, S., Scarani, V.: Phys. Rev. E **99**(4), 042202 (2019). https://doi.org/10.1103/PhysRevE.99.042202

Rotations of Adsorbed Molecules Induced by Tunneling Electrons

N. Lorente and C. Joachim

Abstract The advent of molecular machines is placing special attention on the rotation of a single molecule. Arguably, rotations are central for the diverse movements of a machine during its working dynamics. Here, we consider molecules that are constrained by the surface and effect rotations over an axle. They are then planar molecule-rotors. The excitation of a rotation by tunneling electrons, induced for example by the tip of a scanning tunneling microscope (STM), can be quite efficient as shown by a large body of experimental evidence. These rotations are indeed excited by single tunneling electron effect and are limited by the damping of the rotation by the different degrees of freedom of the substrate. When the molecule inertia momentum is small, quantum effects become apparent and rotation becomes very efficient by the large transfer of angular momentum produced by the transferred electrons through the STM junction. For larger molecules the classical limit is rapidly attained. After several considerations on the electron-induced rotation of a single molecule, we show how to evaluate the rotational dynamics during the tunneling of electrons through a molecule-rotor.

Keywords Molecule-Motor · LT-UHV STM · Inelastic tunneling effect

N. Lorente (✉)
Centro de Física de Materiales CFM/MPC (CSIC-UPV/EHU), Paseo Manuel de Lardizabal 5, 20018 Donostia-San Sebastián, Spain
e-mail: nicolas.lorente@ehu.eus

Donostia International Physics Center (DIPC), Paseo Manuel de Lardizabal 4, 20018 Donostia-San Sebastián, Spain

C. Joachim
GNS, Centre D'Elaboration Des Matériaux et D'Etudes Structurales (CEMES, University of Toulouse, CNRS), 29 Rue J. Marvig, 31055 Toulouse Cedex, France

International Center for Materials Nanoarchitectonics (WPI-MANA), National Institute for Material Sciences (NIMS), 1-1 Namiki, Tsukuba, Ibaraki 305-0044, Japan

C. Joachim (ed.), *Building and Probing Small for Mechanics*, Advances in Atom and Single Molecule Machines, https://doi.org/10.1007/978-3-030-56777-4_12

1 Introduction

Controlling a directed motion of a single molecule is a challenging aspect of chemical physics research with strong impact in nanoscale applications [1]. Among these, the possibility to design molecular motors is enticing as shown by the 2016 Noble prize of chemistry [2]. These molecules can be powered by optical [3–6], chemical [7, 8] together with thermal [9] processes. Rotation in these molecules is a key figure for the design of molecular motors [10–13]. On a surface, the design of a single molecule-motor (i.e. a single molecule able alone to deliver some motive power) is also actively explored. The rotation can be triggered one molecule at a time by thermal or electrical means [14–17]. It is then of great interest to understand intra molecular conformation changes resulting in their rotor rotation and how they can be achieved in a controlled way.

For a single molecule on a surface, a thermal excitation cannot power the one-way rotation of its rotor (essential for such a molecule-motor to function properly) due to the micro-reversibility principle [18]. More explicitly a single molecule in contact only with a single thermostat is not supposed to break the second principle of thermodynamics. Electrical powering using the tip apex of a scanning tunneling microscope (STM) [15–17] (and in the future using crystalline metallic nano-electrodes [19]), allows to attain a great precision both by the amount of energy delivered to the rotating part [20] and by the very fine location of the energy delivery on the molecule [21]. Tunneling electrons are particularly efficient in locally exciting a molecular rotation [15, 17, 22, 23]. The lateral evanescent character of tunneling electrons leads to a great precision in the localization of the STM tunneling current, and hence on the ability to transfer electrons with a picometer spatial precision from the STM tip to the surface through the very molecular rotor under the tip apex.

There are different ways of driving a single molecule rotation by electrical means. Here, we can broadly distinguish its driving by (i) an electric field or by (ii) an electrical current. For (i), the electric field distribution at the end of the STM tip apex is polarizing in a non-uniform way the molecular electronic cloud and the molecule changes its conformation or position on the surface to lower its potential energy in this peculiar electrostatic environment. This kind of energy relaxation can drive a molecule with great control when the STM tip apex is displaced step by step on the surface around the molecule-rotor [20, 24]. For (ii), the elementary electron transfer processes at the origin of the tunneling current passing through the molecule-rotor are providing energy to the nuclear degrees of freedom via inelastic interactions with some of the transferred electrons [25]. Our chapter focuses on this last effect where besides the great spatial accuracy of the rotation excitations, the efficiency of an inelastic interaction is so low that both the time response as well as the integrity of the molecule are under control when using a low bias voltage.

A single molecule can display quantum effects depending on its number of atoms and on the support where it is functioning. In the quantum to classical passage during STM experiments, the supporting surface is the main source of decohence of the nuclear degrees of freedom [26]. While the rotation of a large molecule can be treated quasi-classically [27, 28] on its supporting surface, the electron transfer events building up the tunneling current and the consequent inelastic excitations of this rotation must be described by intra-molecular quantum physics [29].

The frontier between large and small molecules is not very strict. Recent experiments have shown the controlled excitation of H_2 molecule rotations by STM tunneling electrons [30, 31] where the manipulated H_2 molecule is either immersed in an H_2 molecular monolayer [30] or trapped between the STM tip apex end atom and the supporting surface [31]. When the bias voltage is matching the rotational transition energies of an H_2 molecule, the STM tunnel junction conductance changes abruptly as compared with the ones of H_2 vibrational excitation [32]. The observed conductance changes correspond to the H_2 free-molecule rotation $J = 0 \rightarrow J = 2$ transition for homo-nuclear molecules, where J is the total angular momentum of the molecule, while for HD, the $J = 0 \rightarrow J = 1$ transition becomes visible. In another system with the same number of atoms, the STM tunneling current induced random rotation of a single O_2 molecule was described using at least its ground and first electronic excited states Born-Oppenheimer potential energy surfaces [22].

Molecular vibrations are efficiently excited by tunneling electrons. Recording the change of a tunnel junction differential conductance as a function of its applied bias voltage has become a standard in an electronic spectroscopy termed inelastic electron tunneling spectroscopy (IETS) [32]. For the low electron transfer rate at work in a tunneling junction, the excitation of vibrations proceeds via single-electron scattering events [33]. Hence, sizable scattering cross sections are required in order to be able to trigger some excitations.

The excitation of rotations may not seem to be different from the inelastic excitation of vibrations. However, this picture changes while on a surface. There, a molecule is not free, its angular momentum is not defined, and its low-energy excitations proceed via a small perturbation of the molecular nuclear coordinates. As the work on O_2 and H_2 rotations had demonstrated [22, 30, 31], surface rotations of a molecule are different from its vibrations. This is particularly true for molecule-rotors where the quantization of rotations and their excitation selection rules are different from the vibrational ones [34]. Even very light molecules show their nuclear dynamics profoundly affected by the presence of the supporting surface [22, 35, 36]. For rotation to occur, a tunneling electron has first to build up some angular momentum for the molecular rotor which is only a part of the whole molecule-motor chemical structure. Due to the quantization of angular momentum, a large amount of angular momentum can be transferred by a single electron. When allowed by the molecule symmetry (e.g. the above $J = 0 \rightarrow J = 2$ transition for H_2), this leads to a rather efficient excitation process. Rotational excitation becomes very efficient [37]. By comparison, in the case of vibrational excitation, the transfer of linear momentum to the molecule is negligible and the excitation needs to proceed via other mechanisms [38, 39].

In this chapter, we consider single planar molecule-rotors on a supporting solid surface to discuss their rotational excitations by tunneling electrons that are spatially localized on the molecule by the tip apex of an STM. In Sect. 2, we will identify the molecular quantities characterizing a molecule-rotor. In Sect. 3, we will study the excitation of the molecule-rotor by tunneling electrons and the de-excitation processes. The rotation of the molecule by itself will be described in Sect. 4 and the application to molecule-motors (i.e. molecule-rotor expected to display some motive power) will be discussed in conclusion.

2 Planar Molecule-Rotor

A planar free rotor [40, 41] can be mapped to a particle moving along a ring. Hence, the dynamics correspond to a 1-D problem given by the conjugate canonical operators $\hat{\varphi}$, $\hat{L}_z$. In the absence of an angle-dependent interaction potential energy with the surface, the Hamiltonian is given by the free one:

$$\hat{H}_0 = \frac{\hat{L}_z^2}{2I}. \tag{1}$$

where I is the momentum of inertia of the molecule calculated for an axis z perpendicular to the rotation plane. The eigenstates $|L_m\rangle$ of the Hamiltonian are also eigenstates of L_z with eigenvalue $\hbar m$ where m is a $\mathbb{Z}$ number. The energy eigenstates, $\hbar^2 m^2/2I$, are obviously positively defined, unbounded and doubly degenerate. This doubly degeneracy can be associated to the two classical motions of the particle in the ring.

The different interactions between molecule and its supporting surface are going to produce an angle-dependent Hamiltonian that changes the above description. The symmetry is reduced and the 2π periodicity of the free rotor rotation is no longer maintained. As a consequence, and if the rotation motion is still quantified (no decoherence), the Hamiltonian eigenstates are not the eigenstates of L_z any longer. It is natural to use the original free-rotor eigenstates as a canonical basis set as a basis to describe a more complex motion. Using Bloch's theorem, we can write the Hamiltonian eigenstates $|R_{k,n}\rangle$ characterized by two quantum numbers k,n where k is a real number such that $-0.5 < k < 0.5$ and $\hbar k$ corresponds to a pseudo angular momentum eigenvalue. The integer n labels the energy band. The corresponding wavefunction can be written as

$$R_{k,n}(\varphi) = \langle \varphi | R_{k,n} \rangle = e^{ik\varphi} u_{k,n}(\varphi). \tag{2}$$

where $e^{ik\varphi}$ is non-periodic and $u_{k,n}(\varphi)$ is periodic with $u_{k,n}(\varphi) = u_{k,n}(\varphi + 2\pi)$.

The rotation symmetry is increasingly reduced as the symmetries of the molecule-rotor and of the atomic structure of supporting surface are generally completely different. For a large class of molecules and surfaces, certain angular symmetries are maintained and the interaction Hamiltonian can be decomposed in a few Fourier components [27]. When the interaction Hamiltonian is proportional to a single Fourier component, it is straightforward to see that the Hamiltonian matrix only contains one single upper and lower off diagonal when expressed in the free-rotor basis set. More components add more off-diagonal components of the Hamiltonian matrix, reducing the sparsity of the problem.

3 The Moment of Inertia to Classify Molecule-Rotors

In first approximation, the constrained rotation on the surface of a planar molecule-rotor can be reduced to the rotation of a rigid body characterized by its moment of inertia for this rotation (See also chapter "Mechanical Transmission of Rotation for Molecule Gears and Solid-State Gears" of this volume). Hence, very different molecules with extraordinary different chemical structures may actually behave similarly when they are rotating around a fixed axle. This rotation can be considered as the only low-energy mechanical degree of freedom of this molecule-rotor among all its higher-energy mechanical degrees of freedom. The definition of this collective rotation angle requires a transformation from the set of Cartesian coordinates of the atoms of the molecule-rotor to a set of internal coordinates (bond stretching, bond angles and dihedral angles) [27]. As an example, one can compare the small quantum H_2 molecule and the much larger copper phthalocyanine molecule presented in Fig. 1. After re-organizing the H_2 and copper phthalocyanine molecules coordinates, both molecule rotation can be described simply by a rotation angle Θ after calculating in both case the position of their center of mass. For H_2, there are also the H-H molecular stretching mode and a possible tilting relative to the supporting surface

Fig. 1 Adsorbed molecule displaying local rotations of 15° due to the interactions with the substrate. The combination of two different symmetries (D_{4h} for the molecule, and the C_{3v} one for the FCC(111) substrate) leads to the overall reduction of symmetry of the system, but local interactions can lead to other unexpected rotations

during the Θ rotation. The copper phthalocyanine is a π-conjugated molecule with 57 atoms. As shown in Fig. 1, they are configured in a cross-like fashion, with a D_{4h} symmetry in the gas phase. There are (3 × 57 – 3) internal coordinates for this molecule plus the Θ collective rotation angle. This Θ is defined after the determination of its axle of rotation depending on the atomic structure of its supporting surface with the hypothesis that there is also no eccentricity involved for this rotation mode (see also chapter "A Simple Example of a Molecule-Gear Train: PF3 Molecules on a Cu(111)Surface" of this volume) [28].

4 The Classical Limit

It is important to estimate when a molecule-rotor becomes a classical planar rotor first by itself and then when adsorbed on its supporting surface. A classical free planar rotor is characterized by an angular momentum $L_z = I\omega$ where ω is the angular velocity. Quantum effects are still present when $L_z \sim \hbar m$ with m an integer number. A molecule-rotor behaves classically when $L_z/\hbar = I\omega/\hbar$ is so large that it cannot be identified as an integer:

$$I \gg \frac{\hbar}{\omega}. \tag{3}$$

Therefore, large molecules characterized by a large momentum of inertia will easily behave classically even at extremely low angular velocities. Incidentally, we note here that the minimum angular velocity of a planarly rotating molecule is $\omega_{min} = \hbar/I$. However, a molecule-rotor will rotate at angular velocities much larger than $\hbar/I$ in order to be treated classically.

5 Hindered Rotation

The ground state potential energy surface (PES) of a molecule-rotor on its surface has in general many other local minima than the expected one defining its center of rotation for Θ (See the PF3 example, chapter "A Simple Example of a Molecule-Gear Train: PF3 Molecules on a Cu(111)Surface" of this volume). They are at the origin of induced hindered, local rotations or local deformations. They are corresponding to all the others nuclear degrees of freedom not considered when mathematically defining Θ and its associated momentum of inertia. Some of them define local motions similar to bond vibrations in the molecule-rotor. They also show a dependence with the moment of inertia, and can become delocalized rotations when the excitations are high enough in energy to overcome the PES local barrier height.

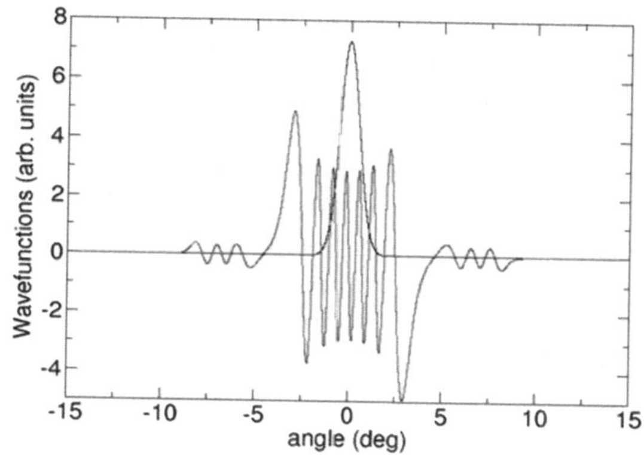

Fig. 2 Short rotation of 15° from a local minimum to the equivalent rotated molecular minimum. R_0 in full line, R_{25} in dashed lines. R_{25} is the 1st quantum state of the series leading to an important rotation probability that can be easily seen by the distribution of the state over the three minima of the molecule when it rotates

An example is shown in Ref. [42, 43]. There, a square-like phthalocyanine molecule on an fcc(111) surface presents two extra local minima off the main mirror-plane of the system (see Fig. 1). Tunneling electrons are shown to induce local rotation between these two local minima [42, 43]. Inspection of the $k = 0$ wavefunctions, in Eq. (2), gives an insight on the rotational process. For values $n \geq 25$, the wave-functions $R_{k=0,n}$ develop weight over the two minima (See Fig. 2), with an important overlap between them. When these states are energetically accessible, the molecule rotates randomly powered by surface thermal excitations.

6 Tunneling Electron Excitation for a Molecule Rotation

As mentioned in the introduction, tunneling electrons present several advantages when trying to power a molecule-rotor. These advantages are the extreme spatial lateral localization of the excitations due to the exponential decay of the tunneling evanescent electronic waves, the high energy resolution thanks to the sharp Fermi edge of the electrodes especially at low temperature and the very small number of electrons transferred per second through the molecule-rotor that controllably interact inelastically with it. This last advantage is coming from the fact that the time interval between two inelastic excitations is very long. For an $I = 1$ nA tunneling current intensity, the average time interval between two electrons transfer events is 0.1 ns with a 10–4 inelastic efficiency [29]. Therefore, one electron per 1 μs is able in average to deliver some energy to the mechanical molecule-motor degrees of freedom. This should suffice to power a rotation by superposing many of such inelastic events depending also on the lifetime of the molecular electronic states involved in those elementary electron transfer events. There is here a balance ideally to be controlled between electronic excitation lifetime, the time interval between two successive excitations and the energy dissipation towards the supporting surface [27]. This superposition is generally not coherent (each electron transfer event is totally independent from the others) and must lead to an accumulation of energy in the targeted Θ collective rotation degrees of freedom.

Through the molecule-rotor, an electron is transferred from an initial electronic state $|\psi_i\rangle$ of the STM's tip apex (energy ε_i) to a final electronic state $|\psi_f\rangle$ of the sample below the surface (energy ε_f). With an efficiency of about 10–4, this process is inelastic with $\varepsilon_i \neq \varepsilon_f$. The excess energy is transferred to the nuclear degree of freedom of the molecule-rotor. With a good positioning of the STM tip apex, it is possible to target the molecular electronic state giving access to the Θ rotation angle. In this case, the molecule-rotor goes from its ro-vibronic state $|R_0\rangle$ to its $|R_n\rangle$ undergoing an excitation of energy $E = E_n - E_0$, where E_0 and E_n are the ro-vibronic initial and final state energies. The total energies is $E_i = E_0 + \varepsilon_i$ for the initial state $|i\rangle = |R_0\rangle \otimes |\psi_i\rangle$ and $E_f = E_n + \varepsilon_f$, for the target state $|f\rangle = |R_n\rangle \otimes |\psi_f\rangle$.

In first approximation, the Fermi's golden rule can be used to calculate the probability that a single electron transfer event through the molecule-rotor excites the Θ rotation collective degrees of freedom. After our reduction (discussed above) of the molecule-rotor motion to a simple planar rotor with its rigid moment of inertia, $|i\rangle$ and $|f\rangle$ are electronically coupled via an interaction V that depends on this collective rotation angle Θ. Using the Bardeen approximation (later used by Tersoff and Hamann), $V_{i,f}$ can be approach by the molecule-rotor wavefunction amplitude calculated at the tip's apex end atom [44] leading to:

$$V_{i,f} = \left\langle \psi_i | \hat{V} | \psi_f \right\rangle \approx C \psi_f(\varphi, \vec{r}_0). \tag{4}$$

The value of the proportionality constant in Eq. (4) can be approached from Eq. (8) of Ref. [44]. Simplifying the sum over electronic states of the tip by its density of states $D_{tip}(\varepsilon)$, the rate of electron-tunneling and rotational excitation is given by:

$$\begin{aligned} \frac{1}{\tau_{ine}} = \frac{2\pi}{\hbar} \sum_{n>N_R, f} & D_{tip}(\varepsilon_f + E)|C|^2 \\ & \times \left|\langle R_0 | \psi_f(\varphi, \vec{r}_0) | R_n \rangle\right|^2 \Theta\left(\varepsilon_F + eV - \varepsilon_f - E\right) \Theta\left(\varepsilon_f - \varepsilon_F\right) \end{aligned} \tag{5}$$

where we have assumed a low surface and STM tip temperatures to substitute the Fermi distribution function by the zero-Kelvin value.

Here, we have supposed that the inelastic energy provided by the tunneling electrons will go directly to the Θ direction on the PES excited and ground electronic state manifold of the molecule-rotor. This is a crucial hypothesis that requires a careful attention in the future if one wants to design the molecule-rotor to be sure that Θ is the only collective mode that the tunneling electrons can be preferentially powered. The manifold topology of a PES can generally open many other outcomes for this inelastic energy. This points out at the care needed concerning the definition of the internal coordinates from the Cartesian ones even for simple molecule like O_2, NH_3 [28] or PF_3.

7 Rotational De-Excitation

The supporting surface is a cold source thermostat for the molecule-motor and is also causing the quantum decoherence of its total nuclear wavefunction. It allows to define the molecule-rotor dynamics induced by the inelastic excitations by a classical molecular dynamic on the ground state PES often mixed by some contributions from the excited states which can also described by some classical molecular dynamics on the corresponding PES [45]. It also allows those molecule-rotor electronic excited states to relax towards the ground state PES while carrying the ro-vibronic excitations during this molecular dynamic process. The Fermi's golden rule introduced in the previous section can be used in first approximation to describe this ro-vibronic de-excitation towards the supporting surface. Metallic substrates possess a continuum of electronic states that encompass efficient de-excitation channels via the surface [45, 46]. Then, $V_{i,f}$ can be considered like a deformation potential [46] for the substrate electrons as the molecule-rotor changes its angle Θ. In the particular case where a local rotation is involved, $R_0(\varphi)$ is going to be a wavefunction very localized at $\varphi = 0$ of the local rotational minima. In this case, we can easily express $V_{i,f}$ as:

$$V_{i,f} = \left\langle \lambda, R_n \left| \frac{\partial H}{\partial \varphi} \varphi \right| \mu, R_0 \right\rangle \tag{6}$$

where the initial state $|i\rangle = |\lambda, R_n\rangle$ is a tensorial product of the substrate electronic states $|\lambda\rangle$ and the excited rotational one $|R_n\rangle$. The final state contains another electronic state and the rotational ground state $|R_0\rangle$. The transition is such that the initial electronic state is below the Fermi energy of the surface. The final state above is producing an electron-hole pair and the energy of this excitation matches the de-excitation energy between $|R_n\rangle$ and $|R_0\rangle$. Written in this form, the potential is separable and becomes:

$$V_{i,f} = \left\langle \lambda \left| \frac{\partial H}{\partial \varphi_0} \right| \mu \right\rangle \langle R_n | \varphi | R_0 \rangle \tag{7}$$

such that rotational transition proceeding towards the ground state PES are given by $\langle R_n | \varphi | R_0 \rangle$. The electronic part can be evaluated from ab initio codes using the variation of the electronic wavefunctions with molecular angle of rotation.

The de-excitation or damping rate, γ, can be easily written now as:

$$\gamma_{n,0} = \frac{2\pi}{\hbar} \sum_{\lambda,\mu} \left| \left\langle \lambda \left| \frac{\partial H}{\partial \varphi} \right| \mu \right\rangle \right|^2 \Theta(\varepsilon_\mu - E_F) \Theta(E_F - \varepsilon_\lambda) \times |\langle R_n | \varphi | R_0 \rangle|^2 \delta(\varepsilon_\mu - \varepsilon_\lambda - E_n - E_0) \tag{8}$$

where we have again assumed a very low surface temperature with E_F the Fermi level, E_0 and E_n the involved rotational energies.

8 Molecular Rotation Under Tunneling Electrons

In the previous sections, we have evaluated the probability of powering the molecule-rotor from its ground state to an excited state and its probability of de-excitation by a coupling to the supporting surface. However, this model of the excitation process is a simplification of the actual intramolecular dynamical process. In order to be more realistic, we need to include the time dependent quantum evolution of the molecule-motor during and after excitation. The correct way to do this would be to consider excitation and evolution on the same footing. This leads to a large number of coupled dynamical equations that are virtually impossible to solve at the quantum level. However, simplifications exist that render such computation feasible.

As shown by the action spectra work [47, 48], molecular motion (generally overcoming some dynamical barrier, for example to start a free-like rotation) can be excited over certain vibrational thresholds. The vibration acts as a resonance that efficiently funnels the tunneling electron energy into the molecular dynamics. To work like this, the normal modes of the vibration need to be coupled to the reaction coordinates. In the case of a rotation, it means an inter mode coupling between the high-energy mode that can easily accept a large amount of energy into the molecular degrees of freedom, and also a hindered rotation (see above the hindered rotation section).

As recalled in the previous section, the probabilities for rotation excitation by a single electron are rather low, and in the concurrent even of the excitation of other degrees of freedom inter mode coupling can help in rendering an unlike reaction, observable [48]. Inter-mode coupling proceeds through the anharmonicities of the normal modes [46], in the vision that a rotation is locally similar to molecular vibration, the theory developed to inter mode coupling [46, 48] can be easily used here.

The dynamics of these processes can be modeled using a term that couples all degrees of freedom of the density matrix [49, 50]. Assuming that the time evolution only depends on the previous time state and has no long-range memory (the so-called Markov approximation), the time-dependent equation for the density matrix can be greatly simplified. If we further assume that a large energy difference between the pairs of states entering the density matrix, it is reasonable to uncouple the diagonal and non-diagonal parts of the density matrix.

The diagonal part gives rise to the so-called master equation. It gives the population of the different states along the time dependent trajectory of the total quantum evolution. Direct access to these quantities is already very interesting because it gives information on the outcome of the intramolecular evolutions. The master equation can be of an increasing complexity. The Linblad form of the master equation is the most general form of a Makovian master equation. It is characterized by including transition operators at the lowest order using the weak coupling to a reservoir that induces transitions in the finite quantum system (here the molecule-rotor).

In the previous section, we have indicated how to calculate the lifetimes of different processes induced by the electron tunneling process through the molecule-rotor.

These act directly on the population of different rotational levels of the molecule when the molecule is small enough for its rotation behavior around its center of mass to be quantified. We can further consider the supporting surface by including the loss of rotational energy by exciting its degrees of freedom. At low-temperatures and for a metal surface, those will be the continuum of metallic electronic excitations. On a semiconductor and more generally an electronic gapped support, phonons degrees of freedom will prevail, which can be rather inefficient leading to long life times.

Let us name $\tau_{n,m}$ the transition time between two levels n and m. The master equation describing the population N_m of a given quantum state |m > can be built up by considering the population all states |n > having a lower energy than |m > and which are excited by the tunneling electrons. Then, one has to consider the decay of the population of this $|m>$ into all those $|n>$ by a damping process to induce a relaxation towards the support degrees of freedom. In this way, one can de-excite the molecule from $|m>$ to $|n>$ with a rate given by $\gamma_{n,m}$. This can be simply expressed as:

$$\frac{dN_m}{dt} = \sum_{n<m} \frac{1}{\tau_{n,m}} N_n - \gamma_{n,m} N_m \tag{9}$$

This describes the balance between excitation and de-excitation from rotational states with the subsidiary condition that there is only one molecule in the calculation, $\sum_m N_m = 1$. However, in experiments describing local rotations, we are interested in the probability of finding the molecule at a given local minimum about a certain angle. This can be accounted for by a linear combination of the lowest-energy rotational states that does not present any conceptual difficulty. Then, we can describe with the simple Eq. (9) rather difficult situations where several degrees of freedom are dynamically involved [48].

In order to compute the probability that a molecule-rotor actually rotates, we need to add the possible paths from the ground state to all excited states with their corresponding occupation probability. But the excitation probability is going to decay exponentially with increasing energy. Then, the most important rotational states to account for delocalized rotations are those just above the rotation barrier, within a small energy window that can be determined numerically. On a metallic supporting surface, the damping is efficient. Therefore, one transferred electron after the other through the molecule-motor, the delocalized excitation will be built up.

From these calculations it is apparent that the decay mechanism needs to be reduced to avoid decay of the rotational level in order to attain delocalized rotations. According to Eq. (9), we can actually compute the rate of rotational excitation needed to maintain a steady delocalized rotation regime. This is the regime of interest for a single molecule-motor, which was also explored recently in a more semi-classical approach [27].

9 Conclusions

We have discussed how to describe the excitation of the molecular rotations of a single molecule-rotor by tunneling electrons. (i) At low energy, the molecule-rotor encounters rotational energy barriers imposed by the supporting surface (or by its stator in case of a molecule-motor) that constrains those rotations to be local. For the case of a molecule-motor, the rotation can be hindered, with no collective rotation of the complete molecular structure. It will result in a net rotation around the molecule-motor center of mass. (ii) At higher energies, the molecule- rotor is not hampered to freely rotate albeit following a certain potential energy landscape that will modify its topology from the free planar rotor energy value $\hbar^2 m^2/2I$. The time dynamics of this rotation can become semi-classical depending on the inertia momentum I, that depends on the value of its molecular rotor weight. All these motions can be classified as rotations because they originate in the angular dependence of the rotor molecular dynamics. Actually, the low-energy type (i) of excitation can lead to a repetitive pattern in the time dependent rotor molecular dynamics. Together with the stochastic nature of the tunnel electron process and with the damping contribution from the supporting substrate, this leads (even at low supporting surface temperature) to a fluctuation-like dynamics [42, 43, 51]. Although depending on the molecule-motor molecular design [23] and on the supporting surface atomic scale structure [52], some directionality can be imposed for this dynamics by playing with the contribution of the electronic excited states to the intrinsic molecular dynamics on the corresponding excited state potential energy surface. One can optimize the tunneling excitation relative lifetime of the electronic excited states by tuning the bias voltage applied. This can be performed even if the contribution of those excited states in the tunneling process is of a relatively small weight relative to the ground state contribution during the intramolecular dynamics.

Acknowledgements The authors acknowledge financial support from the European-Union Integrated Project MEMO "Mechanics with Molecule(s)" (Grant 766864).

References

1. Gimzewski, J.K., Joachim, C.: Science **283**, 1783 (1999). https://doi.org/10.1126/science.283.5408.1683
2. Feringa, B.L.: Angew. Chem. Int. Ed. **56**, 11060 (2017). https://doi.org/10.1002/anie.201702979
3. Vacek, J., Michl, J.: Proc. Natl. Acad. Sci. **98**, 5481 (2001). https://doi.org/10.1073/pnas.091100598
4. Tan, S., Lopez, H.A., Cai, C.W., Zhang, Y.: Nano Lett. **4**, 1415 (2004). https://doi.org/10.1021/nl049347g
5. Saywell, A., Bakker, A., Mielke, J., Kumagai, T., Wolf, M., Garcia-Lopez, V., Chiang, P.-T., Tour, J.M., Grill, L.: ACS Nano **10**, 10945 (2016). https://doi.org/10.1021/acsnano.6b05650
6. van Leeuwen, T., Lubbe, A.S., Stacko, P., Wezenberg, S.J., Feringa, B.L.: Nat. Rev. Chem. **1**, 0096 (2017). https://doi.org/10.1038/s41570-017-0096

7. Kelly, T.R., Silva, H.D., Silva, R.A.: Nature **401**, 150 (1999). https://doi.org/10.1038/43639
8. Hernandez, J.V., Kay, E.R., Leigh, D.A.: Science **306**, 1532 (2004). https://doi.org/10.1126/science.1103949
9. Leigh, D.A., Wong, J.K.Y., Dehez, F., Zerbetto, F.: Nature **424**, 174 (2003). https://doi.org/10.1038/nature01758
10. Balzani, V., Gomez-Lopez, M., Stoddart, J.F.: Acc. Chem. Res. **31**, 405 (1998). https://doi.org/10.1021/ar970340y
11. Sauvage, J.-P.: Acc. Chem. Res. **31**, 611 (1998). https://doi.org/10.1021/ar960263r
12. Michl, J., Sykes, E.C.H.: ACS Nano **3**, 1042 (2009). https://doi.org/10.1021/nn900411n
13. Zhang, R.-Q., Zhao, Y.-L., Qi, F., Hermann, K., Hove, M.A.V.: Phys. Chem. Chem. Phys. **18**, 29665 (2016). https://doi.org/10.1039/C6CP05996A
14. Gimzewski, J.K., Joachim, C., Schlittler, R.R., Langlais, V., Tang, H., Johannsen, I.: Science **281**, 531 (1998). https://doi.org/10.1126/science.281.5376.531
15. Tierney, H.L., Murphy, C.J., Jewell, A.D., Baber, A.E., Iski, E.V., Khodaverdian, H.Y., McGuire, A.F., Klebanov, N., Sykes, E.C.H.: Nat. Nanotechnol. **6**, 625 (2011). https://doi.org/10.1038/nnano.2011.142
16. Kudernac, T., Ruangsupapichat, N., Parschau, M., Maci, B., Katsonis, N., Harutyunyan, S.R., Ernst, K.-H., Feringa, B.L.: Nature **479**, 208 (2011). https://doi.org/10.1038/nature10587
17. Zhang, Y., Kersell, H., Stefak, R., Echeverria, J., Iancu, V., Perera, U.G.E., Li, Y., Deshpande, A., Braun, K.-F., Joachim, C., et al.: Nat. Nanotechnol. **11**, 706 (2016). https://doi.org/10.1038/nnano.2016.69
18. Astumian, R.D.: Nat. Nanotechnol. **7**, 684 (2012). CODEN: NNAABX; ISSN: 1748-3387
19. Joachim, C., Martrou, D., Rezeq, M., Troadec, C., Jie, D., Chandrasekhar, N., Gauthier, S.: J. Phys. CM **22**, 084025 (2010). https://doi.org/10.1088/0953-8984/22/8/084025
20. Wang, B., Vukovic, L., Kral, P.: Phys. Rev. Lett. **101** (2008). https://doi.org/10.1103/physrevlett.101.186808
21. Seldenthuis, J.S., Prins, F., Thijssen, J.M., van der Zant, H.S.J.: ACS Nano **4**, 6681 (2010). https://doi.org/10.1021/nn1021499
22. Stipe, B.C., Rezaei, M.A., Ho, W.: Science **279**, 1907 (1998). https://doi.org/10.1126/science.279.5358.1907
23. Perera, U.G.E., Ample, F., Kersell, H., Zhang, Y., Vives, G., Echeverria, J., Grisolia, M., Rapenne, G., Joachim, C., Hla, S.-W.: Nat. Nanotechnol. **8**, 46 (2013). https://doi.org/10.1038/nnano.2012.218
24. Simpson, G.J., Garcia-Lopez, V., Petermeier, P., Grill, L., Tour, J.M.: Nat. Nanotechnol. **12**, 604 (2017). https://doi.org/10.1038/nnano.2017.137
25. Ohmann, R., Meyer, J., Nickel, A., Echevaria, J., Joachim, C., Moresco, F., Cuniberti, G.: ACS Nano **9**, 8394 (2015). https://doi.org/10.1021/acsnano.5b03131
26. Joachim, C., Gimzewski, J.K.: Springer Book Series: Structure and Bonding, vol. 99, p. 1 (2001)
27. Echeverría, J., Monturet, S., Joachim, C.: Nanoscale **6**, 2793 (2014). https://doi.org/10.1039/C3NR05814J
28. Havighorst, M., Prager, M., Coddens, G.: Chem. Phys. Lett. **259**, 1 (1996). https://doi.org/10.1016/0009-2614(96)00740-3
29. Monturet, S., Kepenekian, M., Lorente, N., Joachim, C.: Chem. Phys. Lett. **573**, 1 (2013). https://doi.org/10.1016/j.cplett.2013.02.057
30. Natterer, F.D., Patthey, F.M.C., Brune, H.: Phys. Rev. Lett. **111**, 175303 (2013). https://doi.org/10.1103/PhysRevLett.111.175303
31. Li, S., Yu, A., Toledo, F., Han, Z., Wang, H., He, H.Y., Wu, R., Ho, W.: Phys. Rev. Lett. **111**, 146102 (2013). https://doi.org/10.1103/PhysRevLett.111.146102
32. Komeda, T.: Prog. Surf. Sci. **78**, 41 (2005). https://doi.org/10.1016/j.progsurf.2005.05.001
33. Godlewski, S., Kawai, H., Kolmer, M., Zuzak, R., Echavarren, A.M., Joachim, C., Szymonski, M., Saeys, M.: ACS Nano **10**, 8499 (2016). https://doi.org/10.1021/acsnano.6b03590
34. Schulz, G.J.: Rev. Mod. Phys. **45**, 378 (1973). https://doi.org/10.1103/RevModPhys.45.378

35. Teillet-Billy, D., Gauyacq, J.P., Persson, M.: Phys. Rev. B **62**, R13306 (2000). https://doi.org/10.1103/PhysRevB.62.R13306
36. Olsson, F.E., Lorente, N., Persson, M.: Surf. Sci. **522**, L27 (2003). https://doi.org/10.1016/S0039-6028(02)02358-0
37. Sugimoto, T., Kunisada, Y., Fukutani, K.: Phys. Rev. B **96**, 241409 (2017). https://doi.org/10.1103/PhysRevB.96.241409
38. Persson, B.N.J., Baratoff, A.: Phys. Rev. Lett. **59**, 339 (1987). https://doi.org/10.1103/PhysRevLett.59.339
39. Lorente, N., Persson, M.: Phys. Rev. Lett. **85**, 2997 (2000). https://doi.org/10.1103/PhysRevLett.85.2997
40. Vazquez, P.C.L., Garcia, A.: Phys. Scr. **91**, 055101 (2016). https://doi.org/10.1088/0031-8949/91/5/055101
41. Seah, S., Nimmrichter, S., Scarani, V.: New J. Phys. **20**, 043045 (2018). https://doi.org/10.1088/1367-2630/aab704
42. Schaffert, J., Cottin, M.C., Sonntag, A., Karacuban, H., Bobisch, C.A., Lorente, N., Gauyacq, J.-P., Mller, R.: Nat. Mater. **12**, 223 (2013). https://doi.org/10.1038/nmat3527
43. Schaffert, J., Cottin, M.C., Sonntag, A., Bobisch, C.A., Mller, R., Gauyacq, J.-P., Lorente, N.: Phys. Rev. B **88**, 075410 (2013). https://doi.org/10.1103/PhysRevB.88.075410
44. Tersoff, J., Hamann, D.R.: Phys. Rev. B **31**, 805 (1985). https://doi.org/10.1103/PhysRevB.31.805
45. Persson, B., Persson, M.: Surf. Sci. **97**, 609 (1980). https://doi.org/10.1016/0039-6028(80)90691-3
46. Lorente, N.: In: Hasselbrink, E., Lundqvist, B. (eds.) Dynamics, (North-Holland, 2008), vol. 3 of Handbook of Surface Science, pp. 575–620 (2008)
47. Motobayashi, K., Kim, Y., Ueba, H., Kawai, M.: Phys. Rev. Lett. **105**, 076101 (2010). https://doi.org/10.1103/PhysRevLett.105.076101
48. Ueba, H., Mii, T., Lorente, N., Persson, B.N.J.: J. Chem. Phys. **123**, 084707 (2005). https://doi.org/10.1063/1.2007631
49. Saalfrank, P.: Chem. Rev. **106**, 4116 (2006). https://doi.org/10.1021/cr0501691
50. Delgado, F., Fernandez-Rossier, J.: Prog. Surf. Sci. **92**, 40 (2017). https://doi.org/10.1016/j.progsurf.2016.12.001
51. Gao, L., Liu, Q., Zhang, Y.Y., Jiang, N., Zhang, H.G., Cheng, Z.H., Qiu, W.F., Du, S.X., Liu, Y.Q., Hofer, W.A., et al.: Phys. Rev. Lett. **101**, 197209 (2008). https://doi.org/10.1103/PhysRevLett.101.197209
52. Mishra, P., Hill, J.P., Vijayaraghavan, S., Rossom, W.V., Yoshizawa, S., Grisolia, M., Echeverria, J., Ono, T., Ariga, K., Nakayama, T., et al.: Nano Lett. **15**, 4793 (2015). https://doi.org/10.1021/acs.nanolett.5b01908

Motion and Nanomechanical Effects in Supramolecular Catalysts

Michael Schmittel, Abir Goswami, Indrajit Paul, and Pronay Kumar Biswas

Abstract In this chapter, we describe the effect of "static" and dynamic nanomechanical actuation at zinc porphyrin sites that are loaded with an organocatalyst. In detail, the actuation embraces several molecular events culminating in a nanometer rearrangement of a toggling arm onto the loaded zinc porphyrin site that finally entails liberation of the catalyst into solution. In this brief review we first describe the development of nanoswitches that upon toggling the switching arm turn ON/OFF catalytic processes. Then, we discuss the prospects of dynamic actuation leading to catalyst liberation in appropriately designed devices. There the catalytic activity is then correlated with the machine speed. A theoretical explanation is presented that identifies the kinetic origin of the phenomenon.

Keywords Catalysis · Allosteric regulation · Dynamic actuation · Dynamic multicomponent devices · Catalytic machinery

1 Introduction

Since mechanical motion is so prominent in our daily life, the enthrallment for nanomechanical aspects in molecular processes is readily understandable [1]. Indeed, if one analyzes the current state of the art, many life-sustaining machine-type movements [2] depend on linear, rotary, reciprocating and oscillating motions in the nanometer range. A large gamut of biological machines, like the ATP-synthase [3], kinesin [4], etc., have been identified and many of them require nanomechanical motions for proper operation [5]. Obviously, it is not only the appealing conception of biological machines but likewise our scientific aspiration for innovative designs of complex structures and functions that is driving the current developments. The groundbreaking and seminal contributions of the Nobel Prize winners, Jean-Pierre Sauvage [6], Fraser Stoddart [7], and Ben Feringa [8] have revealed fascinating

M. Schmittel (✉) · A. Goswami · I. Paul · P. K. Biswas
Department of Chemie – Biologie, Center of Micro and Nanochemistry and Engineering, Universität Siegen, 57068 Siegen, Germany
e-mail: schmittel@chemie.uni-siegen.de

C. Joachim (ed.), *Building and Probing Small for Mechanics*, Advances in Atom and Single Molecule Machines, https://doi.org/10.1007/978-3-030-56777-4_13

advancements in the arena of man-made molecular machines that keep stimulating current new developments.

In the present work we will highlight the potential of "nanomechanical actuation", which should not be confused with conformational changes, for control of catalysis. Conformational analysis, a field strongly linked to the pioneering work of Hassel [9], Barton and others [10], is defined by the IUPAC [11]: "conformation means the spatial arrangement of atoms affording distinction between stereoisomers which can be interconverted by rotations about formally single bonds" [12]. We use the term nanomechanical actuation for nm-wide geometrical changes in molecules that are triggered thermally or by an input (chemical, redox, light) and lead to bond formation. The bond formation step triggers the actuation of an event. Due to the remarkable versatility of the coordinative metal-ligand interaction we and others have used it in dozens of publications for nanomechanical actuation (vide infra) [13–19]. In our opinion, metal coordination has many advantages: (a) the binding strength and metal-ligand kinetics can be readily moderated, and (b) the intrinsic directionality of the metal-ligand interaction facilitates conceiving molecular aspects of the nanodevices. Furthermore, liberation and capture of metal ions, the latter by more strongly binding ligands, provide us with options for the facile reversible actuation of the anticipated nanomechanical motion.

2 Allosteric Regulation of Catalysis at Zinc-Porphyrin Stations (Static Actuation)

2.1 Release of Catalyst at Zinc-Porphyrin-Based Nanoswitches

The concept of allosteric regulation of catalysis as advanced by our group is a lot based on the immobilization of a catalytically active nitrogen heterocycle at a zinc porphyrin (ZnPor) site and on its release from this ZnPor site through a nano-mechanical motion that itself is triggered by a chemical input. Zinc(II) porphyrins bind a variety of nitrogen heterocycles so strongly that under defined conditions the thermal dissociation of the complex is so minor that no catalytic action will result (Table 1).

The principle of "static" allosteric regulation of catalysis may readily be explained by having a look at the operation of nanoswitch **13** [23] (Fig. 1). In State I, nano-switch **13** has the azabipyridine arm bound in an intermolecular heteroleptic copper(I) complex with phenanthroline **14**, while the zinc porphyrin (ZnPor) subunit is free. In the presence of piperidine (**8**), chosen as an organocatalyst for a Knoevenagel reaction between 4-nitrobenzaldehyde (**15**) and diethyl malonate (**16**), the ZnPor unit was expected to immobilize the catalyst. Indeed, when the ensemble of Cu^+, **8**, **13**, **14**, **15** and **16** (1:1:1:1:10:1000) was heated to 55 °C in $CDCl_3$ for 30 min, no conversion of the substrates was detected. However, upon removal of the copper(I) ions by a

Table 1 Binding constants of selected nitrogen compounds to zinc(II) porphyrins

Ligand	Log *K* (solvent)[a]	References
pyridine (**1**)	3.78, 3.89 (DCM)	[20, 21]
2-methylpyridine (**2**)	2.30 (DCM)	[20]
4-methylpyridine (**3**)	4.02 (DCM)	[20]
4-aminopyridine (**4**)	4.65 (DCM)	[20]
1-butylamine (**5**)	4.60 (DCM)	[21]
benzylamine (**6**)	4.08 (DCM)	[21]
DABCO (**7**)	8.60, 2.90 (benzene)	[22]
piperidine (**8**)	5.05 (DCM)	[20]
imidazole (**9**)	4.50 (DCE)	[22]
diethylamine (**10**)	3.15 (toluene)	[22]
triethylamine (**11**)	1.10 (toluene)	[22]
t-butylamine (**12**)	4.15 (DCM)	[21]

[a]DCM: dichloromethane. DCE: 1,2-dichloroethane

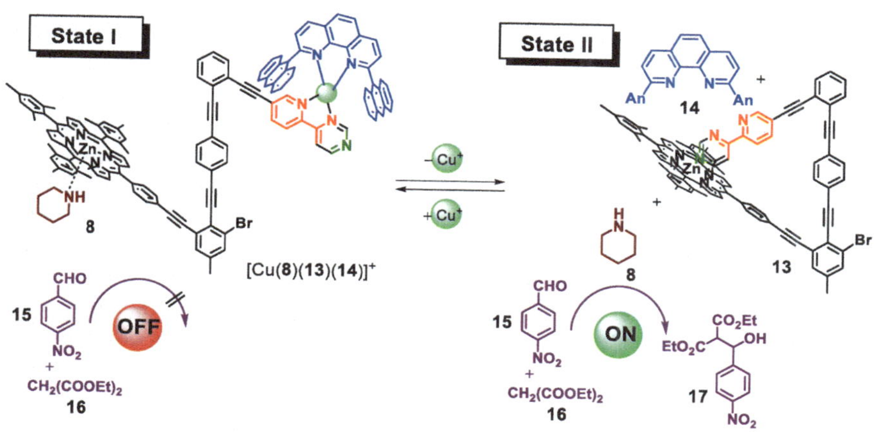

Fig. 1 Allosteric regulation at the ZnPor site of nanoswitch **13**. Depending on the switching state the ZnPor site either strongly binds the organocatalyst **8** or releases it into solution

trapping agent, the intermolecular copper complex was broken up and the azabipyridine arm moved onto the ZnPor unit. Since the ZnPor unit is not able to tightly harbor six donor atoms/ligands [24], the piperidine was released into solution and as a result the Knoevenagel reaction was turned ON (23% in 30 min). Upon addition of copper(I) ions, phenanthroline **14**—due to the bulk of both anthryl groups—regenerated the intermolecular complex and forced the arm to detach from the ZnPor station (State I). In this state, catalysis was again turned OFF due to the allosteric binding of the organocatalyst. The reversible ON/OFF regulation was accomplished over three cycles without loss of catalytic activity [23]. This mode of action mimics the

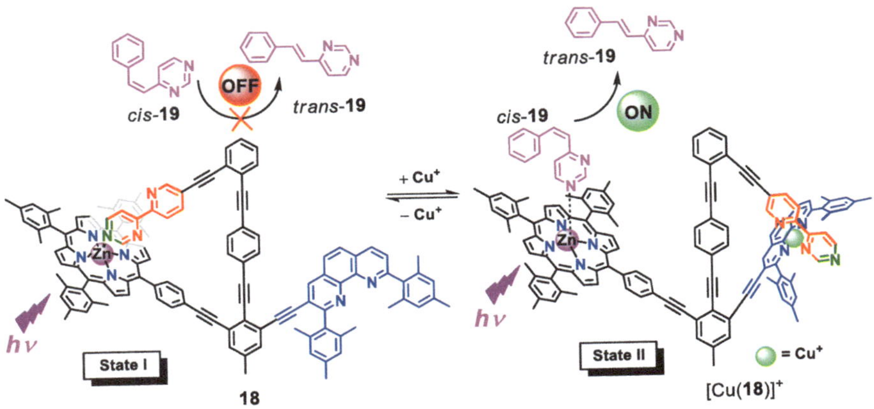

Fig. 2 Allosteric regulation at the ZnPor site of nanoswitch **18**. Depending on the switching state the ZnPor site either strongly binds the substrate **19** or releases it into solution. Triplet photosensitization is only possible in the bound state (State II)

operating principle of the calcium/calmodulin-dependent protein kinase II (CaMKII) [25].

The same principle provided a reliable strategy for controlling a photoisomerization reaction [26]. Here, the intramolecularly locked conformation with the azabipyridine arm bound to the ZnPor unit is inactive (OFF), whereas the ON state was associated with freeing the ZnPor site. For the required two-state switching a new versatile design was chosen by adding a further arm to **13**, so that two intramolecular switching states became possible. In the thus created nanoswitch **18** (Fig. 2), the azabipyridine arm was tightly locked on the ZnPor unit, which did not allow diazastilbene **19** to attach to this binding site. As a result, photochemical irradiation would generate the triplet state of the ZnPor unit, but since transfer of triplet energy is only possible by an exchange mechanism requiring contact of ^{3}ZnPor and *cis*-**19**, no *cis* → *trans* isomerization of **19** was detected [26]. Upon addition of copper(I), formation of the heteroleptic copper complex required intramolecular toggling of the bipyridine arm to the shielded [1,10]-phenanthroline. In [Cu(**18**)]$^+$ (State II) the ZnPor site is now available for the Dexter exchange mechanism. Indeed, upon irradiation at 419 nm, the *cis*- → *trans*-isomerization of diazastilbene **19** was promoted (Fig. 2), actually in more or less the same yield as in the reference reaction with zinc 5,10,15,20-tetraphenylporphyrin (ZnTPP). As in the former example, addition of a trapping agent formed a strong copper(I) chelate complex, removing the copper ions from [Cu(**18**)]$^+$ regenerating **18** and pushing back the bipyridine arm to the ZnPor binding site (State I). After this allosteric actuation, photoisomerization proved to be OFF even when irradiation was upheld.

Multiple switching was realized in a system designed for controlling dual catalysis that was based on stoichiometry-dependent self-sorting [27]. The decision-making in this complex cascade of switching events ultimately implicated three self-sortings [28–30] of a mixture of altogether ten different components thus demanding perfect

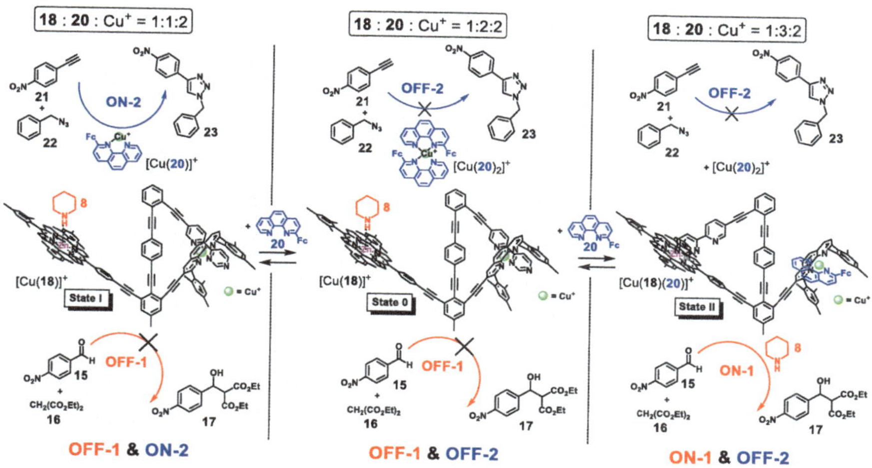

Fig. 3 Stoichiometry-dependent toggling in a ten-component catalytic machinery allows control over two catalytic reactions

orthogonality between the switching states (Fig. 3). At the heart of the dual switching of catalysis, nanoswitch **18** was toggled between its intramolecular copper complex (in States 0,I) and an intermolecular copper complex with the azapyridine arm on the ZnPor unit (see State II). Depending on the situation about the ZnPor unit, a catalyzed Knoevenagel reaction was regulated by binding and release of piperidine (**8**) as described for nanoswitch **13** (Fig. 1). Parallel a click reaction was ON/OFF controlled by switching between $[Cu(\mathbf{20})]^+$ (active catalyst) and $[Cu(\mathbf{20})_2]^+$ (inactive). The first switching led to State I (Fig. 3). It required a two-fold completive [31] self-sorting of **18**, **20** and $Cu^+ = 1{:}1{:}2$ providing the two complexes $[Cu(\mathbf{18})]^+$ and $[Cu(\mathbf{20})]^+$ (1:1). Thus click catalysis was ON, but due to the firm binding of **8** to $[Cu(\mathbf{18})]^+$ organocatalysis was OFF. Formation of State 0 required a self-sorting of **18**, **20** and Cu^+ (1:2:2) furnishing $[Cu(\mathbf{18})]^+ + [Cu(\mathbf{20})_2]^+$. In this state, click catalysis was turned OFF due to formation of $[Cu(\mathbf{20})_2]^+$. Finally, the Knoevenagel reaction was turned ON in State II by a self-sorting of **18**, **20** and Cu^+ (1:3:2) generating $[Cu(\mathbf{18})(\mathbf{20})]^+ + [Cu(\mathbf{20})_2]^+$. In $[Cu(\mathbf{18})(\mathbf{20})]^+$, the aza-bipyridine arm was located on the ZnPor unit thus liberating the organocatalyst **8** into solution. The cycle was restarted (State II → I) by addition of one equivalent of copper(I) ions and continued (State I → 0 → II) by further addition of **20**.

A higher level of complexity, i.e., remote allosteric actuation, was realized when the toggling of nanoswitch **18** was controlled in a cybernetic network with nanoswitch **24** (Fig. 4) [32]. Here, the switching states of both nanoswitches need to be closely coupled via chemical communication [33–36].

A proper starting point of regulating the catalytic activity of *N*-methylpyrrolidine in a cybernetic network was complex $[Cu(\mathbf{18})]^+$, which served to bind the organocatalyst **27** so that no activity was obtained, in combination with $[Fe(\mathbf{24})_2]^{2+}$ (NetState I). As a result, in NetState I catalysis of the conjugate addition was shut down. ON&OFF

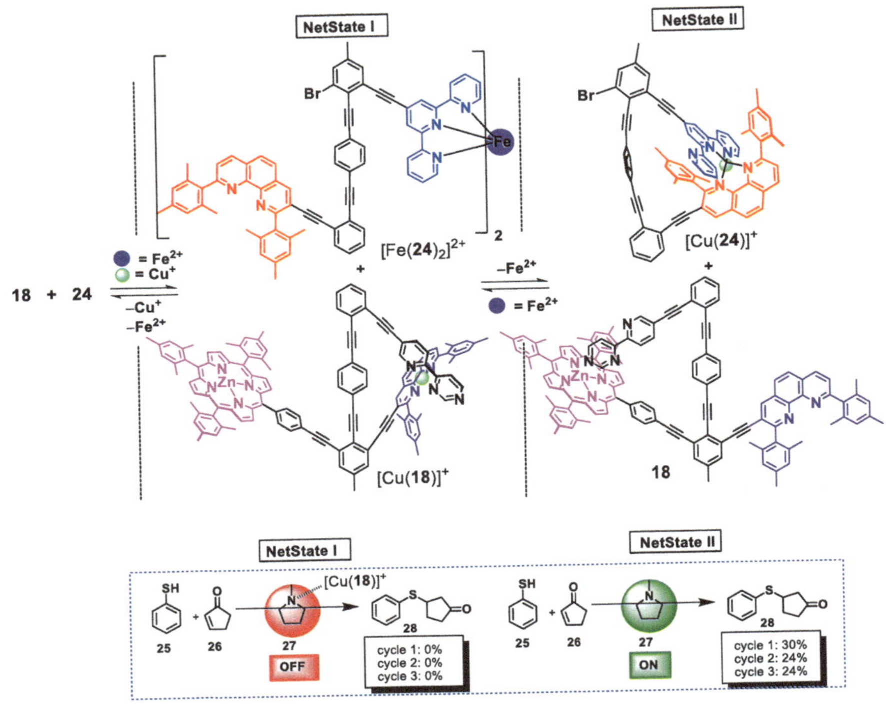

Fig. 4 Allosteric regulation at the ZnPor site of nanoswitch **18** in a cybernetic network with nanoswitch **24**

regulation of organocatalysis was accomplished by removal and addition of iron(II) ions. When Fe^{2+} was removed, the high stability of complex [Cu(**24**)]$^+$ claimed the copper(I) ions to depart from [Cu(**18**)]$^+$ (NetState II). In this signaling between two switches, copper(I) ion loss from [Cu(**18**)]$^+$ was facilitated by gaining binding energy through the toggling of the azabipyridine arm onto the ZnPor platform. The latter process, though, released *N*-methylpyrrolidine (**27**) into solution enabling catalysis of the conjugate addition. The cycle was continued by addition of Fe^{2+} regenerating the highly stable iron bis-terpyridine complex [Fe(**24**)$_2$]$^{2+}$. The thus required decomposition of [Cu(**24**)]$^+$ led to liberation of copper(I) ions that translocated to switch **18** where the copper complex [Cu(**18**)]$^+$ was furnished. Formation of this complex led to the departure of the bipyridine arm from the ZnPor and thus strong binding of the organocatalyst and suppression of catalysis.

Multistate allosteric regulation of catalysis was observed in a three-state supramolecular walker (Figs. 5, 6, 7) [37]. The assembly [**29**·**30**] was fabricated from biped **30** with its two picoline terminals and the tetrahedral ligand **29**. In this assembly (State III), both ZnPor units were occupied by the biped and thus the added organocatalyst, *N*-methylpyrrolidine (**27**) was available for catalysis. Upon addition of one equiv. of

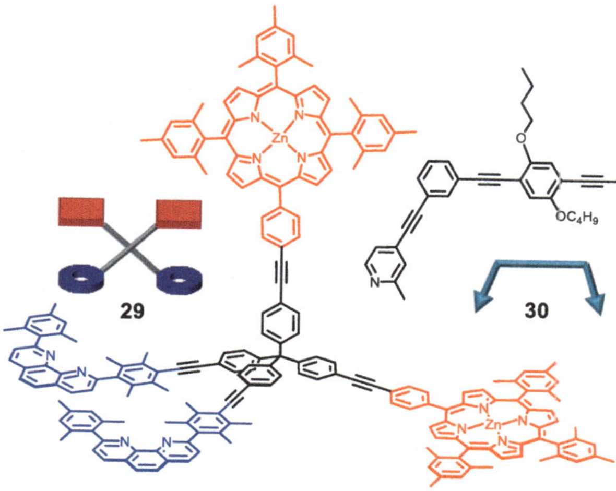

Fig. 5 Components used for the walker assembly [**29·30**] (see Fig. 6). Adapted with permission from ref [1], Copyright 2020 by the American Chemical Society

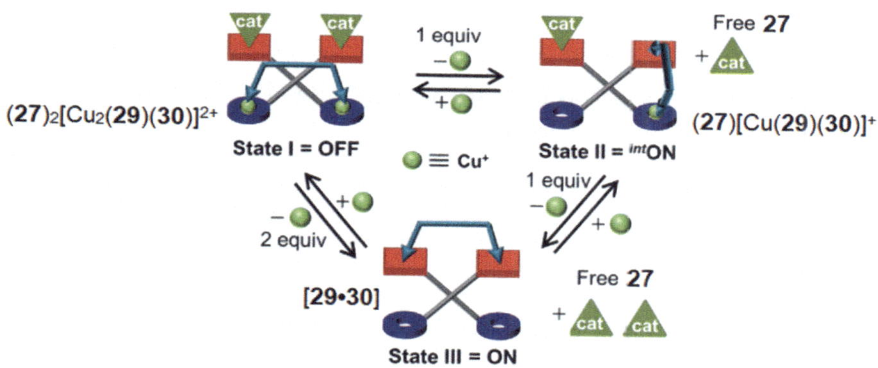

Fig. 6 Reversibly walking the assembly of [**29·30**] that is built from two picoline → ZnPor interactions (see State III) by addition/removal of copper(I) ions. The copper(I) ions fill the phenanthroline voids and generate picoline → [Cu(phenAr$_2$)]$^+$ complexes that make the biped walk. Adapted with permission from ref [1], Copyright 2020 by the American Chemical Society

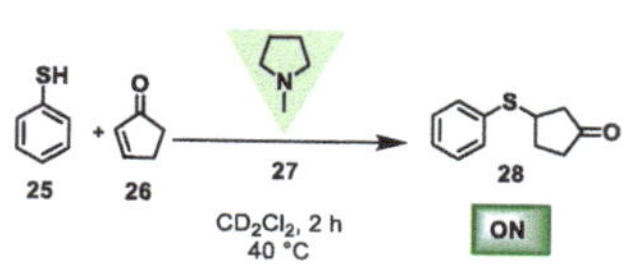

Fig. 7 Catalysis controlled by the three-state supramolecular walker (see Fig. 6). Adapted with permission from ref [1], Copyright 2020 by the American Chemical Society

copper(I) ions, one foot of the biped walked to the newly filled copper(I) phenanthroline site of **29** due to the formation of a picoline → [Cu(phenAr$_2$)]$^+$ complex. As a result, one ZnPor site is now free for binding the organocatalyst. Finally, addition of one more equiv. of copper(I) ions afforded the assembly [Cu$_2$(**29**)(**30**)]$^{2+}$ (State I). Due to the availability of both ZnPor units for binding the organocatalyst, organocatalysis should be turned OFF.

Indeed, when we heated the walker assembly [Cu$_2$(**29**)(**30**)]$^{2+}$, catalyst **27** and the substrates of a conjugate addition (**25** and **26**), there was no conjugate addition reaction observed. Addition of the electron-rich 2-ferrocenyl-9-mesityl-1,10-phenanthroline to sequentially remove the Cu$^+$ ions afforded first the three-component assembly [Cu(**29**)(**30**)]$^+$ and then the two-component assembly [**29**·**30**]. Upon removal of the first equiv of copper(I) ions, one picoline foot of the biped departed from its phenanthroline site and moved over to the ZnPor unit where the formerly attached catalyst **27** was released into solution. Heating now generated product **28** of the conjugate addition in 12% yield (State II). In State III all of the organocatalyst **27** was freed into solution. Heating State III afforded 24% yield of **28**, i.e. double the yield as that obtained in State II. The process turned out to be reversible.

3 Allosteric Regulation of Catalysis at Zinc-Porphyrin Stations (Dynamic Actuation)

Research over the past decades has demonstrated that nanomechanical motion is at the heart of some of the most amazing biological processes, such as ATP synthase or kinesin [3, 4]. Despite immense efforts in designing artificial machines, most devices so far have been created via covalent or topological methods [38–43], whereas nature assembles its highly sophisticated nanomechanical machinery through cooperative supramolecular interactions.

In the present chapter we will have a look at some multicomponent dynamic devices. Our strategy uses the principle of self-sorting, in which carefully chosen orthogonal binding motifs are combined in order to avoid interference. For a sophisticated example of three orthogonal complexation motifs [44], see Fig. 8.

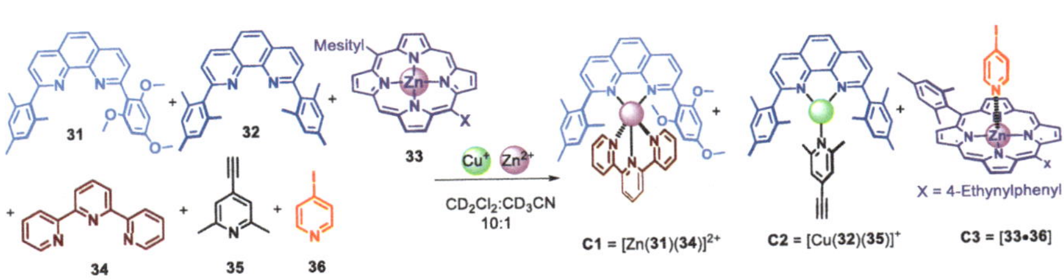

Fig. 8 Three orthogonal complexation motifs developed with simple ligands

Elaborating allosteric regulation of catalysis by dynamic effects required an understanding of the dynamics in zinc-porphyrin based rotor and slider systems. For this purpose, we have chosen zinc porphyrins with hydrogen atoms at *meso* positions because these proton sites are ^{1}H NMR sensitive to the formation of complexes at the zinc center.

3.1 Dynamic Devices: Zinc-Porphyrin-Based Rotors/Sliders

The three-component rotor [Zn(**37**)(**38**)]$^{2+}$ (Fig. 9) is held together by the strong HETTAP complexation motif (Heteroleptic Terpyridine And Phenanthroline, **C1**: $\log \beta = 15.1$, $\log K_1 \geq 6$) and the $N_{py} \rightarrow$ ZnPor (**C3**: $\log K = 4.45$) interaction [45]. Due to the geometric fitting of the components in [Zn(**37**)(**38**)]$^{2+}$, the pyridyl arm of the phenanthroline ligand may exchange between both degenerate ZnPor stations. Considering the weakness of the $N_{py} \rightarrow$ ZnPor bond, it was unsurprising to see that the exchange happened at 24 kHz (298 K) as determined by variable temperature (VT) ^{1}H NMR. Activation data were determined as $\Delta H^{\neq} = 52.5$ kJ mol^{-1}, $\Delta S^{\neq} = 16.3$ J mol^{-1} K^{-1} and at 298 K the activation free energy as $\Delta G^{\neq}_{298} = 47.6$ kJ mol^{-1}. Replacement of the zinc(II) by copper(I) ion led to another rotor with just slightly different data: $\Delta H^{\neq} = 52.5$ kJ mol^{-1}, $\Delta S^{\neq} = 19.8$ J mol^{-1} K^{-1} and $\Delta G^{\neq}_{298} = 46.6$ kJ mol^{-1} and actually a higher speed at 298 K: 34.0 kHz. These data may be compared with the dynamics ($\Delta G^{\neq}_{298} = 32.4$ kJ mol^{-1}) about the pyridine **36** $\rightarrow$ ZnPor (in **33**) interaction [45] suggesting cooperative effects in the three-component rotors!

Fabrication of the five-component oscillator [CuZn$_2$(**39**)(**40**)$_2$(**41**)]$^{5+}$ convincingly demonstrated the utility of cooperative stabilization (Fig. 10) [44]. Initially, the three-component precursor [Zn$_2$(**39**)(**40**)$_2$]$^{4+}$ had only formed in ca. 95% (incomplete self-sorting [31]) because both HETTAP binding sites showed some dissociation at low concentration. However, when the oscillator arm **41** and copper(I) ion were added the resultant 5-component device **43** = [CuZn$_2$(**39**)(**40**)$_2$(**41**)]$^{5+}$ was afforded quantitatively, most likely due to cooperative stabilization. In **43**, the three orthogonal interactions (see Fig. 8) mutually stabilized the five-component device. A thermal

Fig. 9 The three-component rotor [Zn(**37**)(**38**)]$^{2+}$ operating at 24 kHz at room temperature

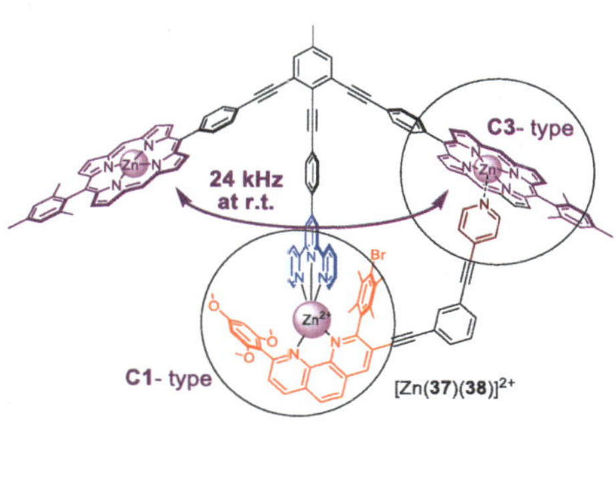

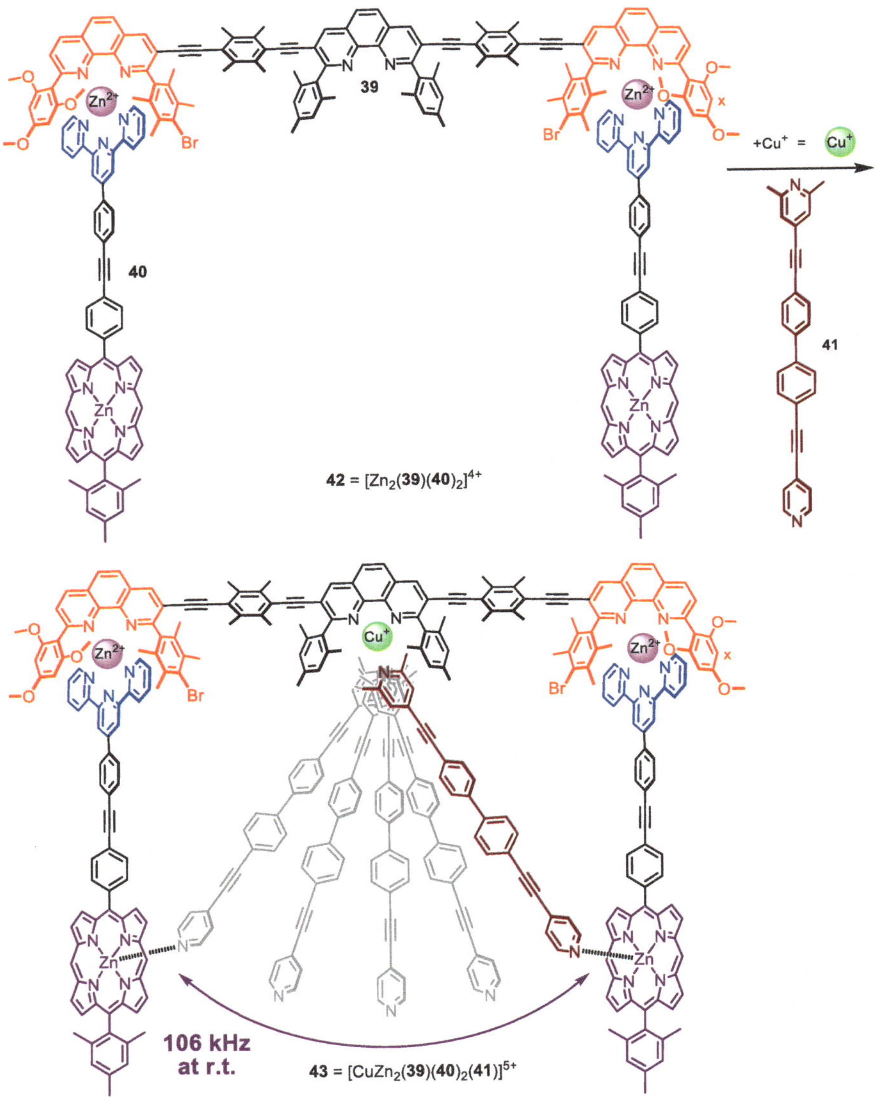

Fig. 10 A five-component oscillator built from three orthogonal interactions, see Fig. 8

oscillatory motion of 106 kHz, as determined by VT-^{1}H NMR, was observed and as expected the weak interaction in **21** was cleaved {log $K = 4.45$ (**C3**)}. The measured activation data amounted to $\Delta H^{\neq} = 45.0$ kJ mol^{-1}, $\Delta S^{\neq} = 6.9$ J mol^{-1} K^{-1} and $\Delta G^{\neq}_{298} = 43.6$ kJ mol^{-1}. Additional experimental and computational data indicated that the oscillator arm **41** is rather strained once its pyridine foot is attached to one of the ZnPor units (experimental strain energy: 4.7 kJ mol^{-1}) providing a reasonable

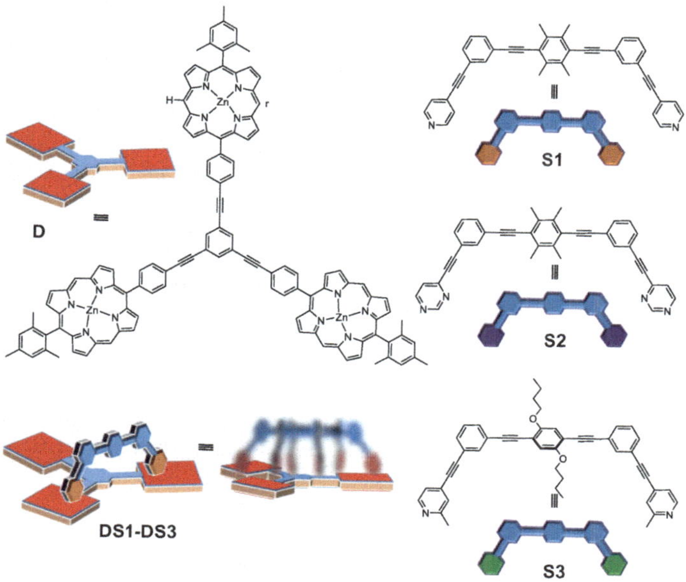

Fig. 11 Two-component slider-on-deck systems **DS1–DS3**. Adapted with permission from ref [46], Copyright 2018 WILEY-VCH Verlag GmbH & Co. KGaA, Weinheim

rationale why the experimental free activation energy of 43.6 kJ mol^{-1} at 298 K was lower than that of system [Zn(**37**)(**38**)]$^{2+}$.

The facile oscillation/rotation of pyridine-terminated arms between ZnPor units in [$CuZn_2$(**39**)(**40**)$_2$(**41**)]$^{5+}$ and [Zn(**37**)(**38**)]$^{2+}$ raised our interest to develop and investigate slider-on-deck systems.

The two-component slider-on-deck systems **DS1–DS3** (Fig. 11) were generated by mixing the tris-ZnPor deck **D** with one of the bipeds **S1–S3** in a 1:1 ratio [46]. Obviously, since the bipeds carry only two nitrogen feet and since the $N_{py} \rightarrow$ ZnPor dissociation is facile [47], we expected that the biped would slide over all three ZnPor units. Varying the feet of the biped allows addressing the speed of sliding in a rational manner, since lower binding to ZnPor should lead to faster sliding. The sliders feet, pyridine, pyrimidine and methylpyridine were selected because of their vastly different binding affinities. We determined the data with regard to a reference zinc porphyrin mimicking the ZnPor site in **D**: for 4-iodopyridine (py), 4-iodopyrimidine (pym) and 4-bromo-2-methylpyridine (pic) we measured log $K_{py} = 4.31 \pm 0.12$, log $K_{pym} = 3.35 \pm 0.15$ and log $K_{pic} = 2.72 \pm 0.04$.

First pieces of information on the dynamics of the slider-on-deck systems **DS1–DS3** were extracted from the ^{1}H NMR spectra at room temperature, because in all systems alike a single ^{1}H NMR signal was detected for protons r-H at all three ZnPor units. If a static situation would prevail in **DS1–DS3** two sets of signals were expected at a ratio 2:1. Indeed, upon cooling, the signals of r-H in **DS1–DS3** started to first

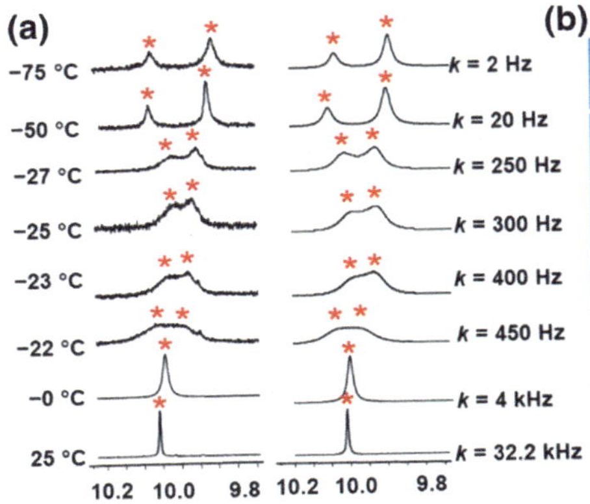

(b)

Slider-on-deck	DS1	DS2	DS3
$\Delta H^{\ddagger}$ (kJ mol^{-1})	54.7 ± 0.5	45.5 ± 0.6	42.9 ± 0.6
$\Delta S^{\ddagger}$ (J mol^{-1} K^{-1})	24.8 ± 2.2	10.1 ± 2.5	7.5 ± 2.5
$\Delta G^{\ddagger}$ (kJ mol^{-1})	47.3	42.5	40.7
k_{25} (s^{-1})	32200	220000	440000
Coalescence temp (°C)	−22	−35	−50

Fig. 12 **a** Partial VT-^{1}H NMR of proton signal r-H in **DS1**. **b** Kinetic data for **DS1–DS3**. Adapted with permission from ref [46], Copyright 2018 WILEY-VCH Verlag GmbH & Co. KGaA, Weinheim

broaden and then separated in two singlets (2:1). The downfield signal was assigned to the free ZnPor and the upfield one to the ZnPor units loaded with biped.

The activation data are depicted in Fig. 12b. As expected, the weaker $N_{\text{pym}} \rightarrow$ ZnPor interaction ($\log K_{2\cdot 4} = 3.35 \pm 0.15$) in **DS2** showed an exchange frequency of $k_{25} = 220$ kHz thus representing a faster sliding process than that in **DS1** ($k_{25} = 32.2$ kHz) with its stronger $N_{\text{py}} \rightarrow$ ZnPor binding ($\log K_{\text{py}} = 4.31 \pm 0.12$). Accordingly, the very weak $N_{\text{pic}} \rightarrow$ ZnPor interaction in **DS3** ($\log K_{\text{pic}} = 2.72 \pm 0.04$) resulted in the fastest exchange frequency (440 kHz at 25 °C).

For the motion in **DS1–DS3**, one may consider two different mechanisms: (1) full dissociation into **D** and **S1–S3** followed by reassociation, or (2) sliding from one ZnPor unit to the other through disconnection/reconnection of just one foot of the biped. The first possibility demanding the complete dissociation of **S1–S3** and **D** proved to be energetically inadequate with thermodynamic data. Moreover, the activation barrier of [Zn(**37**)(**38**)]$^{2+}$ serving as a reference operating definitely on a single $N_{\text{py}} \rightarrow$ ZnPor cleavage is basically identical to that of **DS1**.

3.2 Dynamic Allosteric Regulation of Catalysis

Many heterocycles that are known for their action as organocatalyst bind strongly to ZnPor units and are in such complexes unavailable for catalysis. For instance, see the model conjugate addition of thiophenol (**25**) with cyclopentenone **26** catalyzed by *N*-methylpyrrolidine (**27**). In the range from 25 to 60 °C, the yields of product **28** after 4 h showed the expected increase with higher temperature (Fig. 13). In presence of ZnTTP and catalyst **27** (both 10 mol%), though, the very same reaction afforded no addition product **28** at temperatures below 50 °C. Obviously, product formation is impeded due to the strong binding of *N*-methylpyrrolidine (**27**) to zinc

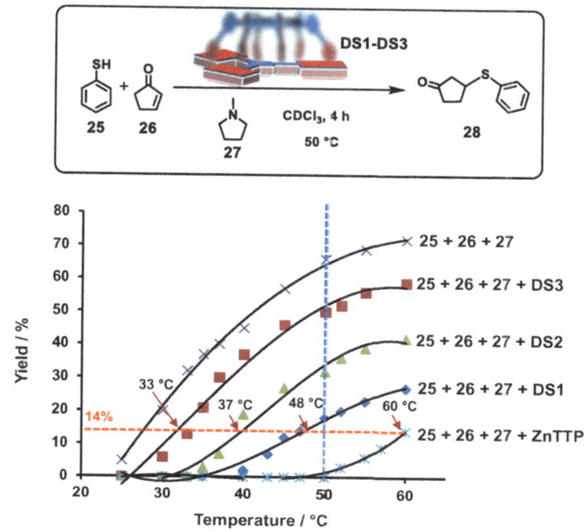

Fig. 13 The test reaction (Top). Temperature dependence of the yield of product **28** (Bottom). Adapted with permission from ref [46], Copyright 2018 WILEY-VCH Verlag GmbH & Co. KGaA, Weinheim

tetraphenylporphyrin ZnTTP (log $K = 4.01 \pm 0.12$). It is clear that static binding of **27** to ZnTTP reduces the availability of the free organocatalyst in solution for catalysis. However, at temperatures above 50 °C the entropy of dissociation shifts the equilibrium so that released **27** is able to perceptibly generate product formation.

The facile sliding of the biped across the surface in **DS1–DS3** raised the question whether a 1:1 mixture of **DS1** and **27** would impede catalysis in the same manner as a 1:1 mixture of ZnTTP and **27**. Our expectation, triggered by the release seen in the systems described in chapter "From the Pascaline to a 5-Digits Metallic and Planar Miniature Mechanical Calculator 1 cm in Lateral Size", was that the dynamics in the slider-on-deck systems would liberate the organocatalyst into solution with each movement of the biped. However, the catalyst would have the chance to reassociate to **DS1–DS3** at the freed ZnPor unit (dynamic actuation). How big will this effect be?

The catalytic activity of the three-component ensemble **27**·(**DS1–DS3**) was assessed at different temperatures using the same conditions as mentioned above (same concentrations, 4 h reaction time). We observed distinct yields for each individual slider-on-deck system with regard to the addition reaction (Fig. 13). In comparison with the impeded reference reaction (no yield at 50 °C), the system **27·DS1** catalyzed the conjugate addition discernibly better since it furnished (18 ± 2)% of **8** at 50 °C. The onset of some catalytic activity was seen for **27·DS2** already at 35 °C; at 50 °C it afforded thiol adduct **28** in (32 ± 2)%. Most impressively, the ensemble **27·DS3** led to an even more pronounced acceleration of the conjugate addition. There was a notable catalytic activity already at a temperature of 30 °C. The yield of **28** at 50 °C was (50 ± 2%). Control experiments revealed that product formation under these conditions was kinetically controlled and that the slider-on-deck systems **DS1–DS3** themselves were catalytically inactive.

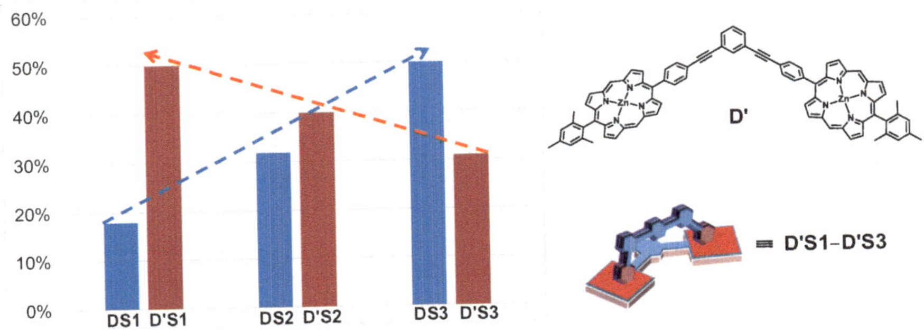

Fig. 14 Yields of **25** + **26** → **28** in presence of **27**·(**DS1–DS3**) and **27**·(**D'S1–D'S3**). Adapted with permission from ref [46], Copyright 2018 WILEY-VCH Verlag GmbH & Co. KGaA, Weinheim

To understand the effect of the dynamics on the catalysis, we evaluated how much of the organocatalyst was liberated from the three-component ensembles **27**·(**DS1–DS3**) at 50 °C. The analysis revealed that **27**·**DS3** exposed the highest catalytic activity due to 76% of catalyst **27** being liberated from its ZnPor binding sites. The catalytic machinery **27**·**DS2** and **27**·**DS1** showed a release of 48% and 29% of catalyst **27**. Clearly, the liberation of catalyst is more noticeable, the higher the sliding rate and the weaker the thermodynamic binding within the slider-on-deck systems.

An interesting detail on the side is that the data allow prediction of the presumed outcome at zero and infinite speed of the catalytic machinery. At infinite sliding speed the system should liberate 100% of **28** thus providing 66% of **28**, i.e., the same yield as the unimpeded model reaction at 50 °C. Contrastingly, at zero sliding speed, the ZnPor unit would act as a static binding site and the yield should be the same as that in the model reaction with ZnTPP acting as inhibitor, i.e., 0% at 50 °C.

For further clarification of the underlying mechanisms in the catalytic machinery, the corresponding static systems **D'S1–D'S3** were studied (Fig. 14). In **D'S1** to **D'S3** the same bipeds were combined with the bis(zinc porphyrin) platform **D'**. Obviously in these systems the feet of the bipeds can equally detach from the ZnPor platforms but due to the lack of alternative binding sites the feet will usually return to the site from which they departed. As long as the feet are detached from the ZnPor platforms, any added organocatalyst may bind and thus may be inactive during the time it forms a complex.

In Fig. 14, the data clear demonstrate that the effects on catalysis are opposite for **D'S1–D'S3** as compared to **DS1–DS3**. When the binding of the biped to the platform **D'** becomes weaker in the series **D'S1** > **D'S2** > **D'S3** the ZnPor units will be more often freed from the intrasupramolecular binding of the biped with the result that the organocatalyst is now bound. As a consequence, the external organocatalyst is immobilized at the ZnPor unit and the catalytic yield drops. With the dynamic systems **DS1** to **DS3** the trend is opposite.

3.3 *Thermodynamic Versus Kinetic Control in Dynamic Device*

Since the exchange frequency in the slider-on-deck systems **DS1-DS3** increased with decreasing binding ability of the biped's feet, the findings in chapter "Photo-Microlithography Fabrication of the Parts of a Micro-Mechanical Calculator" raised the question whether the higher catalytic activity with higher speed is a thermodynamic or a kinetic effect.

For an exact numerical analysis it is advised to analyze how the rate of liberation and reassociation of the catalyst develops at the "third" site that is not occupied by the biped of the individual sliders **DS1–DS3**. Here we will use a model very competently developed for a divalent host and a divalent guest [48] and described by Huskens to evaluate the dissociation in the slider-on-deck systems. It will use the concept of the effective molarity.

3.3.1 Rate of Liberating the Catalyst

The starting point of our explanation is the description of the simple systems of host (H) and guest (G) as depicted in Fig. 15. K_x are association constants, $k_{a,x}$, $k_{d,x}$ are association and dissociation rate constants with x denoting either the intrinsic (=i) binding site or binding sites 1 and 2 of the divalent system.

The situation for the monovalent host-guest system (Fig. 15a) is textbook knowledge and reminds us that the equilibrium constant can be described as a quotient of the association and dissociation rate constants ($K_i = k_{a,i}/k_{d,i}$). For the divalent host - divalent guest system (Fig. 15b) the first binding constant K_1 can be expressed alike that of K_i after taking in account a statistical factor of four due to the presence of two binding sites in both host and guest. The second binding constant K_2 has to be unitless because both $k_{a,2}$ and $k_{d,2}$ the same unit. In combination with the intrinsic rate constants, K_2 is best assessed using the concept of the effective molarity (EM)

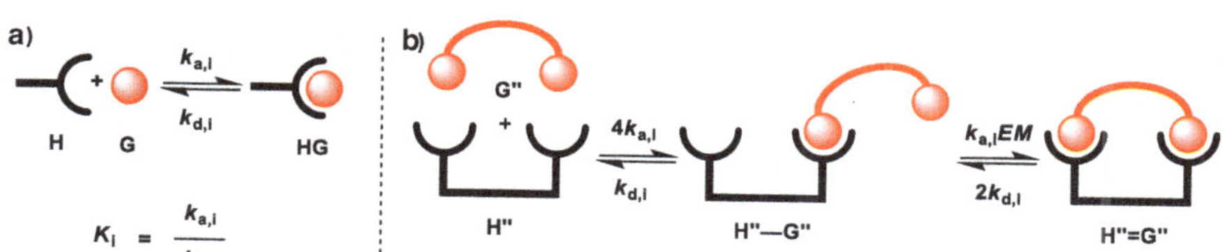

$$K_i = \frac{k_{a,i}}{k_{d,i}} \qquad K_i = \frac{[HG]}{[H][G]}$$

$$K_1 = \frac{k_{a,1}}{k_{d,1}} = \frac{4k_{a,i}}{k_{d,i}} = 4K_i \qquad K_2 = \frac{k_{a,2}}{k_{d,2}} = \frac{k_{a,i}EM}{2k_{d,i}} = \frac{1}{2}K_iEM$$

$$K_1 = \frac{[H''\text{—}G'']}{[H''][G'']} \qquad K_2 = \frac{[H''\text{=}G'']}{[H''\text{—}G'']} \qquad K_{ov} = K_1K_2 = 2K_i^2EM$$

Fig. 15 Numerical treatment of the equilibrium of divalent host and divalent guest molecules and the importance of the effective molarity (EM), see reference [48]

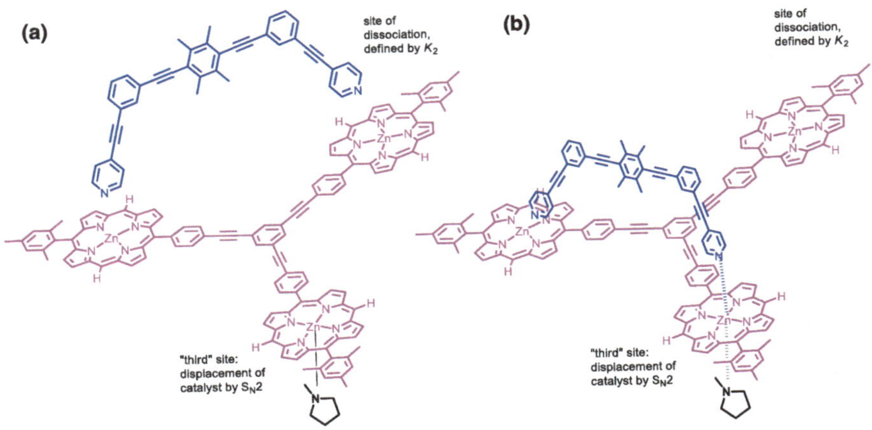

Fig. 16 Depiction of decisive intermediates in the dynamic liberation of catalyst molecules

the latter having M(olarity) as unit. After formation of complex H″ − G″ the intra-complex association rate constant $k_{a,2}$ is described as product of $k_{a,i}$ and *EM* whereas the dissociation rate constant $k_{d,2}$ is twice that of $k_{d,i}$ as there are two sites that may dissociate. With these relationships in mind one can write down

$$K_2 = {}^1/_2 K_i EM \tag{1}$$

$$EM = 2K_2/K_i = 2K_2 k_{d,i}/k_{a,i} \tag{2}$$

With these expressions from reference [48] in mind we can now address the situation in the slider-on-deck systems. Since the sliding barriers indicate that the rate determining step is the dissociation of one foot (of the biped) from the ZnPor site (Fig. 16a), we expect that this step is described by the rate constant $k_{d,2}$ of the divalent host—divalent guest system. A static snapshot at time $t = 0$ would reveal that two of the three ZnPor sites are occupied by the biped, the third one is coordinated to the *N*-methylpyrrolidine. Such ensemble, if static, would be catalytically inactive in the reference reaction described in chapter "Photo-Microlithography Fabrication of the Parts of a Micro-Mechanical Calculator".

Let's have a look at the rate of liberation of the catalyst assuming a nucleophilic substitution of the S_N2-type at the third ZnPor site (Fig. 16b). In this step the dissociated foot of the biped would displace the catalyst into solution. The rate of liberation, $v_{cat,lib}$, would be the product of the association rate constant at ZnPor 3, i.e. $k_{a,3}$, multiplied by *EM* and a statistical factor. The factor ½ is derived from the fact that the foot can return to two different ZnPor sites, either site 2 from which it departed or site 3.

$$v_{cat,lib} = 0.5k_{a,3}[\text{slider foot}] = 0.5k_{a,3} \cdot EM \tag{3}$$

If one replaces *EM* by the expression in Eq. (2) and considers two possible scenarios then the outcome is:

(a) $k_{a,3} = k_{a,i}$ (same barrier of association of the free foot to ZnPor with and without catalyst attached; i.e. $k_{a,3} = k_{a,i}$), then

$$v_{cat,lib} = 0.5k_{a,3} \cdot EM = k_{a,3} \cdot K_2 \cdot k_{d,i}/k_{a,i} = K_2 \cdot k_{d,i} \tag{4}$$

(b) $k_{a,3} < k_{a,i}$ {the rate constant for association to the catalystloaded ZnPor site is smaller than that for the free ZnPor, hence, $k_{a,3} = \mathrm{f} \cdot k_{a,i}$ ($\mathrm{f} < 1$)}

$$v_{cat,lib} = 0.5k_{a,3} \cdot EM = k_{a,3} \cdot K_2 \cdot k_{d,i}/k_{a,i} = \mathrm{f} \cdot K_2 \cdot k_{d,i} \tag{5}$$

In both scenarios, a higher dissociation rate constant $k_{d,i}$ and higher association equilibrium constant K_2 lead to a higher rate of liberation of catalyst. Which factor is more important, the thermodynamics represented by K_2 or the kinetics embodied by $k_{d,i}$? Here, the analysis of the three sliders with their different feet (pyridine = py, pyrimidine = pym, picoline = pic) provides a crisp clue. Since K_2 (py) > K_2 (pym) > K_2 (pic) (see the experimental binding data), the outcome under thermochemical control for $v_{cat,lib}$ should follow the sequence py > pym > pic. As demonstrated, however, the experimental data on catalyst liberation measured by catalytic activity showed the opposite picture: pic > pym > py. This finding suggests that the rate constant $k_{d,i}$ is by far the more important contributor (for $k_{d,i}$ the result is pic > pym > py, see Fig. 12b), in agreement with the postulate that the "machine" speed determines the liberation of the catalyst and thus the catalytic conversion.

3.3.2 Equilibrium Amount of Liberated Catalyst

For the catalysis obviously the equilibrium amount of free catalyst is the important factor, not the rate of liberation. For a more complete picture we thus have to consider not only $v_{cat,lib}$ but also $v_{cat,ass}$ (the rate of the catalyst associating to the ZnPor units). In the equilibrium both will need to be identical!

For the rebound of the catalyst to the slider-on-deck, there are several options to be considered. Evidently, the displacement of a bound slider foot by the catalyst {see pathway (b) in Fig. 17A} should be kinetically more difficult than the simple association with an empty ZnPor site {see pathway (a)}. Case (b) thus should play a minor role and is neglected in the following discussion. Anyhow, in the S_N2 displacement of a bound slider foot by the catalyst along pathway (b) the rate $v_{cat,ass} = k_{cat,ass,SN2} \cdot$ [free cat] should develop along pic > pym > py as judged by the binding data of the foot pieces, and thus opposite to the experimental finding.

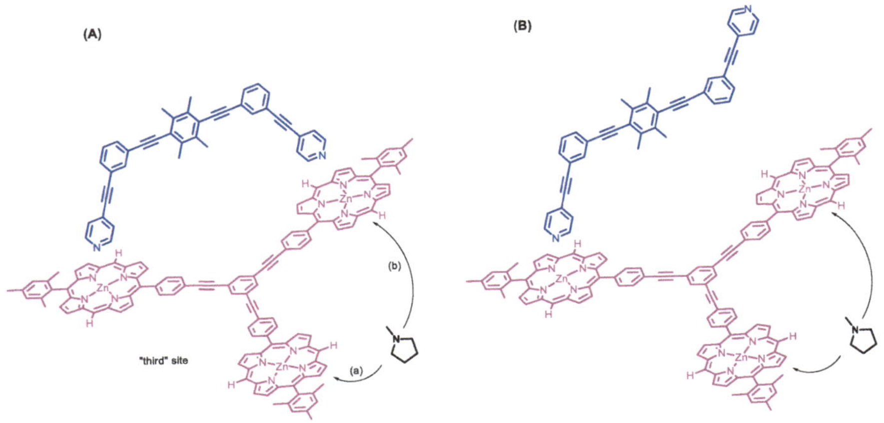

Fig. 17 Depiction of options for the rebound of catalyst molecules

Considering now pathway (a) (Fig. 17A) we assume that the free ZnPor site in all three slider-on-deck systems should have the same rate of association with the catalyst $v_{cat,ass} = k_{cat,ass} \cdot$ [free cat], as the biped's interaction with the free ZnPor site should be negligible. The rate should thus develop along: py = pym = pic.

An even more in-depth analysis would evaluate the rebound of the catalyst to three different species; i.e. the free zinc porphyrin site(s) in (i) the slider-on-deck with both feet on the deck (Fig. 17A), (ii) the slider-on-deck with only one foot on the deck (Fig. 17B) and the finally (iii) the deck with the slider being fully dissociated. The amounts of the entities (i)–(iii) may be derived from thermodynamic considerations. To account for the different amounts of ZnPor sites available in the three species, the rates are multiplied with factors 1/3, 2/3 and 1. The total concentration of the slider-on-deck is [DS]. The rate constant reflects the intrinsic rate constant $k_{cat,ass}$ (unknown) of the *N*-methylpyrrolidine (=cat) to the zinc porphyrin unit.

$$v_{cat,ass} = \underbrace{(\tfrac{1}{3} k_{cat,ass}}_{\text{Term for (i)}} + \underbrace{\tfrac{2}{3} k_{cat,ass} K_{-2}}_{\text{Term for (ii)}} + \underbrace{k_{cat,ass} K_{-1} K_{-2}}_{\text{Term for (iii)}}) \, [DS] \bullet [\text{free cat}] \quad (6)$$

Term for (i) Term for (ii) Term for (iii)

For a simple approximation only the first term is relevant, because K_{-1} and K_{-2} (dissociation constants, see Fig. 15) are much smaller than the increase in the factor when going from 1/3 to 2/3 and 1. Finally, we obtain an expression that is independent of the rate and thermodynamics of the slider:

$$v_{cat,ass} = 1/3 k_{cat,ass} [DS] \cdot [\text{free cat}] \quad (7)$$

The message from this subchapter is that the theoretical models either claim independence of the association rate from the rate of sliding or they suggest the sequence $v_{cat,ass:}$ pic > pym > py, and thus opposite to the experimental finding.

Summarizing chapter "Photo-Microlithography Fabrication of the Parts of a Micro-Mechanical Calculator", the association rate of the catalyst onto the ZnPor station(s) is not able to explain the experimental finding, whereas the dissociation rate constant of the slider (as derived in Sect. 3.3.1), which also determines the sliding speed, is the key factor as it increases the *EM* of the dissociated biped's foot. The *EM* is directly correlated with the dissociation rate of the catalyst. The higher the machine speed the higher the catalytic activity in the catalytic machinery **27**·(**DS1–DS3**).

3.4 Thermodynamic Versus Kinetic Control in Static Devices

As described in chapter "Photo-Microlithography Fabrication of the Parts of a Micro-Mechanical Calculator" the behavior of the dynamic catalytic machineries **27**·(**DS1–DS3**) was opposite to that of the "static" catalytic machineries **27**·(**D'S1–D'S3**). A theoretical treatment of **27**·(**D'S1–D'S3**) recognizes that dissociation of the biped occurs at the same site were the catalyst molecule will associate. One can still apply the concept and the binding constants (K_2) of the divalent host—divalent guest system (Fig. 15), but now it follows the classic scheme.

For the **D'S**, the intramolecular dissociation of the biped from the deck should be in equilibrium with the association of the free catalyst onto the same ZnPor site:

$$K_2 = K_{\mathrm{cat,ass}}[\text{free cat}] \tag{8}$$

The stronger the biped is bound to the deck the higher the concentration of the free catalyst. Hence, [free cat] and the catalytic activity will decrease along py > pym > pic! This yield development is opposite to that for **27**·(**DS1–DS3**) but exactly in line with that in our control experiments with **27**·(**D'S1–D'S3**).

4 Perspectives for Other Systems

While the theoretical treatment in Sects. 3.2–3.4 rationalizes the finding that the liberation of catalyst from the machinery **27**·(**DS1–DS3**) is more the faster the machine speed, the legitimate question arises whether this feature may be exploited for other modes of molecular machines as well. Indeed, it turned out that the same theoretical treatment predicts that in catalytic machines impeded by product inhibition faster machine speed should lead to an increased turn-over in the catalyst. The concept appreciates that nanomechanical actuation started at a remote site by dissociation of a moving arm should be able to "kick out" the product alike the catalyst was liberated in the slider-on-deck systems (Fig. 11).

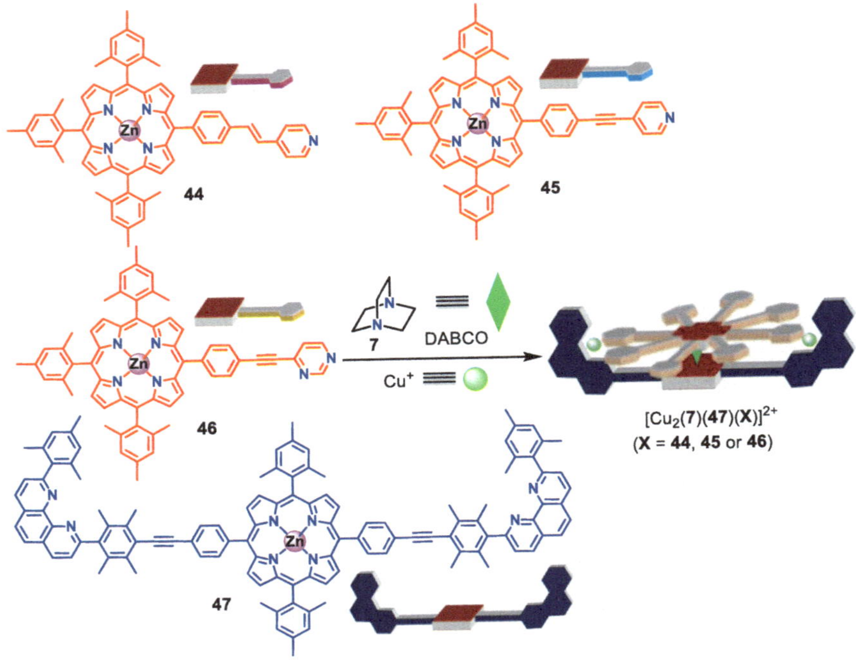

Fig. 18 Components of the four-component nanorotors $[Cu_2(\mathbf{7})(\mathbf{47})(\mathbf{X})]^{2+}$. DABCO serves as hinge between both zinc porphyrins in this hetero-sandwich assembly. Adapted with permission from ref [54]. Copyright 2018 by the American Chemical Society

4.1 Nanomechanical Reduction of Product Inhibition

Recently, the rotating four-component catalyst $[Cu_2(\mathbf{7})(\mathbf{47})(\mathbf{X})]^{2+}$ (Fig. 18), self-assembled from copper(I) ions, DABCO (**7**), stator **47** and various rotators (where **X** is **44**, **45** or **46**), were studied with regard to their rotational speed and catalytic activity in a click reaction [54]. Since in any given moment there should be one copper(I) phenanthroline site not occupied by the monodentate rotator **X** in $[Cu_2(\mathbf{7})(\mathbf{47})(\mathbf{X})]^{2+}$, we expected that some catalytic activity would unfold in a click reaction.

The ability of the nanorotors to effect click catalysis via the free copper(I) phenanthroline centers was probed for all the nanorotors listed above (Table 2).

When nanorotors $[Cu_2(\mathbf{7})(\mathbf{47})(\mathbf{X})]^{2+}$ were utilized as catalyst (10 mol%) for the click reaction of 9-(azidomethyl)anthracene (**48**) and (prop-2-yn-1-yloxy)benzene (**49**) at 55 °C (4 h), the rotor $[Cu_2(\mathbf{7})(\mathbf{47})(\mathbf{46})]^{2+}$ furnished the highest yield of click product **50** (62%) followed by rotor $[Cu_2(\mathbf{7})(\mathbf{47})(\mathbf{45})]^{2+}$ (44%) and $[Cu_2(\mathbf{7})(\mathbf{47})(\mathbf{44})]^{2+}$ (20%) (Table 2). The same trend was recognized in the reaction with another set of compounds, i.e. **51** and **52** furnishing **53** (Table 2). Remarkably, for both transformations the yield was linearly correlated with the exchange

Table 2 Activation data of nanorotors and catalytic reaction yields. Adapted with permission from ref [54], Copyright 2018 by the American Chemical Society

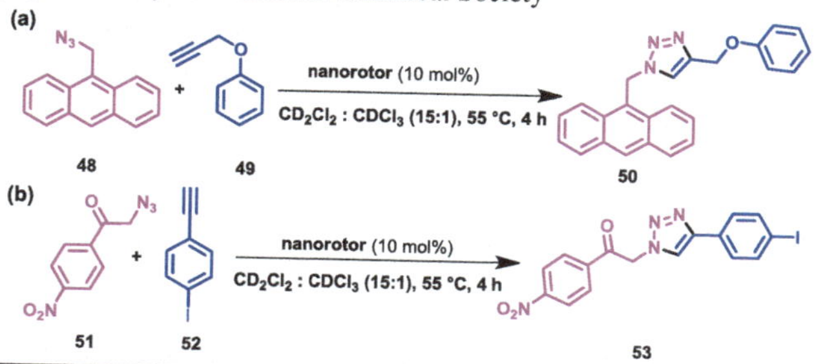

Nanorotor	$\Delta G^{\ddagger}_{25}$ (kJ mol^{-1})	k_{25} (s^{-1})	Yield of **50**	Yield of **53**
[Cu$_2$(**7**)(**47**)(**44**)]$^{2+}$	57.7 ± 0.7	0.50×10^3	20±1%	27±2%
[Cu$_2$(**7**)(**47**)(**45**)]$^{2+}$	48.6 ± 0.2	20.0×10^3	44±2%	42±2%
[Cu$_2$(**7**)(**47**)(**46**)]$^{2+}$	46.6 ± 0.3	42.0×10^3	62±2%	56±3%

frequency of the nanorotors; the faster the rotational exchange at the two copper(I)-loaded phenanthroline stations the higher the catalytic activity (Fig. 19). Even more convincing, for reaction (a) yielding **50** the measured rate of catalysis at zero time, v_0, was linearly correlated with the yield.

The hypothesis for the experimental correlations depicted in Fig. 19 is that higher rotational speed leads to a reduction of product inhibition by kicking out the product from the catalytic site. The catalytic site for all nanorotors is the copper(I) phenanthroline unit not occupied by the rotator head. When the rotator dissociates from the geometrically opposite copper(I) phenanthroline site of the stator it will be able to coordinate to the product-filled site after kicking out the product.

Indeed, it was independently confirmed that increased rotator motion liberates more of the product into solution. Moreover, when the yield of the click reaction was probed with copper(I) phenanthroline [Cu(**32**)]$^+$ as a "static" reference catalyst, the

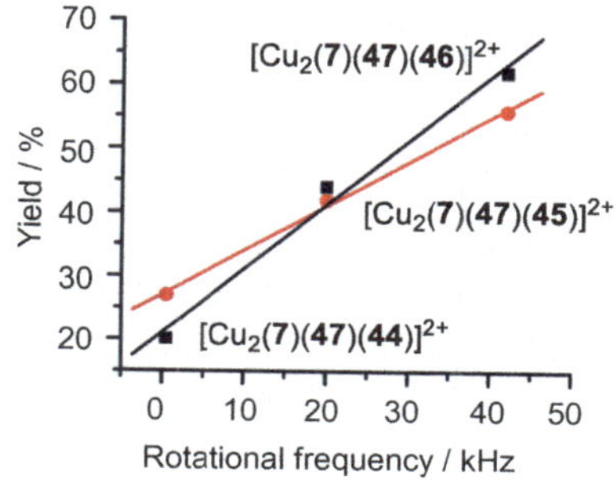

Fig. 19 Linear correlation between the yield of the click reaction and the rotational exchange frequency of the nanorotors. Data points (see Table 2) for reactions (**a**) and (**b**) are given in black squares and red dots, respectively. Adapted with permission from ref [54], Copyright 2018 by the American Chemical Society

conversion stopped basically at a turn-over number of 1. Clearly, product inhibition is a problem with "static" copper(I) phenanthrolines for the used test reaction.

The concept that higher rotating rates in rotational copper(I) catalysts lead to higher yields was noted in follow-up publications on the side [49]. It seems to be a general phenomenon.

5 Conclusion

The relation of protein conformational dynamics and enzymatic activity is fiercely debated [50, 51], since both accurate geometric arrangement and high dynamics need to function synergistically for high turnover rates in enzymes [52, 53]. It is thus a fascinating challenge to methodically use dynamic effects in artificial catalysts for increasing catalytic turnover rates and numbers.

The present chapter describes dynamic catalysts and catalytic machinery in which the catalytic activity is correlated with the machine speed. A rationale is given based on the numerical treatment of multivalent host-guest systems that the above described phenomenological observations are indeed caused by a kinetic effect.

Acknowledgements The authors are highly obliged to a gifted and highly motivated group of colleagues and coworkers, whose names are given in the cited references. Moreover, we would like to express our gratefulness to the Deutsche Forschungsgemeinschaft for continued funding of our research (Schm 647/19-1, 19-2 and 20-1, 20-2) and the University of Siegen.

References

1. Goswami, A., Saha, S., Biswas, P.K., Schmittel, M.: (Nano)mechanical motion triggered by metal coordination: from functional devices to networked multicomponent catalytic machinery. Chem. Rev. **120**, 125 (2020)
2. Noji, H., Yasuda, R., Yoshida, M., Kinosita Jr., K.: Direct observation of the rotation of F_1-ATPase. Nature **386**, 299 (1997)
3. Walker, J.E.: The ATP synthase: the understood, the uncertain and the unknown. Biochem. Soc. Trans. **41**, 1 (2013)
4. Hirokawa, N.: Kinesin and dynein superfamily proteins and the mechanism of organelle transport. Science **279**, 519 (1998)
5. Vale, R.D., Milligan, R.A.: The way things move: looking under the hood of molecular motor proteins. Science **288**, 88 (2000)
6. Sauvage, J.-P.: From chemical topology to molecular machines (Nobel Lecture). Angew. Chem. Int. Ed. **56**, 11080 (2017)
7. Stoddart, J.F.: Mechanically interlocked molecules (MIMs)-molecular shuttles, switches, and machines (Nobel Lecture). Angew. Chem. Int. Ed. **56**, 11094 (2017)
8. Feringa, B.L.: The art of building small: from molecular switches to motors (Nobel Lecture). Angew. Chem. Int. Ed. **56**, 11060 (2017)
9. Davis, M., Hassel, O., Arro, L., Kvande, P.C., Meisingseth, E.: Electron diffraction investigation of molecules containing a cyclohexane type six-membered ring. Acta Chem. Scand. **17**, 1181 (1963)

10. Barton, D.H.R., Cookson, R.C.: The principles of conformational analysis. Q. Rev., Chem. Soc. **10**, 44 (1956)
11. https://goldbook.iupac.org/terms/view/C01258
12. Anslyn, E.V., Dougherty, D.A.: Modern Physical Organic Chemistry. University Science Books (2006) Chapter 2.3
13. Barboiu, M., Lehn, J.-M.: Dynamic chemical devices: Modulation of contraction/extension molecular motion by coupled-ion binding/pH change-induced structural switching. Proc. Natl. Acad. Sci. **99**, 5201 (2002)
14. Linke-Schaetzel, M., Anson, C.E., Powell, A.K., Buth, G., Palomares, E., Durrant, J.D., Balaban, T.S., Lehn, J.-M.: Dynamic chemical devices: photoinduced electron transfer and its ion-triggered switching in nanomechanical butterfly-type bis(porphyrin)terpyridines. Chem. Eur. J. **12**, 1931 (2006)
15. Samanta, S.K., Rana, A., Schmittel, M.: Conformational slippage determines rotational frequency in five-component nanorotors. Angew. Chem. Int. Ed. **55**, 2267 (2016)
16. Barboiu, M., Stadler, A.-M., Lehn, J.-M.: Controlled folding, motional, and constitutional dynamic processes of polyheterocyclic molecular strands. Angew. Chem. Int. Ed. **55**, 4130 (2016)
17. Gaikwad, S., Goswami, A., De, S., Schmittel, M.: A metalloregulated four-state nanoswitch controls two-step sequential catalysis in an eleven-component system. Angew. Chem. Int. Ed. **55**, 10512 (2016)
18. Goswami, A., Schmittel, M.: Heteroleptic copper phenanthroline complexes in motion: from stand-alone devices to multi-component machinery. Coord. Chem. Rev. **376**, 478 (2018)
19. Lee, E., Ju, H., Park, I.-H., Jung, J.H., Ikeda, M., Kuwahara, S., Habata, Y., Lee, S.S.: Pseudo[1]catenane-type pillar[5]thiacrown whose planar chiral inversion is triggered by metal cation and controlled by anion. J. Am. Chem. Soc. **140**, 9669 (2018)
20. Kirksey, C.H., Hambright, P., Storm, C.B.: Stability constants and proton magnetic resonance studies of zinc α, β, γ, δ-tetraphenylporphin and substituted pyridines. Inorg. Chem. **8**, 2141 (1969)
21. Mizutani, T., Wada, K., Kitagawa, S.: Molecular recognition of amines and amino esters by zinc porphyrin receptors: binding mechanisms and solvent effects. J. Org. Chem. **65**, 6097 (2000)
22. Satake, A., Yoshiaki Kobuke, Y.: Dynamic supramolecular porphyrin systems. Tetrahedron **61**, 13 (2005)
23. Schmittel, M., De, S., Pramanik, S.: Reversible ON/OFF nanoswitch for organocatalysis: mimicking the locking and unlocking operation of CaMKII. Angew. Chem. Int. Ed. **51**, 3832 (2012)
24. Favereau, L., Cnossen, A., Kelber, J.B., Gong, J.Q., Oetterli, R.M., Cremers, J., Herz, L.M., Anderson, H.L.: Six-coordinate zinc porphyrins for template-directed synthesis of spiro-fused nanorings. J. Am. Chem. Soc. **137**, 14256 (2015)
25. Schulman, H.: The multifunctional Ca^{2+}/calmodulin-dependent protein kinases. Curr. Opin. Chem. Biol. **5**, 247 (1993)
26. Schmittel, M., Pramanik, S., De, S.: A reversible nanoswitch as an ON-OFF photocatalyst. Chem. Commun. **48**, 11730 (2012)
27. De, S., Pramanik, S., Schmittel, M.: A toggle nanoswitch alternately controlling two catalytic reactions. Angew. Chem. Int. Ed. **53**, 14255 (2014)
28. Mukhopadhyay, P., Wu, A., Isaacs, L.: Social self-sorting in aqueous solution. J. Org. Chem. **69**, 6157 (2004)
29. Samanta, D., Mukherjee, P.S.: Multicomponent self-sorting of a Pd7 molecular boat and its use in catalytic Knoevenagel condensation. Chem. Commun. **49**, 4307 (2013)
30. Chervyachkova, E., Wegner, S.V.: Reversible social self-sorting of colloidal cell-mimics with blue light switchable proteins. ACS Synth. Biol. **7**, 1817 (2018)
31. Saha, M.L., Schmittel, M.: Degree of molecular self-sorting in multicomponent systems. Org. Biomol. Chem. **10**, 4651 (2012)

32. Mittal, N., Pramanik, S., Paul, I., De, S., Schmittel, M.: Networking nanoswitches for ON/OFF control of catalysis. J. Am. Chem. Soc. **139**, 4270 (2017)
33. Schmittel, M.: Dynamic functional molecular systems: from supramolecular structures to multi-component machinery and to molecular cybernetics. Isr. J. Chem. **59**, 197 (2019)
34. Remón, P., Pischel, U.: Chemical communication between molecules. Chem Phys Chem **18**, 1667 (2017)
35. Ashkenasy, G., Hermans, T.M., Otto, S., Taylor, A.F.: Systems chemistry. Chem. Soc. Rev. **46**, 2543 (2017)
36. Ludlow, R.F., Otto, S.: Systems chemistry. Chem. Soc. Rev. **37**, 101 (2008)
37. Mittal, N., Özer, M.S., Schmittel, M.: Four-component catalytic machinery: reversible three-state control of organocatalysis by walking back and forth on a track. Inorg. Chem. **57**, 3579 (2018)
38. McConnell, A.J., Wood, C.S., Neelakandan, P.P., Nitschke, J.R.: Stimuli-responsive metal-ligand assemblies. Chem. Rev. **115**, 7729 (2015)
39. Erbas-Cakmak, S., Leigh, D.A., McTernan, C.T., Nussbaumer, A.L.: Artificial molecular machines. Chem. Rev. **115**, 10081 (2015)
40. Cheng, C., Stoddart, J.F.: Wholly synthetic molecular machines. Chem. Phys. Chem. **17**, 1780 (2016)
41. Kassem, S., van Leeuwen, T., Lubbe, A.S., Wilson, M.R., Feringa, B.L., Leigh, D.A.: Artificial molecular motors. Chem. Soc. Rev. **46**, 2592 (2017)
42. van Dijk, L., Tilby, M.J., Szpera, R., Smith, O.A., Bunce, H.A.P., Fletcher, S.P.: Molecular machines for catalysis. Nat. Rev. Chem. **2**, 117 (2018)
43. Baroncini, M., Silvi, S., Credi, A.: Photo- and redox-driven artificial molecular motors. Chem. Rev. **120**, 200 (2020)
44. Samanta, D., Paul, I., Schmittel, M.: Supramolecular five-component nano-oscillator. Chem. Commun. **53**, 9709 (2017)
45. Goswami, A., Paul, I., Schmittel, M.: Three-component nanorotors generated from fusion of complexes and post-fusion metal–metal exchange. Chem. Commun. **53**, 5186 (2017)
46. Paul, I., Goswami, A., Mittal, N., Schmittel, M.: Catalytic three-component machinery: control of catalytic activity by machine speed. Angew. Chem. Int. Ed. **57**, 354 (2018)
47. Liu, S., Kondratuk, D.V., Rousseaux, S.A.L., Gil-Ramirez, G., O'Sullivan, M.C., Cremers, J., Claridge, T.D.W., Anderson, H.L.: Caterpillar track complexes in template-directed synthesis and correlated molecular motion. Angew. Chem. Int. Ed. **54**, 5355 (2015)
48. Huskens, J.: In: Huskens, J., Prins, L.J., Haag, R., Ravoo, B.J. (eds) Multivalency, chapter 2, pp. 23–52. Wiley, Hoboken (2018)
49. Goswami, A., Schmittel, M.: Double Rotors with Fluxional Axles: Domino Rotation and Azide– Alkyne Huisgen Cycloaddition Catalysis. Angew. Chem. Int. Ed. **59**, 12362 (2020)
50. Kim, T.H., Mehrabi, P., Ren, Z., Sljoka, A., Ing, C., Bezginov, A., Ye, L., Pomès, R., Prosser, R.S., Pai, E.F.: The role of dimer asymmetry and protomer dynamics in enzyme catalysis. Science **355** (2017). https://doi.org/10.1126/science.aag2355
51. Luk, L.Y.P., Loveridge, E.J., Allemann, R.K.: Protein motions and dynamic effects in enzyme catalysis. Phys. Chem. Chem. Phys. **17**, 30817 (2015)
52. Larion, M., Hansen, A.L., Zhang, F., Bruschweiler-Li, L., Tugarinov, V., Miller, B.G., Brüschweiler, R.: Kinetic cooperativity in human pancreatic glucokinase originates from millisecond dynamics of the small domain. Angew. Chem. Int. Ed. **54**, 8129 (2015)
53. Warshel, A., Bora, R.P.: Perspective: defining and quantifying the role of dynamics in enzyme catalysis. J. Chem. Phys. **144**, 180901 (2016)
54. Biswas, P.K., Saha, S., Paululat, T., Schmittel, M.: Rotating catalysts are superior: suppressing product Inhibition by anchimeric assistance in four-component catalytic machinery. J. Am. Chem. Soc. **140**, 9038 (2018)

Five Minutes in the Life of a Molecular Shuttle: Near-Equilibrium Measurements of Shuttling Dynamics Using Optical Tweezers

Kateryna M. Lemishko, Teresa Naranjo, Emilio M. Pérez, and Borja Ibarra

Abstract Molecular shuttles are prototypes of most complex synthetic molecular machines. In a molecular shuttle, a ring-shaped molecule or macrocycle is threaded onto a molecular axle. One of the most prominent features of these devices is that the macrocycle can shuttle reversibly between different recognition sites on the axle as a reaction to external stimuli. Molecular shuttles are currently of great interest to researchers due to their potential applications in various fields. Although kinetics and thermodynamics of these systems in bulk are well understood, the mechanistic principles of operation of molecular shuttles and their dynamics are not quantified yet. Here, we show how to use the single-molecule manipulation technique of optical tweezers to probe mechanically and perform near-equilibrium measurements on single molecular shuttles in near-physiological conditions. The method described in this chapter can be used to study the mechanical strength and the real-time operation of other artificial systems at the single-molecule level in near-physiological conditions.

Keywords Rotaxane · Single molecule · Optical tweezers · DNA · Nanodevice · Equilibrium

---

K. M. Lemishko · T. Naranjo · E. M. Pérez (✉) · B. Ibarra (✉)
Instituto IMDEA Nanociencia, Faraday 9, Campus de Excelencia UAM-CSIC, Cantoblanco, 28049 Madrid, Spain
e-mail: emilio.perez@imdea.org

B. Ibarra
e-mail: borja.ibarra@imdea.org

B. Ibarra
Nanobiotecnología (IMDEA-Nanociencia, Unidad Asociada al Centro Nacional de Biotecnología (CSIC)), 28049 Madrid, Spain

T. Naranjo
Instituto IMDEA Energía, Avda. Ramón de La Sagra, Parque Tecnológico de Móstoles 3, 28935 Móstoles, Madrid, Spain

C. Joachim (ed.), *Building and Probing Small for Mechanics*, Advances in Atom and Single Molecule Machines, https://doi.org/10.1007/978-3-030-56777-4_14

1 Introduction

Over the last few decades, researchers working in the field of supramolecular chemistry have shown that supramolecular systems, due to their structural organization and functional integration of their components, can perform useful tasks, thus demonstrating that the concept of a device can be transferred to the molecular level [1]. In such molecular devices, different components of the molecule can change their relative positions as a response to external stimuli [2, 3]. Prominent examples of such synthetic supramolecular systems are rotaxane-based molecular shuttles. These devices are constructed of a molecular chain encircled by a macrocycle that can move between different recognition sites (two or more) on the thread as a response to some external stimulus. Molecular shuttles are currently of great interest to researchers due to their potential applications in different fields, from molecular machinery to biomedicine [4–6]. Making the step from molecules to molecular machinery, which incorporates various molecular components to generate work, will require a precise control of the dynamics and mechano-chemical processes (conversion of thermal/chemical energy to motion and work) governing the operation of these systems at the nanoscale. Although the thermodynamics and kinetics of rotaxanes in bulk are well-understood [7], a good understanding of their operational dynamics and mechano-chemical properties at the single-molecule level is still missing. To date, Atomic Force Microscopy (AFM) is the most common tool to probe the mechanical properties of mechanically interlocked macromolecular systems [8–13] at the single-molecule level. However, classical AFM-based techniques present some drawbacks hindering the full comprehension of the dynamics of the non-covalent interactions ruling out the operation of these systems: (i) the stiffness of the cantilever limits force resolution to a range above ~20 picoNewtons (pN), which is higher than the strength of non-covalent interactions responsible for remodeling events at the molecular level [14] and higher than the forces characteristic of biological molecular motors [15, 16]; (ii) limited force precision and stability makes challenging to perform near-equilibrium measurements and (iii) non-specific adsorption of the sample to the cantilever and, in many cases, the lack of a proper reporter, has hindered the unambiguous identification of single-molecule events. We note that recent technological advances in the development of ultra-stable AFM devices are now starting to overcome these issues [17, 18].

In this chapter, we show how to use the single-molecule manipulation technique of optical tweezers to probe mechanically and perform near-equilibrium measurements on single molecular shuttles under aqueous conditions. Optical tweezers rely on forces imparted to matter by light [19]. Very generally, by focusing a laser beam into a fluidics chamber, objects in the solution like a micron-sized polystyrene bead can be optically trapped on the focus. An essential feature of the optical trap is that near the focus, it behaves as an extremely sensitive linear 'Hookean' spring with stiffness typically in the picoNewton/nanometer range providing high resolution force (1–100 pN) and position (1–10 nm) measurements. The extraordinary force stability and resolution of optical tweezers have been proven highly relevant for the study of the near-equilibrium dynamics of non-covalent interactions governing the structure of biological systems [20–23]. To interface individual molecular shuttles with optical tweezers, we connected a rotaxane molecule to two double-stranded DNA (dsDNA) molecules, which serve as handles for manipulation. In addition, such coupling enabled us to (*i*) probe the operation of the shuttle in aqueous (near-physiological) conditions, (ii) identify individual rotaxane-DNA constructs, and (iii) avoid non-specific interactions between the rotaxane and the surface of the beads.

The experimental design described in this chapter permitted us to measure the mechanical stability as well as real-time shuttling events of tens of individual rotaxane molecular shuttles. We hope this method could open new avenues to investigate the real-time operation of other synthetic devices at the single-molecule level.

2 Chemical Synthesis of a Rotaxane Suitable for Isolation and Mechanical Manipulation

We synthetized a rotaxane molecule containing an oligoethyleneglycol molecular thread with two binding stations: a fumaramide station and a succinic amide-ester station, Fig. 1. Fumaramide, due to the trans-orientation of its two amide carbonyls, binds a tetraamide macrocycle strongly by four hydrogen bonds, Fig. 2. Succinic amide-ester presents less affinity to the macrocycle than fumaramide, because of the substitution of one of the amides with an ester, a weaker hydrogen bond acceptor. Therefore, the fumaramide: succinic amide-ester occupancy ratio is biased towards the fumaramide station. Two spacious diphenylethyl groups on the two ends of the thread serve as stoppers to prevent the unlacing of the macrocycle, Fig. 2.

The axle for the shuttle was synthesized in 12 steps as described in Fig. 1 [24]. Briefly, and in reverse order of synthesis, the final thread (compound 12) was obtained from the (*Z/E* 55/45)-thread, labelled as compound 11 in Fig. 1. The (*Z/E* 55/45)-thread was synthesized from two constructs: compound 6 and compound 10. Compound 10 was obtained in 3 steps from *O,O*´-bis (2-aminoethyl) hexacosaethylene glycol, and compound 6 was synthesized in 3 steps from 2,2-diphenylethyl 6-([((9H-fluoren-9-yl)methoxy)carbonyl]amino)-2-aminohexanoate and *N*-Hydroxysuccinimidobiotin and purified by flash chromatography. The axle included additional functionalization for the manipulation of the shuttle with the optical tweezers. The diphenylethyl group close to the fumaramide station was functionalized with a biotin group, which will later be used to immobilize the rotaxane on a polystyrene bead functionalized with streptavidin (see below).

The macrocycle encircling the axle at the fumaramide station was synthesized out of 5-azidoisophthaloyl dichloride and N2,N6-bis(4-(aminomethyl)benzyl)pyridine-2,6-dicarboxamide, see Fig. 2. The former leaves an azide group that will be used as an attachment point to connect the macrocycle to the optically trapped bead via DNA (see below).

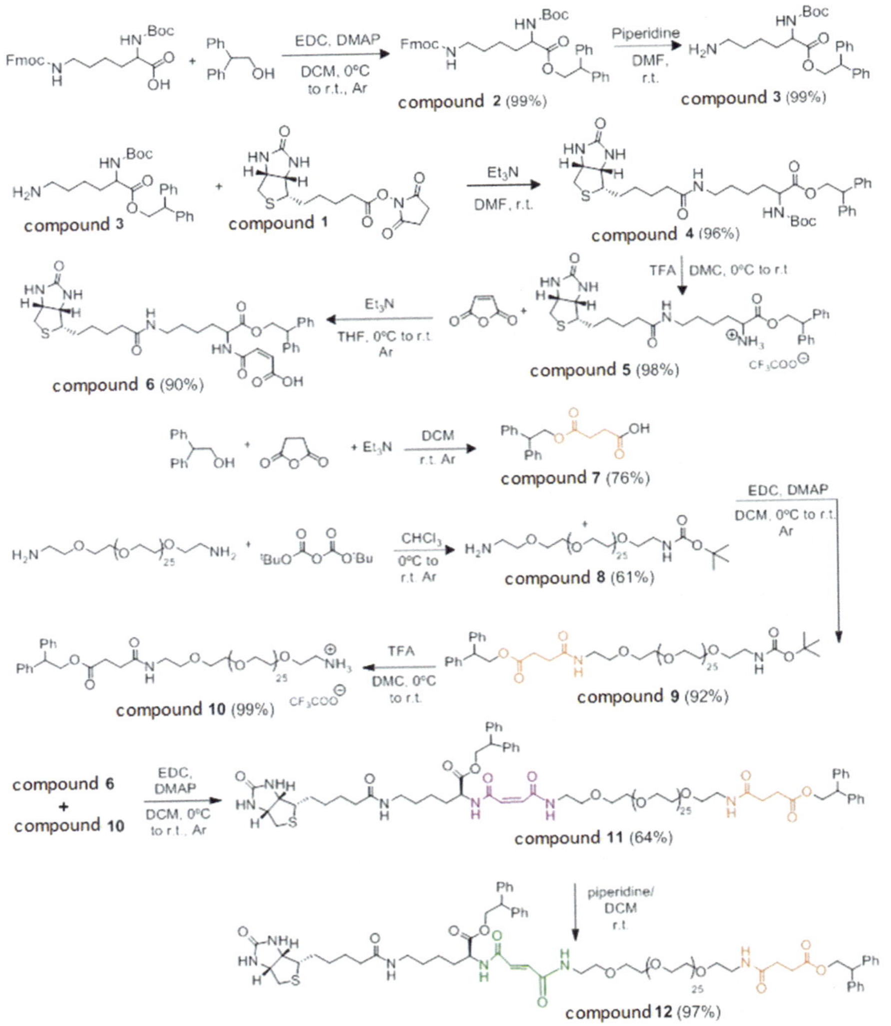

Fig. 1 **Synthesis of the axle.** Schematic representation of the synthesis pathway of the rotaxane' axle (compound 12). Each step is indicated by an arrow specifying the reaction conditions. Adapted with permission from [24] (https://creativecommons.org/licenses/by/4.0/)

3 Coupling of the Rotaxane with Optical Tweezers

Coupling the rotaxane with the optical tweezers involved the following sequential steps:

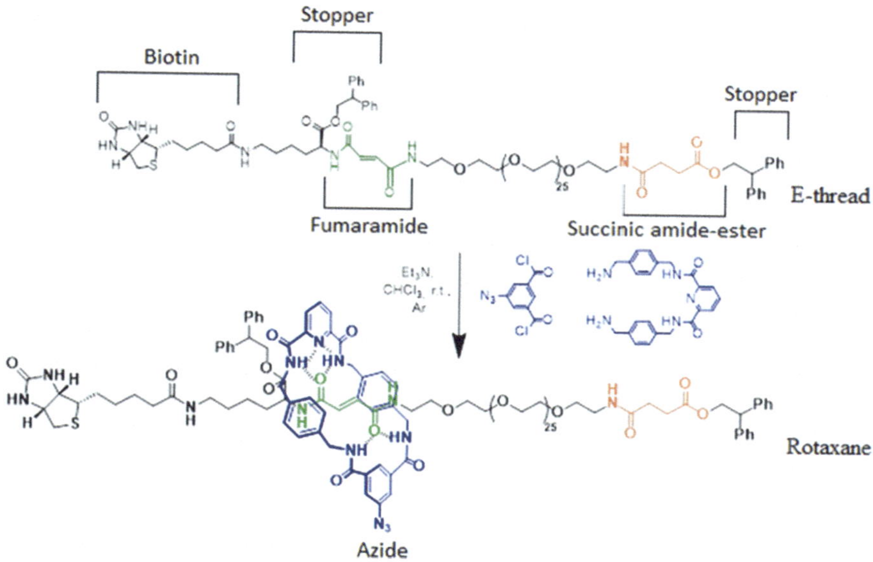

Fig. 2 Synthesis of the macrocycle. 5-azidoisophtaloyl and N^2,N^6-bis(4-(aminomethyl)benzyl)pyridine-2,6-dicarboxamide react at room temperature in chlorophorm to form a macrocycle around the fumaramide station of the E-thread. Triethylamine (Et_3N) is used in this reaction as a base to remove HCl. The azide at the macrocycle and the biotin at the stopper close to the fumaramide station work as attachment points to interface the rotaxane with the optical tweezers. Adapted with permission from [24] (https://creativecommons.org/licenses/by/4.0/)

3.1 Coupling of the Macrocycle with a ssDNA Oligonucleotide

Two single-stranded DNA (ssDNA) oligonucleotides, 5'TTTTTTTTTTTTTTT-TAGCT ($(dT)_{15}$AGCT) and 3'AAAAAAAAAAAAAAA5'($(dA)_{15}$) were synthetized with a MerMade4 synthesizer using phosphoramidite methodology, Fig. 3. This methodology is based on the use of DNA phosphoramidite nucleosides, modified with a 4,4'-dimethoxytrityl (DMTr) group, protecting the 5'-OH, a ß-cyanoethyl, protecting the 3'-phosphite and appropriate conventional protecting groups on the reactive primary amines in the heterocyclic nucleobase. The 5' end of $(dT)_{15}$AGCT was functionalized with a cyclooctyne-containing reacting group (R), which will later react with the azide located at the macrocycle (Fig. 3a). Therefore, in the last step of the solid supported oligonucleotide synthesis, we added a phosphoramidate modified with a cyclooctyne reacting group. The columns for synthesis of the oligonucleotides were filled with Controlled Pore Glass (CPG) solid support and anhydrous acetonitrile was used as solvent. The synthesis cycle (from 3' to 5') included the following steps: deblocking, activation, capping and oxidation. The 5'-DMTr group of the 5'-terminal base was removed by brief exposure to the 3% CCl_3COOH in anhydrous DCM, which later was removed via purging with

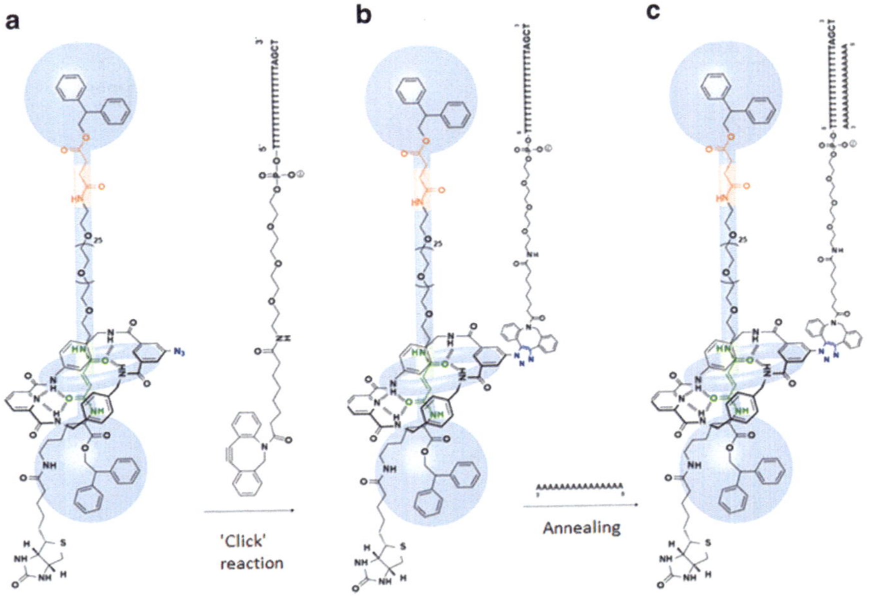

Fig. 3 **Coupling the rotaxane with DNA**. **a** The azide group of the macrocycle (N_3, blue) interacts in a copper-free 'click' reaction with the cyclooctyne reacting group linked to the 5'-end of the R-$(dT)_{15}$AGCT oligonucleotide. **b** After the 'click reaction', the R-$(dT)_{15}$AGCT oligonucleotide is annealed to a complementary $(dA)_{15}$ oligonucleotide. **c** The final product of the reaction is a double stranded DNA segment containing an overhang -AGCT 3' end, and a 5' end bound covalently to the macrocycle of the rotaxane

anhydrous acetonitrile. The phosphoramidite functionality was activated with the addition of the 0.25 M solution of benzylthiotetrazole in anhydrous acetonitrile. The activated species was then reacted with the 5'-OH to give a trivalent phosphite triester. The standard phosphoramidites coupling time was 2 min, and for the cyclooctyne derivative, it was 5 min. The P(III)-species were oxidized using 20 mM iodine in THF/Py/water. The unreacted 5'-OH-groups were capped with a mixture of two solutions: the first solution was composed of 10% pyridine, 10% Ac_2O, and 80% THF, and the second one was a 10% solution of 1-methylimidazole in THF. Upon synthesis completion, the obtained oligonucleotides were cleaved from the solid support and the Fmoc and β-cyanoethyl groups were removed by adding 28% aqueous NH_3. After 20 h, the obtained solution was filtered and concentrated in vacuum. The concentrate was then dissolved in water. To purify the synthesized oligonucleotides, we used gel electrophoresis. The segments of the gel containing oligonucleotides were visualized with ultraviolet light and isolated from the gel. The oligonucleotides were then extracted with elutrap system (for 3 h at 200 V).

The R-(dT)15AGCT oligonucleotide was bound to the macrocycle of the rotaxane via a copper-free 'click' reaction. The term 'click chemistry' was introduced to describe a family of chemical reactions, that are high yielding, simple to perform and allow to quickly obtain chemical compounds by joining small molecular units. Typically, this kind of reactions should lead only to byproducts that can be easily removed, be stereospecific, can be conducted in physiological conditions, be thermodynamically-favored and lead to only one product. This family of reactions includes the sterically promoted alkyne-azide cycloaddition (SPAAC), in which azides and sterically strained cycloalkynes are cobbled together. The driving force of the reaction is the removal of steric stress in cycloalkyne. We used this reaction to connect the azide of the rotaxane macrocycle to the dibenzoazacyclooctyne derivative (R), attached to the 5' end of the $(dT)_{15}$AGCT oligonucleotide (Fig. 3). To carry out this reaction, the oligonucleotide in H2O/DMF (50 μM) was mixed with the rotaxane in H2O/DMF (60 μM), and stirred for several hours at room temperature. The product of the reaction was purified with gel electrophoresis (20% polyacrylamide). After purification, the product of the 'click' reaction was mixed (1:1 molar ratio) and annealed with the complementary oligonucleotide $(dA)_{15}$ (1:1 molar ratio). The annealing yielded a rotaxane-oligo construct with a protruding, sticky AGCT 3' end, which will be used for subsequent ligations with DNA handles, see below (Fig. 3c).

3.2 Coupling the Rotaxane-Oligo Construct with dsDNA

To manipulate individual rotaxane molecules with optical tweezers, we coupled the rotaxane with two dsDNA molecules: one, used as 'handle' to monitor and manipulate the movement of the macrocycle (*DIG-DNA handle*) and the other, used as a 'spacer' to separate the rotaxane from the surface of the bead (*spacer-DNA*).

The *DIG-DNA handle* is a dsDNA molecule of 2,686 base pairs (bp) bearing multiple digoxigenines at one end and a protruding -TCGA 5' at the other end (note that the protruding 5' end is complementary to the AGCT 3' end of the rotaxane-DNA construct). This molecule was synthetized as described in [25]. The *DIG-DNA handle* was later ligated to the rotaxane-oligo construct using the T4 DNA ligase (NEB).

The *spacer-DNA* is dsDNA molecule of 827 bp functionalized with biotin at one end and digoxigenin at the other end. This DNA molecule was obtained via PCR amplification of the polylinker segment of the pUC19 plasmid using two DNA primers labeled either with digoxigenin or biotin at their 5'-ends. The *spacer-DNA* was mixed with an excessive amount of streptavidin and purified with standard DNA purification columns (Promega). Therefore, the final product of these reactions is a dsDNA molecule labeled with digoxigenin at one end and streptavidin at the other.

3.3 Interfacing the Rotaxane with the Optical Tweezers

To interface the rotaxane with the optical tweezers, we first incubated the rotaxane-*DIG DNA handle* construct and the *spacer-DNA* molecules separately with polystyrene beads (2 μm diameter, Spherotech) functionalized with antibodies against digoxigenin for 20 min at room temperature. In the optical tweezers, we introduced the two types of beads separately into the fluidic chamber using two different channels connected to a central reaction channel, where the optical trap and the glass micropipette are located (Fig. 4a). To isolate a single rotaxane molecule, we used the laser trap to first place a bead coated with the *spacer-DNA* on top of the micropipette. Then, we trapped a bead functionalized with the rotaxane-*DIG-DNA handle* and placed it close to the bead held on top of the micropipette. The optical trap is then moved in the Y-direction to approach the two beads. An attachment is made when the biotin group of the rotaxane binds to the streptavidin molecule at the end of the *spacer-DNA* attached to the bead on top of the pipette (Fig. 4b). Upon attachment, moving the trap away from the micropipette in the Y-direction provided the force-extension curve of the resulting DNA-rotaxane construct (Fig. 5a). At low tensions (< 10 pN, see below), the obtained force-extension curves resembled the elastic behavior of a polymer with a persistence length of 50 nm found from the fit to the Worm-Like Chain model (WLC), a typical persistence length value of an individual dsDNA molecule. A characteristic rip in the force-extension curves at forces < 10 pN revealed the presence of the rotaxane in our constructs (see below and Fig. 5a). Therefore, the force-extension curves help us to identify the attachment of single rotaxane-DNA constructs unequivocally.

4 Determination of the Mechanical Strength of the Interactions Between the Macrocycle and the Two Stations

Upon isolation of a single rotaxane-DNA construct, mechanical force was applied to the system by moving the optical trap relatively to the glass micropipette at a constant rate (200 nm s^{-1}), Fig 4b and Fig. 5. In this way, we took multiple pulling and relaxing cycles on single rotaxane-DNA construct, which enable us to determine the mechanical stability of the interactions of the tetraamide macrocycle with the fumaramide and succinic amide-ester stations. Typically, at forces < 9pN, the obtained force-extension curves are in a good agreement with the WLC model for a dsDNA molecule of the same length as the total length of the dsDNA handles used in our set up. Remarkably, a sudden increase of extension was observed when the applied force surpasses the strength of the four hydrogen bonds connecting the macrocycle to the fumaramide station (Fig. 5a). This extension rip is due to the macrocycle transitioning from the fumaramide station to the succinic amide-ester station. Pulling-relaxing data was

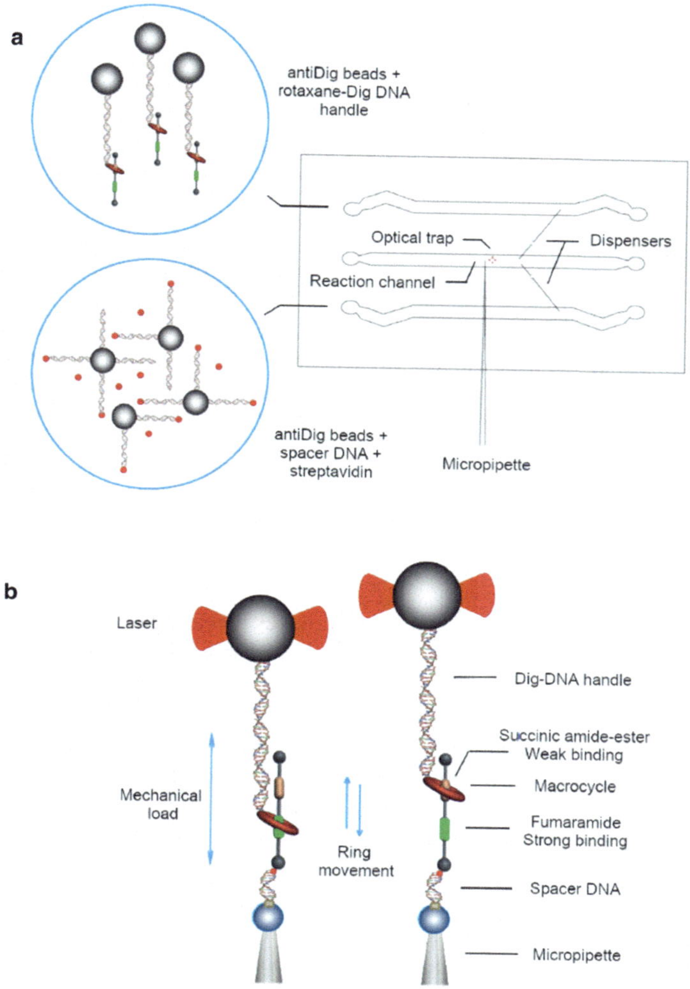

Fig. 4 Interfacing the rotaxane with optical tweezers. a Polystyrene beads functionalized with anti-digoxigenin are incubated either with dsDNA-rotaxane constructs (top) or with *spacer-DNA* molecules bearing streptavidin at one end (bottom). After incubations, each set of functionalized beads is introduced to a microfluidic chamber through separate channels, connected to the central reaction channel by means of glass dispensers. Once the polystyrene beads arrive to the central reaction channel, a bead functionalized with *spacer-DNA* is placed on top of the micropipette and a bead containing a dsDNA-rotaxane construct is brought with the optical trap close to the bead on top of the micropipette to make an attachment. **b** Schematic representation of the experimental setup. The Dig-DNA handle and the spacer DNA molecules are connected to the polystyrene beads through digoxigenin-antidigoxigenin interactions. The *spacer-DNA* connect the rotaxane through biotin-streptavidin interactions (red dot). The resulting construct can be pulled and relaxed multiple times by moving along the Y-axis the optical trap relatively to the glass micropipette at a constant rate (200 nm s^{-1})

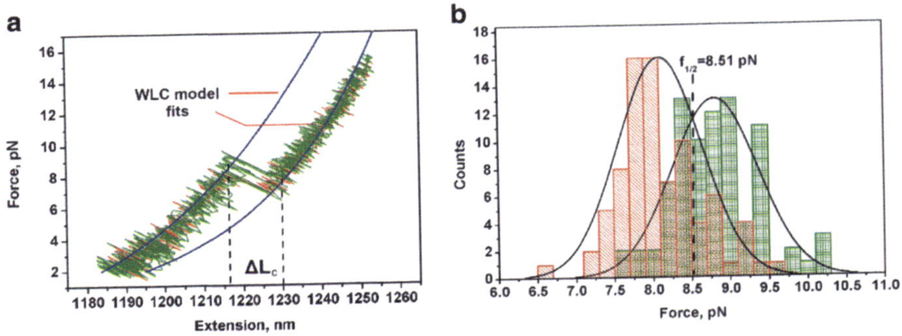

Fig. 5 Quantification of the mechanical strength of the macrocycle at each station. a Force-extension curves characterizing mechanical stretching (green) and relaxing (magenta) of a single dsDNA-rotaxane complex at constant pulling rate (200 nm s^{-1}). Blue lines represent fits of the experimental curves with the WLC model (persistence length ~50 nm)). ΔL_c is an increment of the contour length of the dsDNA-rotaxane complex after a shuttling event, ΔL_c ~15 nm. **b** Distribution of breaking forces at each station. The crossing point of the distributions of breaking forces at fumaramide (green) and succinic amide-ester (magenta) stations yielded an average coexistence force of, $f_{1/2} = 8.51$ pN ($N = 450$)

collected for different rotaxane-dsDNA complexes revealing an increment in extension, ΔLc, of 15 ± 2 nm (450 pulling-relaxing cycles in total). This value corresponds to the contour length of the oligoethyleneglycol thread separating the two stations at forces where the macrocycle shuttles in between stations. The average rupture force of the non-covalent interactions between the macrocycle and the fumaramide station was 8.8 ± 0.6 pN (Fig. 5b). Interestingly, as force decreases on the rotaxane-dsDNA molecule (relaxation cycle), an abrupt decrease of the molecule's length can be clearly observed, and the original extension is recovered. This sudden decrease in extension is due to the macrocycle returning from the succinic amide ester station to the thermodynamically favorable fumaramide station. The average force at which the original extension is recovered was 8.1 ± 0.5 pN (Fig. 5b). The crossing point of the rupture and recovery force distributions corresponds to the force at which the probability of the macrocycle residing over either of the two stations is equal ($f_{1/2} = 8.51$ pN, Fig. 5b). This force is the so-called coexistence force. Using Jarzynski equality [26, 27], the total free energy of shuttling (ΔG_T) was calculated as the product of the coexistence force ($f_{1/2}$) by the contour length of the thread separating the two stations (ΔLc), as $\Delta G_T = 31 \pm 2$ k_BT. Note that this value corresponds to the sum of the actual energy of shuttling from the fumaramide station to the succinic amide-ester station at f = 0 pN plus the free energy of stretching of the rotaxane-dsDNA complex from f = 0 pN to $f = f_{1/2}$. The latter was calculated as 11.8 k_BT therefore, the actual free energy of shuttling at f = 0 pN was 19 ± 2 k_BT [24].

5 Determination of Tension-Dependent Shuttling Rates

We next studied the dynamics and bistability of the rotaxane system by maintaining constant forces on single rotaxane-dsDNA complexes (Fig. 6a). For every single rotaxane-dsDNA complex held at a constant force close to the coexistence force ($f_{1/2} = 8.51$ pN), continuous shuttling events between the fumaramide and succinic amide-ester stations were followed for several minutes by recording extension traces over time. The obtained traces presented precise residence times at the two stations (Fig. 6a). At forces close to the coexistence force, the macrocycle spent approximately equal amounts of time at the two stations. The occupancy of one of the two stations was favored by applying tensions either slightly higher or lower than the coexistence force (±0.5 pN). At forces higher than the coexistence force, the macrocycle resided at the succinic amide-ester station preferentially. In contrast, at forces lower than the coexistence force, the occupancy of the fumaramide station was favored. The extension histograms obtained from the recorded traces showed two peaks that fit well two Gaussian distributions separated by $\Delta x = 15.5 \pm 2.5$ nm, which is the distance separating the two stations found in the pulling-relaxing experiment. The energetic profiles of the shuttling at different forces were obtained directly from the

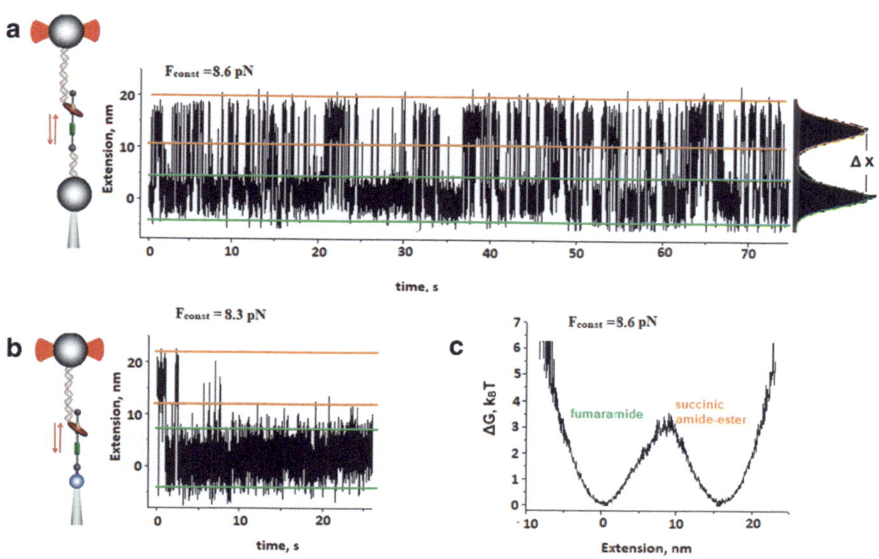

Fig. 6 **Shuttling dynamics at a constant force.** **a** Representative shuttling dynamics of a system in which the rotaxane-DNA construct is separated from the surface of the streptavidin bead by a dsDNA spacer ($F_{const} = 8.6$ pN). **b** Representative shuttling dynamics of a system in which the biotin-end of the rotaxane is attached directly to a bead functionalized with streptavidin ($F_{const} = 8.3$ pN). Direct attachment of the rotaxane to the beads interfered with the shuttling dynamics drastically. **c** Example of an energy profile obtained by using the Boltzmann distribution directly from the extension distributions

extension distributions by using the Boltzmann distribution (Fig. 6c) [28, 29]. The energetic profiles display a clear transition state separating two minima, with the relative height varying with force.

We used the Bell-Evans theory to extract from these data kinetic and energetic parameters of shuttling [30, 31], such as: (i) the coexistence force ($f_{1/2} = 8.83$ pN), (ii) the free energy of shuttling ($\Delta G_T = f_{1/2} \cdot \Delta x = 33 \pm 2\ k_BT$) and (iii) the position of the transition state relative to the fumaramide and succinic amide ester stations. Interestingly, at the coexistence force the transition state was 10 ± 2 nm away from the fumaramide station and 6 ± 2 nm from the succinic amide-ester station, revealing an asymmetric location. Note that the values of $f_{1/2}$ and ΔG_T were identical to those obtained from the pulling-relaxing data.

It is important to point out that separating the rotaxane from the bead with the *spacer-DNA* greatly improved the quality of the data. Figure 6b shows that when the rotaxane was directly attached to streptavidin covered beads, less and noisier transition events were observed, which in addition, attenuated in time quickly. In contrast, coupling the rotaxane to a *dsDNA-spacer* that separates the end of the rotaxane from the surface of a polystyrene bead (Fig. 6a), enabled us to measure several hundreds of shuttling events for each molecule. These results strongly suggest that separation of the rotaxane from the bead diminished or eliminated non-specific interactions between the macrocycle and the surface of the bead, which may affect the shuttling process.

6 Conclusions

Synthetic switchable molecular shuttles have attracted interest from the research community due to their potential applications in biomedicine and nanotechnology [4–6]. The idea of controlling the motion of these molecular devices has become a major research challenge. Although the kinetics and thermodynamics of these systems in bulk are well understood [4], there is no clear understanding of the operational dynamics of these synthetic devices. Here, we have described how to obtain this elusive information from individual rotaxane molecules by taking advantage of the precise force control and force resolution of the single-molecule manipulation technique of optical tweezers. We have shown that the unique biochemical and mechanical properties of dsDNA can be readily exploited to study and manipulate the operation in real-time of synthetic devices. Coupling rotaxane molecules with dsDNA, allowed us to solubilize and study them under near-physiological conditions. In addition, dsDNA was used as (i) a handle to manipulate the macrocycle, (ii) as a single-molecule reporter, and (iii) as a spacer to separate the device from the surface of the beads. It is noteworthy that in the experimental setup we described here, it is possible to change the reaction conditions *in situ*. This possibility will enable us to study the combined effect of force and other additional factors relevant to the operation of the rotaxane, such as the ionic strength, chemical reagents, light and temperature. These studies will shed light into the inner thermo-mechanical and mechano-chemical processes that govern the operation of these synthetic devices. The method described here could be applied to study the mechanical strength and the real-time operation of other artificial systems at the single-molecule level, which will represent a valuable contribution to the advancement of the fascinating field of single-molecule supramolecular chemistry.

Acknowledgements This work was supported by the European Research Council (ERC-StG-MINT 307609), Ministerio de Economía y Competitividad (grants BFU2015-63714-R, CTQ2014-60541-P, CTQ2017-86060-P, SAF2014-56763-R, FIS2016-80458-P). IMDEA Nanociencia acknowledges support from the "Severo Ochoa" Programme for Centers of Excellence in R&D (MINECO, Grant SEV-2016-0686) and Comunidad de Madrid (NANOMAGCOST P2018 INMT-4321). K.M.L was supported by the Ministerio de Educación Cultura y Deporte (FPU2014/06867).

References

1. Shibasaki, M., Stoddart, J.F., Vogtle, F.: Stimulating Concepts in Chemistry. Wiley-VCH, Weinheim (2000)
2. Balzani, V., Credi, A., Raymo, F.M., Stoddart, J.F.: Artificial molecular machines. Angew. Chem. Int. Ed. **39**, 3348–3391 (2000)
3. Pease, A.R., Jeppesen, J.O., Stoddart, J.F., Luo, Y., Collier, C.P., Heath, J.R.: Switching devices based on interlocked molecules. Acc. Chem. Res. **34**, 433–444 (2001)
4. van Dongen, S.F.M., Cantekin, S., Elemans, J., Rowan, A.E., Nolte, R.J.M.: Functional interlocked systems. Chem. Soc. Rev. **43**, 99 (2014)
5. Dvornikovs, D., House, B.E., Kaetzel, M., Dedman, J.R., Smithrud, D.B.: Host-[2]rotaxanes as cellular transport agents. J. Am. Chem. Soc. **125**, 8290 (2003)
6. Nishimura, D., Takashima, Y., Aoki, H., Takahashi, T., Yamaguchi, H., Ito, S., Harada, A.: Single-molecule imaging of rotaxanes immobilized on glass substrates: observation of rotary movement. Angew. Chem. Int. Ed. **47**, 6077 (2008)
7. Kay, E.R., Leigh, D.A., Zerbetto, F: Synthetic molecular motors and mechanical machines. Angew. Chem. Int. Ed. **46**, 72–191 (2007)
8. Brough, B., Northrop, B.H., Schmidt, J.J., Tseng, H.R., Houk, K.N., Stoddart, J.F., Ho, C.M.: Evaluation of synthetic linear motor-molecule actuation energetics. Proc. Natl. Acad. Sci. U.S.A. **103**, 8583 (2006)
9. Lussis, P., Svaldo-Lanero, T., Bertocco, A., Fustin, C.-A., Leigh, D.A., Duwez, A.-S.: A single synthetic small molecule that generates force against a load. Nat. Nanotechnol. **6**, 553–557 (2011)
10. Van Quaethem, A., Lussis, P., Leigh, D.A., Duwez, A.-S., Fustin, C.-A.: Probing the mobility of catenane rings in single molecules. Chem. Sci. **5**, 1449–1452 (2014)
11. Hanozin, E., Mignolet, B., Morsa, D., Sluysmans, D., Duwez, A.-S., Stoddart, J.F., Remacle, F., De Pauw, E.: Where ion mobility and molecular dynamics meet to unravel the (un)folding mechanisms of an oligorotaxane molecular switch. ACS Nano **11**, 10253–10263 (2017)
12. Sluysmans, D., Devaux, F., Bruns, C.J., Stoddart, J.F., Duwez, A.-S.: Dynamic force spectroscopy of synthetic oligorotaxane foldamers. Proc. Natl. Acad. Sci. U.S.A. **115**, 9362–9366 (2018)
13. Sluysmans, D., Hubert, S., Bruns, C.J., Zhu, Z., Stoddart, J.F., Duwez, A.S: Synthetic oligorotaxanes exert high forces when folding under mechanical load. Nat. Nanotechnol. **13**, 209–213 (2018)

14. Biedermann, F., Schneider, H.-J.: Experimental binding energies in supramolecular complexes. Chem. Rev. **116**(9), 5216–5300 (2016)
15. Wang, M.D., Schnitzer, M.J., Yin, H., Landick, R., Gelles, J., Block, S.M.: Force and velocity measured for single molecules of RNA polymerase. Science **282**(5390), 902–907 (1998)
16. Meyhöfer, E., Howard, J.: The force generated by a single kinesin molecule against an elastic load. Proc. Natl. Acad. Sci. U.S.A. **92**(2), 574–578 (1995)
17. Edwards, D.T., Faulk, J.K., LeBlanc, M.A., Perkins, T.T.: Force spectroscopy with 9-μs resolution and Sub-pN stability by tailoring AFM cantilever geometry. Biophys. J. **113**(12), 2595–2600 (2017)
18. Walder, R., Van Patten, W.J., Adhikari, A., Perkins, T.T.: Going vertical to improve the accuracy of atomic force microscopy based single-molecule force spectroscopy. ACS Nano **12**(1), 198–207 (2018)
19. Ashkin, A., Dziedzic, J.M., Bjorkholm, J.E., Chu, S.: Observation of a single-beam gradient force optical trap for dielectric particles. Optical Letters **11**, 288–290 (1986)
20. Li, P.T., Collin, D. Smith, S.B., Bustamante, C., Tinoco, Jr., I.: Probing the mechanical folding kinetics of TAR RNA by hopping, force-jump, and force-ramp methods. Biophys. J. **90**, 250–260 (2006)
21. Wen, j.-D., Manosas, M., Li, P.T.X., Smith, S.B., Bustamante, C., Ritort, F., Tinoco Jr., I.: Force unfolding kinetics of RNA using optical tweezers. I. Effects of experimental variables on measured results. Biophys. J. **92**, 2996–3009 (2007)
22. Alemany, A., Ritort, F.: Force-dependent folding and unfolding kinetics in DNA hairpins reveals transition-state displacements along a single pathway. J. Phys. Chem. Lett. **8**(5), 895–900 (2017)
23. Jiao, J., Rebane, A.A., Ma, L., Zhang, Y.: Single-molecule protein folding experiments using high-precision optical tweezers. Methods Mol. Biol. **1486**, 357–390 (2017)
24. Naranjo, T., Lemishko, K.M., de Lorenzo, S., Somoza, A., Ritort, F., Pérez, E.M., Ibarra, B.: Dynamics of individual molecular shuttles under mechanical force. Nat. Commun. **9**, 4512 (2018)
25. Cerrón, F., de Lorenzo, S., Lemishko, K.M., Cieselski, G.L., Kaguni, L.S., Cao, F.J., Ibarra, B.: Replicative DNA polymerases promote active displacement of SSB proteins during lagging strand synthesis. NAR **47**(11), 5723–5734 (2019)
26. Jarzynski, C.: Nonequilibrium equality for free energy differences. Phys. Rev. Lett. **78**, 2690 (1997)
27. Jarzynski, C.: Equilibrium free-energy differences from nonequilibrium measurements: a master-equation approach. Phys. Rev. E **56**, 5018 (1997)
28. Maxwell, J.C.: Illustrations of the dynamical theory of gases. Part I. On the motions and collisions of perfectly elastic spheres. Lond., Edinb., Dublin Philos. Mag. J. Sci. 4th Series, **19**, 19–32 (1860)
29. Boltzmann, L.: Weitere studien über das Wärmegleichgewicht unter Gasmolekülen. Sitzungsberichte der Kaiserlichen Akademie der Wissenschaften in Wien, mathematisch-naturwissenschaftliche Classe, **66**, 275–370 (1872)
30. Bell, G.: Models for the specific adhesion of cells to cells. Science **200**(513), 25 (1978)
31. Evans, E., Leung, A., Hammer, D., Simon, C.: Chemically distinct transition states govern rapid dissociation of single L-selectin bonds under force. Proc. Natl. Acad. Sci. U.S.A. **98**(7), 3784 (2001)

The manufacturer's authorised representative in the EU is Springer Nature Customer Service Centre GmbH, Europaplatz 3, 69115 Heidelberg, Germany. If you have any concerns regarding our products, please contact ProductSafety@springernature.com

Printed and bound by CPI Group (UK) Ltd, Croydon, CR0 4YY
22/09/2026
02221060-0006